NEW DEVELOPMENTS IN MEDICAL RESEARCH

PHYTOCHEMICALS FROM MEXICAN MEDICINAL PLANTS

POTENTIAL BIOPHARMACEUTICALS AGAINST NONCOMMUNICABLE DISEASES

New Developments in Medical Research

Additional books and e-books in this series can be found on Nova's website under the Series tab.

NEW DEVELOPMENTS IN MEDICAL RESEARCH

PHYTOCHEMICALS FROM MEXICAN MEDICINAL PLANTS

POTENTIAL BIOPHARMACEUTICALS AGAINST NONCOMMUNICABLE DISEASES

J. BASILIO HEREDIA
ERICK P. GUTIÉRREZ-GRIJALVA
AND
LETICIA X. LÓPEZ-MARTÍNEZ
EDITORS

nova
Medicine & Health
New York

Library of Congress Cataloging-in-Publication Data

ISBN: 978-1-53617-397-0

Published by Nova Science Publishers, Inc. † New York

CONTENTS

PREFACE

Noncommunicable diseases are the leading cause of death around the world, accounting for nearly 70% of all deaths. Several factos, such as unhealthy lifestyles, physical inactivity, exposure to UV-light, harmful use of alcohol, among others, cause diseases such as cancer, diabetes, obesity, and cardiovascular diseases. The current conventional pharmacological approach of treatment for these diseases is often costly, which represents an economic burden for most people, mainly from the low- and middle-income countries, who cannot always afford to buy medication. Besides, with or without medical approval and surveillance, people often turn to incorporate preparations from local medicinal plants when they are already under prescribed medication for their particular ailments.

On the other hand, medicinal plants are a rich source of phytochemicals such as alkaloids, terpenes, and polyphenols, which are potential biopharmaceutical agents against several noncommunicable diseases. Several studies have been conducted to isolate, identify and evaluate such compounds with promising results. In this book, we comprehensively review the information that shows the potential of phytochemicals from selected Mexican medicinal plants against noncommunicable diseases such as obesity, diabetes, cancer, cardiovascular diseases, and metabolic syndrome. This work also summarizes the toxicological studies and risks of the use of medicinal plants as well as perspectives in the future use of phytochemicals as biopharmaceutical agents.

Chapter 1, “Mexican Medicinal Plants: Phytochemistry and Pharmacognosy” and Chapter 2, “Regulation and Safety of Medicinal Plants” are introductory chapters to the book. In these review chapters, a brief introduction to the ethnobotanical use of medicinal plants is assessed, also phytochemicals are defined. The safety use of plant species is also assessed with special remarks on toxicology studies in some plants.

Chapters 3 to 7 focus on studies regarding the potential of phytochemicals from plants from the Mexican herbal pharmacopeia against the main noncommunicable diseases obesity, the metabolic syndrome, cancer, and cardiovascular diseases. Special focus is given to studies considering the metabolomic characterization of phytochemicals and their mode/mechanism of action.

Chapter 8, “Bioactive Compounds from Medicinal Plants: Green Extraction Technologies”, focuses on recent studies on the obtention of phytochemicals from plants listed in the Mexican herbal pharmacopeia, using green technologies aimed to reduce the use of extraction with solvents that may be contaminants after use.

In: Phytochemicals from Mexican …
Editors: J. Basilio Heredia et al.
ISBN: 978-1-53617-397-0

Chapter 1

MEXICAN MEDICINAL PLANTS: PHYTOCHEMISTRY AND PHARMACOGNOSY

***Erick P. Gutiérrez-Grijalva*[1,*], *Karla M. Baez-Parra*[2] *and J. Basilio Heredia*[3]**

[1]Cátedras CONACYT - Centro de Investigación en Alimentación y Desarrollo, A. C., Culiacán, México

[2]Facultad de Ciencias Químico Biológicas, Universidad Autónoma de Sinaloa, Culiacán, Sinaloa, Mexico

[3]Centro de Investigación en Alimentación y Desarrollo, A. C., Culiacán, México

ABSTRACT

Noncommunicable diseases (NCD) like diabetes, cancer, and cardiovascular diseases are the leading cause of death around the world. NCD represent around 70% of the total world deaths. Even though the first line of treatment is through the adminstration of pharmaceutical drugs, an increasing number of people from low and middle-income countries use plants as primary or complementary source of treatment.

* Corresponding Author's Email: erick.gutierrez@ciad.mx.

The use of medicinal plants is documented in many countries in pharmacopoeias like the the Herbal Mexican Pharmacopoeia. This review is an introductory chapter for this book and briefly defines the main phytochemical groups found in Mexican medicinal plants which are the metabolites associated with their potential therapeutic effect, as well as some pharmacognostic aspects of these plants.

Keywords: ethnopharmacology, pharmacognosy, medicinal plants, natural products, phytochemicals

1. Traditional Use of Medicinal Plants

Medicinal plants have been traditionally used around the world to treat a myriad of diseases. Such is the importance of medicinal plants worldwide that the World Health Organization has published monographs on selected medicinal plants to provide current scientific information on the safe use of these species. Furthermore, many countries have well established official pharmacopoeias, which are national monographs of the medicinal plants, which aim to provide basic pharmacological basis [1].

The study of medicinal plants is a field of study of ethnopharmacology, which is contained within the concepts of pharmacognosy, phytotherapy and phytomedicine. The differences among these terms lay between the fact that pharmacognosy incorporates the study of medicinal plants and their natural products, phytotherapy aims on plants to treat diseases, and phytomedicine is the use of plants and herbs to mitigate human diseases [2-4].

Even when synthetic drugs occupy most of the market of the pharmaceutical industry, there is a "renewed" and growing interest in the discovery of new natural products from plants used in traditional medicine. It is important to mention that the use of medicinal plants is still an important part of the traditional medicine in many countries, where the use of traditional medicine is the mainstay of health care delivery or a complement to it, which is why in some regions of the world, traditional medicine is often known as complementary medicine [5, 6].

Noncommunicable diseases (NCD) like diabetes, cancer, and cardiovascular diseases are the leading cause of death around the world, causing around the 70%, representing nearly 41 million people each year, of the total deaths worldwide. NCD arise from multifactorial conditions such as unhealthy diet, sedentary lifestyle, pollution, genetic predisposition, etc. Thus, as a result of the increasing incidence of NCD, efforts are being made to identify new molecules as potential preventive or treatment drugs for those conditions [7, 8].

The ethnobotanical use of plants in traditional and complementary medicine is well documented in many countries around the world. The use of medicinal plants is increasing its credibility in many industrialized countries; however, the percentage of people using traditional medicine in developed countries is decreasing [9]. As a result of the increasing use of plants in traditional and complementary medicine, the market of herbs and herbal products is growing rapidly. As of 2002, the global sales of herbal products reached around US$ 60,000 million. Moreover, derived from medicinal plants, there is another increasing market regarding the extraction and isolation of their phytochemical compounds. The isolated phytochemicals can be used in the pharmaceutical and food industries, in the latter, phytochemicals can be used as active functional ingredients in the production of energy drinks, capsules, health supplements, energy boosters, etc. Among the most common plants used for these purposes are turmeric, cumin, garlic, ginger, clove, ginseng, aloe vera, neem, etc. [10]. Moreover, some phytochemicals such as alkaloids, terpenes, and phenolic compunds have been used in the pharmaceutical industry as drugs against diseases. And nowadays, many studies are being conducted for this purpose [11, 12].

The study of medicinal plants for whatever the purpose mentioned above involves a very complicated process and requires time and resources. Before the selection of plants, a thorough study of the availabile literature must be done in order to identify which plants are used ethnobotanically for the treatment of the disease/ailment of interest. After that, plants must be collected and a preliminary screening of their phytochemical content is performed. Later the main compounds are extracted, isolated and evaluated [10]. A step that is commonly neglected is the evaluation of the bioavailability and toxicity of these compounds, which is a pivotal step in the systemic study of the pharmaceutical potential use of phytochemicals [13, 14].

2. Medicinal Plants Used in Mexico

There are some reports of the medicinal plants that are commonly used ethnobotanically by people in different regions of Mexico. Due to the great extension of Mexico, a wide diversity of plants are commonly used throughout the different regions of the country. Ethnobotanic records of medicinal plants used in Mexico are not common, and survey studies [15-17].

In this regard, the Herbal Mexican Pharmacopoeia, which is where this book took the guidance of plant species, contains monographs of plants commonly used in Mexico and some methods of analysis. Briefly, the plants assessed in this manuscript are those that both have been traditionally used in herbal medicine and studied for their potential to treat obesity, metabolic syndrome, diabetes, cancer and cardiovascular diseases.

3. Mexican Medicinal Plants and Noncommunicable Diseases

Noncommunicable diseases (NCDs), also known as chronic diseases, are diseases that are not transmissible directly from one person to another. Noncommunicable diseases arise from the combination of genetic, physiological, environmental and behaviour factors. The most common NCDs are cardiovascular diseases, cancers, chronic respiratory diseases and diabetes. Modifiable lifestyles such as tobacco use, sedentarism, unhealthy diets and harmful use of alcohol increase the risk of NCDs [8].

Table 1. Some medicinal plants listed in the Herbal Mexican Pharmacopoeia that have been included in this book

Scientific name	Common name
Agrimonia eupatoria L.	Agrimony
Allium sativum L.	Garlic
Cynara scolymus L.	Artichoke
Aloe vera L. Burm	Aloe vera
Vaccinium myrtillus L.	Bilberry
Calendula officinalis L.	Common marigold
Cinnamomum verum J Presi	Cinnamon
Elettaria cardamomum L. Maton	Cardamom
Aesculus hippocastanum L.	Horse-chestnut
Turnera diffusa Wild	Damiana
Echinacea angustifolia	Blacksamson echinacea
Echinacea pallida	Pale purple coneflower
Ginkgo biloba	Ginkgo
Psidium guajava L.	Common guava
Hypericum perforatum L.	St John's wort
Hibiscus sabdariffa L.	Roselle
Zingiber officinale Roscoe	Ginger root
Humulus lupulus L.	Hemp
Melissa officinalis L.	Balm
Mentha x piperita L.	Peppermint
Citrus aurantium L.	Bitter Orange
Origanum vulgare L.	Oregano
Rosmarinus officinalis L.	Rosemary
Thymus vulgaris L.	Thyme

Cardiovascular diseases (CVDs) are the leading cause of death in the world, each year more people die from CVDs than any other cause. Around 17.9 million people died from CVDs in 2016 which represents around 31% of deaths worldwide [18].

Furthermore, obesity and the metabolic syndrome (MetS) are risk factors of diabetes, cardiovascular diseases and recent studies have shown that obesity is related with the onset of some cancers. The metabolic syndrome is a cluster of factors closely related with the onset of cardiovascular diseases. The metabolic syndrome is the combination of obesity, hyperglycemia, hypertension and dyslipidemia. [19, 20]. Around 20-25% of the world population have the metabolic syndrome, the incidence of MetS is often similar to the incidence of obesity and incidence of diabetes. The prevalence of MetS may vary depending each country, age of population and even among certain ethnic groups [21].

On the other hand, cancer is a major health problem around the world, representing the second leading cause of death worldwide due to its increasing incidence and high mortality. According to data from the World Health Organization cancer is responsible for an estimated 9.6 million deaths in 2018, and the most common cancers are lung, breast, colorectal, prostate, skin and stomach cancer [18].

The first line of treatment against noncommunicable diseases is the administration of pharmacological drugs. However, these treatments are often costly and can't be easily accessed by people from low- and middle-income countries. Mexican medicinal plants have been used in traditional medicine in Mexican culture to treat different types of cancer, hyperglycemia, diabetes, inflammation-related diseases, among others [11, 5, 1]. Thus, scientific studies have focused on the search of plants with potential effect against noncommunicable diseases. These studies have focused on the ethnobotanical studies of these plants, and on the extraction, identification and evaluation of their phytochemicals which are the compounds associated with the potential pharmacological effect. The interest in medicinal plants, as part of the traditional and complementary medicine is increasingly growing, both for the health and economic implications. In this regard, the World Health Organization has published a manuscript on the Traditional Medicine Strategy 2014-2023 [5].

4. Phytochemicals from Mexican Medicinal Plants

Phytochemicals are a widespread group of compounds derived from the secondary metabolism of plants. Their main function is to act as a defense against pathogens, herbivores, UV-light, water stress, etc. Phytochemicals can be classified for their study in alkaloids, terpenes, and phenolic compounds. These groups of phytochemicals are of current interest in ethnopharmacology due to their potential bioactive properties against noncommunicable diseases, which are the leading cause of death worldwide [22].

Terpenes

Terpenes are phytochemicals that share a common biosynthetic pathway from the C5 substrates isopentenyl pyrophosphate (IPP) and the dimethylallyl diphosphate [23]. Terpenes can be found in many life forms, and in plants they form the largest group of phytochemicals with more than 15,000-20,000 structures characterized. Terpenes are categorized into hemiterpenes (1 isoprene unit, C5), monoterpenes (2 isoprene units, C10), sesquiterpenes (3 isoprene units, C15); diterpenes (4 isoprene units, C20), triterpenes (6 isoprene units, C30), tetraterpenes (8 isoprene units, C40) and polyterpenes (8 isoprene units, C5n) [24].

Terpenes are also responsible for the aroma of some medicinal herbs like rosemary, thyme, oregano and basil, and can be found mainly in the trichomes of plants [25]. Moreover, terpenes are also partially responsible for the antioxidant and bioactive properties of these plants [26, 27]. Some of the most common terpenes have been of interest due to their potential bioactive effects are D-limonene (monoterpene), 1,8-cineole or eucalyptol (monoterpene), boswellic acid (triterpene), betulinic acid (triterpene), ursolic acid (triterpene) (Figure 1) [28].

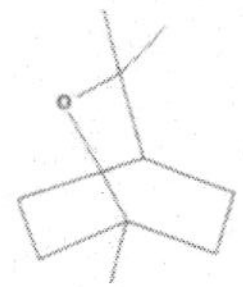

Eucalyptol

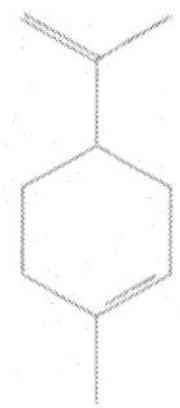

Limonene

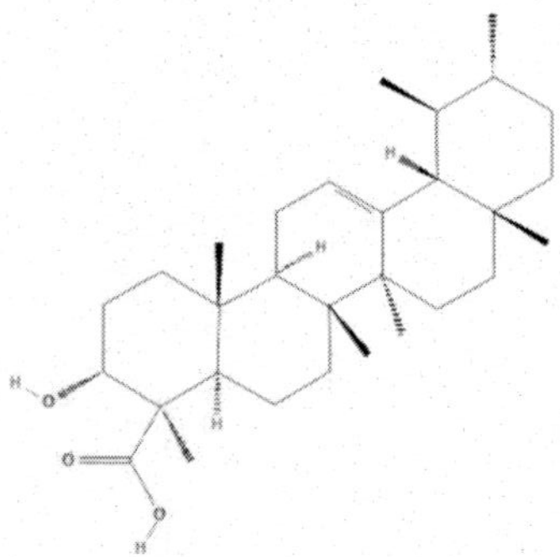

Boswellic acid

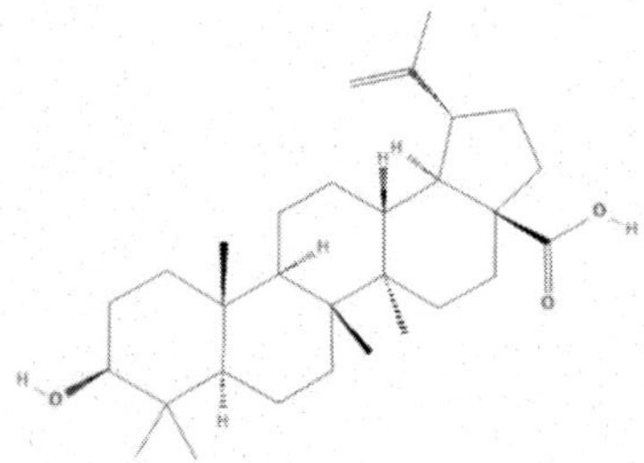

Betulinic acid

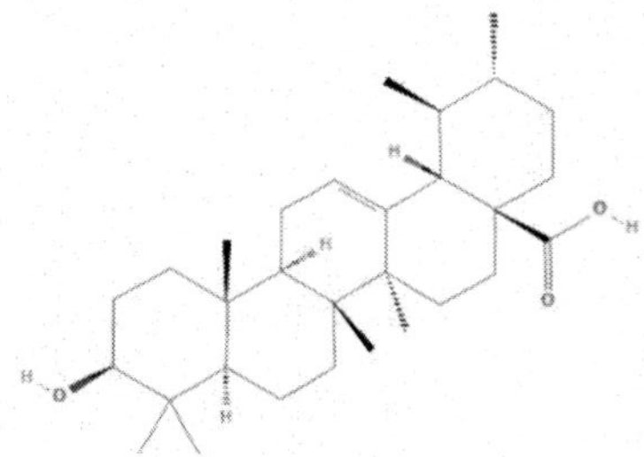

Ursolic acid

Figure 1. Graphical representation of the structures of common terpenes found in plants with potential bioactive potential.

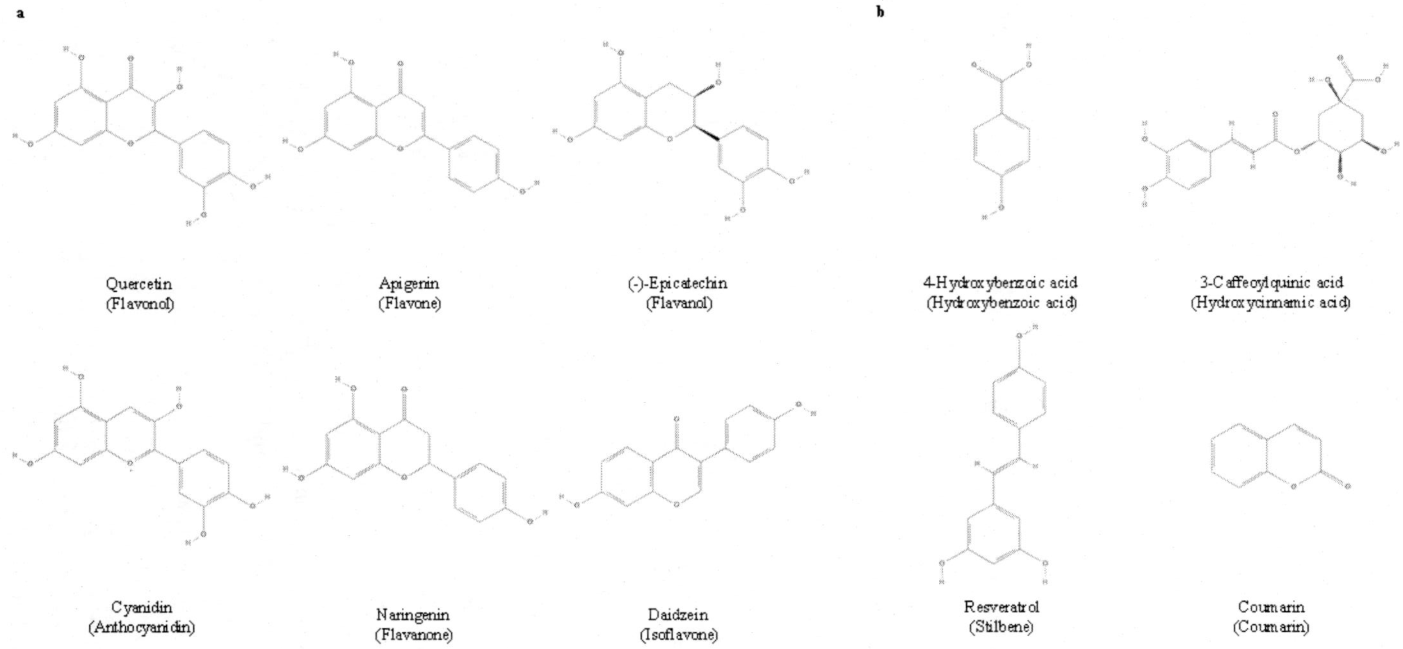

Figure 2. Graphical representation of the structure of common phenolic compounds found in plants with bioactive potential.

Phenolic Compounds

Phenolic compounds are a large group of phytochemicals with a basic core of at least an aromatic ring with one or more hydroxyl radicals (-OH) attached.

Phenolic compounds are metabolites derived from the shikimate pathway; the aromatic ring B and the chromane ring are are derived from the amino acid phenylalanine, which is also a product of the shikimate pathway, whilst the ring A from the three units of malonyl-CoA [29]. Moreover, phenylalanine ammonia lyase is the key enzyme in the phenylpropanoid pathway which catalyzes the conversion of phenylalanine to cinnamate and C6-C3 structures.

Phenolic compounds can be sub-classified in flavonoids (Figure 2a) such as flavonols, flavones, flavan-3-ols, anthocyanidins, flavanones, isoflavones and others; and non-flavonoids such as phenolic acids (Figure 2b), hydroxycinnamic acids, hydroxybenzoic acids, stilbenes, coumarins, and others [30, 31].

Alkaloids

Alkaloids are a widespread group of chemicals characterized by containing cyclic structrures with at least one nitrogen atom [22, 32]. Alkaloids are considered one of the first molecules used by humans in ethnopharmacology, and they are used nowadays as drugs to treat diseases like cancer. Alkaloids of pharmacological use are commonly of plant origin, however, they can also be found in animals, insects, some marine animals and microorganisms. There is not a unique way for alkaloid classification, since these compounds can be classified by ther chemical characteristis, pharmaceutical functions or botanical aspects [33]. Furthermore, in plants alkaloids can occur in any part of the plants, though some specific alkaloids can be found only in certain parts of the plants [34]. Chemically, alkaloids can be classified in: heterocyclic alkaloids and non-heterocyclic alkaloids. Heterocyclic alkaloids are those with nitrogen in the heterocycle, whilst non-hterocyclic alkloids contain the nitrogen atom in a side molecule [35, 34].

Moreover, alkaloids are biosynthesized from the secondary metabolism of plants. Their metabolic route startst from the same compounds as the other phytochemicals, alkaloid biosynthesis starts from acetyl-coenzyme A, shikimic acid, mevalonic acid, and 1-deoxyxylulose 5-phosphate [36].

5. Pharmacognosy of Mexican Medicinal Plants

Medicinal plants have not only had an impact on the discovery of drugs, but they also play an important role for human health care as a base for herbal preparations, also called phytopharmaceuticals or botanicals [37]. According to the World Health Organization [5], herbal medicines or remedies are determined as plant-derived materials with therapeutic benefits or other benefits for human health, which may contain raw or processed ingredients of one or more plants. In Mexico, herbal medicine is made from a standardized extract of leaf, root or bark and is presented in pharmaceutical form. This presentation has a standardized concentration of the active ingredients in each dose. In addition, the herbal medicine must also have supporting studies that have proven the pharmacological effects through experiments, and its clinical efficacy must be demonstrated in clinical studies and in medical practice [38]. The preparation of herbal medicines requires compliance with Good Manufacturing Practices, from their origin place to their use for medicinal purposes [39].

Since the medicinal plant is the raw material to obtain the extract for the herbal medicine, the first step to ensure a quality product is the proper identification of these plants, as well as the part of the plant where the active ingredients are located and the quality of the material. As part of the chemical-technological process, standardization and control of chemical extraction and the process itself are very important [38, 39, 1].

In Mexico, the Secretariat of Health is the organization that regulates these procedures. The general methods of analysis and technical specifications that both, plants and those derived from them, that are used in the elaboration of medicines and herbal remedies are published in the Herbal Pharmacopoeia of the United Mexican States, for 51 medicinal plants that are regulated [40].

The Herbal Mexican Pharmacopoeia, in addition to establishing the official guidelines related to the quality control of the plants used, contributes to the improvement of the quality of the products and their proper use, as well as to disseminate and promote the therapeutic potential of the flora of Mexico [41]. However, the plants reported in this document only correspond to the 1% of the flora of Mexico. So pharmacognosy studies are needed in order to improve the pharmacopoeia with more monographs of medicinal plants used in Mexico.

To ensure the quality of the raw material and avoid contaminants, identity tests of medicinal plants are performed, such as the description by means of botanical, organoleptic, macroscopic and microscopic characteristics, as well as the chromatographic profile, identification reactions and purity parameters, including moisture, ash and yield [42, 40, 5]. These studies are not only useful for the identification that guarantees the use of the appropriate raw material, but also for the control of quality standards, required by health legislation in Western countries [43, 40].

In addition to the identification of the raw material, the standardization and control of the chemical extraction and the elaboration process are very important [38], because the active principles present in the plants can only be found in some parts of the plant (leaves, stems or roots), its concentration and quality can vary depending on factors such as the weather, the time of the year, the type of soil, the humidity, in addition to the extraction procedures to obtain them [41]. That is why, to obtain a quality product, good practices are followed from the cultivation and collection of plants, as well as good practices during the procedure for the final obtaining of herbal medicines [5].

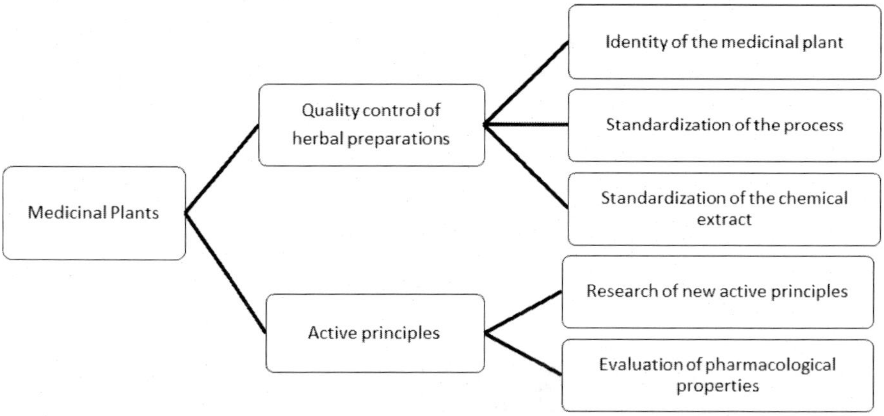

Figure 3. General scheme of the medicinal plants analysis to obtain the quality, safety and efficacy of a medicinal herbal preparation.

In addition to ensuring the quality of herbal preparations, evidence of safety and efficacy of these is required, for this, studies are carried out on the active ingredients that may have a therapeutic potential [44], where, new active principles are sought, as well as the evaluation of their pharmacological properties, either the isolated active principle, or as a mixture of these in a crude extract.

In Mexico about 15% of the total flora (4,000 species of plants) have medicinal attributes, that is, one in seven species has healing property. And it is estimated that the pharmacognostic validation of the active principles they contain has been carried out in approximately 5% of these species [41]. Therefore, there is still a diversity of medicinal plants, whose therapeutic action has not yet been confirmed, that is why, Mexican herbalism represents a viable resource to find new treatments against diseases that affect the world population and for which, there are no effective treatments yet.

CONCLUSION

Phytochemicals from medicinal plants have been associated with ethnopharmacological properties against noncommunicable diseases, which are the leading cause of death worldwide. Mexico is a country with a wide biodiversity and history of use of medicinal plants and herbal medicine. This knowledge is partially summarized in the Herbal Mexican Pharmacopoeia, which also aims a safe use of these medicinal plants. Medicinal plants used in Mexico have been studied for their potential as pharmaceutical agents against NCDs, and these studies are summarized in this book.

REFERENCES

[1] World Health Organization. 1999. *WHO monographs on selected medicinal plants.*

[2] Heinrich, M. 2015. "Ethnopharmacology: A Short History of a Multidisciplinary Field of Research." In: *Ethnopharmacology*, edited by Michael Heinrich and K. Jäger. NJ, USA: John Wiley & Sons Ltd.

[3] Tripathi, P., Gupta, G., and Chauhan, P. S. 2019. "Chapter 16 - Role of Phytomedicine in Diabetes and Cardiovascular Diseases." In: *New Look to Phytomedicine*, edited by Mohd Sajjad Ahmad Khan, Iqbal Ahmad and Debprasad Chattopadhyay, 409-433. Academic Press.

[4] Nagalingam, A. 2017. "Chapter 15 - Drug Delivery Aspects of Herbal Medicines." In: *Japanese Kampo Medicines for the Treatment of Common Diseases: Focus on Inflammation*, edited by Somasundaram Arumugam and Kenichi Watanabe, 143-164. Academic Press.

[5] World Health Organization. 2013. *WHO Traditional Medicine Strategy: 2014-2023*. Accessed September, 2019.

[6] Heinrich, M., Barnes, J., Gibbons, S., and Williamson, E. M. 2012. *Fundamentals of Pharmcognosy and Phytotherapy*. Edited by Michael Heinrich, Joanne Barnes, Simon Gibbons and Elizabeth M. Williamson. 2nd ed: Elsevier Ltd.

[7] World Health Organization. 2018. *Global Health Observatory Data*. accessed 27 de Enero. http://www.who.int/gho/en/.

[8] World Health Organization. 2018. *Noncommunicable diseases.* accessed September 19th. https://www.who.int/news-room/fact-sheets /detail/noncommunicable-diseases.

[9] FAO. 2005. *Trade in Medicinal Plants.* edited by Food and Agriculture Organization of the United Nations. Rome, Italy: Food and Agriculture Organization of the United Nations.

[10] Patra, J. K., Das, G., Lee, S., Kang, S. S., and Shin, H. S. (2018). Selected commercial plants: A review of extraction and isolation of bioactive compounds and their pharmacological market value. *Trends in Food Science & Technology* 82:89-109. doi: https://doi.org/10. 1016/j.tifs.2018.10.001.

[11] Alonso-Castro, A. J., Villarreal, M. L., Salazar-Olivo, L. A., Gomez-Sanchez, M., Dominguez, F., and Garcia-Carranca, A. (2011). Mexican medicinal plants used for cancer treatment: Pharmacological, phytochemical and ethnobotanical studies. *Journal of Ethnopharmacology* 133 (3):945-972. doi: https://doi.org/10.1016/ j.jep.2010.11.055.

[12] Dutta, S., Mahalanobish, S., Saha, S., Ghosh, S., and Sil, P. C. (2019). Natural products: An upcoming therapeutic approach to cancer. *Food and Chemical Toxicology* 128:240-255. doi: https://doi. org/10.1016/j.fct.2019.04.012.

[13] Heleno, S. A., Martins, A., Queiroz, M. J. R. P., and Ferreira, I. C. F. R. (2015). Bioactivity of phenolic acids: Metabolites versus parent compounds: A review. *Food Chemistry* 173:501-513. doi: 10.1016/j. foodchem.2014.10.057.

[14] Rein, M. J., Renouf, M., Cruz-Hernandez, C., Actis-Goretta, L., Thakkar, S. K., and da Silva Pinto, M. (2013). Bioavailability of bioactive food compounds: a challenging journey to bioefficacy. *British Journal of Clinical Pharmacology* 75 (3):588-602.

[15] Gómez-Castellanos, J. R. (2009). El ambiente regulatorio de los medicamentos herbolarios en México. Antecedentes, situación actual y perspectivas al año 2025 [The regulatory environment of herbal medicines in Mexico. Background, current situation and perspectives

to the year 2025]. *Boletín Latinoamericano y del Caribe de Plantas Medicinales y Aromáticas* 8 (1):33-40.

[16] Velázquez-Vázquez, G., Pérez-Armendáriz, B., Ortega-Martínez, L. D., and Nelly-Juarez, Z. (2019). Conocimiento etnobotánico sobre el uso de plantas medicinales en la Sierra Negra de Puebla, México [Ethnobotanical knowledge about the use of medicinal plants in the Sierra Negra de Puebla, Mexico]. *Boletín Latinoamericano y del Caribe de Plantas Medicinales y Aromáticas* 18 (3):265-276.

[17] Esquivel-García, R., Pérez-Calix, E., Ochoa-Zarzosa, A., and García-Pérez, M. E. (2018). Ethnomedicinal plants used for the treatment of dermatological affections on the Purépecha Plateau, Michoacán, Mexico. *J Acta botánica mexicana.*

[18] World Health Organization. 2018. *Cancer - Key facts.* accessed September 29. https://www.who.int/news-room/fact-sheets/detail/cancer.

[19] Alberti, K. G. M. M., Zimmet, P., and Shaw, J. (2005). The metabolic syndrom - a new worldwide definition. *The Lancet* 366 (9491):1059-1062. doi: 10.1016/S0140-6736(05)67402-8.

[20] International Diabetes Federation. 2006. *The IDF consensus worldwide definition of the metabolic syndrome.* Accessed October 2nd, 2019.

[21] Saklayen, M. G. J. C. H. R. (2018). *The Global Epidemic of the Metabolic Syndrome.* 20 (2):12. doi: 10.1007/s11906-018-0812-z.

[22] Croteau, R., Kutchan, T. M., and Lewis, N. G. 2015. "Natural Products (Secondary Metabolites)." In: *Biochemistry & Molecular Biology of Plants*, edited by B. Buchanan, W. Gruissem and R. Jones, 1250-1318. Rockville, Maryland: American Society of Plants.

[23] Oldfield, E., and Lin, F. Y. (2012). Terpene Biosynthesis: Modularity Rules. *Angewandte Chemie International Edition* 51 (5):1124-1137. doi: 10.1002/anie.201103110.

[24] Gonzalez-Burgos, E., and Gomez-Serranillos, M. P. (2012). Terpene Compounds in Nature: A Review of Their Potential Antioxidant Activity. *Current Medicinal Chemistry* 19 (31):5319-5341. doi: 10. 2174/092986712803833335.

[25] Russo, E. B. (2011). Taming THC: potential cannabis synergy and phytocannabinoid-terpenoid entourage effects. *British Journal of Pharmacology* 163 (7):1344-1364. doi: 10.1111/j.1476-5381.2011.01238.x.

[26] Embuscado, M. E. (2015). Spices and herbs: Natural sources of antioxidants - a mini review. *Journal of Functional Foods* 18:811-819. doi: https://doi.org/10.1016/j.jff.2015.03.005.

[27] Vázquez-Fresno, R., Rosana, A. R. R., Sajed, T., Onookome-Okome, T., Wishart, N. A., and Wishart, D. S. (2019). Herbs and Spices-Biomarkers of Intake Based on Human Intervention Studies – A Systematic Review. *Genes & Nutrition* 14 (1):18. doi: 10.1186/s12263-019-0636-8.

[28] Singh, B., and Sharma, R. A. (2015). Plant terpenes: defense responses, phylogenetic analysis, regulation and clinical applications. *3 Biotech* 5 (2):129-151. doi: 10.1007/s13205-014-0220-2.

[29] Tsao, R. (2010). Chemistry and Biochemistry of Dietary Polyphenols. 2 (12):1231-1246.

[30] Gutiérrez-Grijalva, E. P., Picos-Salas, M. A., Leyva-López, N., Criollo-Mendoza, M. S., Vazquez-Olivo, G., and Heredia, J. B. (2018). Flavonoids and phenolic acids from Oregano: Occurrence, biological activity and health benefits. *Plants* 7 (1). doi: 10.3390/plants7010002.

[31] Vermerris, W., and Nicholson, R. 2006. "Families of Phenolic Compounds and Means of Classification." In: *Phenolic Compound Biochemistry*, edited by Wilfred Vermerris and Ralph Nicholson, 1-34. The Netherlands: Springer.

[32] Mondal, A., Gandhi, A., Fimognari, C., Atanasov, A. G., and Bishayee, A. (2019). Alkaloids for cancer prevention and therapy: Current progress and future perspectives. *European Journal of Pharmacology* 858:172472. doi: https://doi.org/10.1016/j.ejphar.2019.172472.

[33] Talapatra, S. K., and Talapatra, B. 2015. "Alkaloids. General Introduction." In: *Chemistry of Plant Natural Products:*

Stereochemistry, Conformation, Synthesis, Biology, and Medicine, 717-724. Berlin, Heidelberg: Springer Berlin Heidelberg.

[34] Cushnie, T. P. T., Cushnie, B., and Lamb, A. J. (2014). Alkaloids: An overview of their antibacterial, antibiotic-enhancing and antivirulence activities. *International Journal of Antimicrobial Agents* 44 (5):377-386. doi: https://doi.org/10.1016/j.ijantimicag.2014.06.001.

[35] Evans, W. C., and Evans, D. 2009. "Chapter 26 - Alkaloids." In: *Trease and Evans' Pharmacognosy (Sixteenth Edition)*, edited by William Charles Evans and Daphne Evans, 353-415. W. B. Saunders.

[36] Aniszewiski, T. 2015. "Alkaloid chemistry." In: *Alkaloids - Chemistry, Biology, Ecology and Applications*, edited by Tadeusz Aniszewiski, 99-193. MA, USA: Elsevier.

[37] David, B., Wolfender, J. L., and Dias, D. A. (2015). The pharmaceutical industry and natural products: historical status and new trends. *Phytochemistry Reviews* 14 (2):299-315. doi: 10.1007/s11101-014-9367-z.

[38] Dominguez, F. (2015). La biotecnología y las plantas medicinales [Biotechnology and medicinal plants]. *Revista de la Academia Mexicana de Ciencias* 66 (1):1-8.

[39] Cuassolo, F., Ladio, A., and Ezcurra, C. (2010). Aspectos de la comercialización y control de calidad de las plantas medicinales más vendidas en una comunidad urbana del NO de la Patagonia Argentina [Aspects of the commercialization and quality control of the best-selling medicinal plants in an urban community of the NO of the Argentine Patagonia]. *Boletín Latinoamericano y del Caribe de Plantas Medicinales y Aromáticas* 9 (3):165-176.

[40] Secretaría de Salud. 2013. *Farmacopea Herbolaria de los Estados Unidos Mexicanos*. 2nd ed [*Pharmacopoeia Herbolaria of the United Mexican States*. 2nd ed]. D.F., México: Secretaría de Salud.

[41] Ocegueda, S., Moreno, E., and Koleff, P. (2005). Plantas utilizadas en la medicina tradicional y su identificación científica [Plants used in traditional medicine and their scientific identification]. CONABIO. *Biodiversitas* 62:12-15.

[42] Garrido, B. M., Polanco, S. C., Gaitán, I. C., Cáceres, A., and Paredes, M. E. (2013). Caracteres farmacognósticos para el control de calidad de *Petiveria alliacea*, *Lippia graveolens* y *Tagetes lucida* [Pharmacognostic characters for quality control of *Petiveria alliacea, Lippia graveolens* and *Tagetes lucida*]. *Dominguezia* 29 (2):25-39.

[43] Di Sapio, O., Bueno, M., Busilacchi, H., Quiroga, M., and Severin, C. (2012). Caracterización Morfoanatómica de Hoja, Tallo, Fruto y Semilla de Salvia hispanica L. (Lamiaceae) [Caracterización Morfoanatómica de Hoja, Tallo, Fruto y Semilla de Salvia hispanica L. (Lamiaceae)]. *Boletín Latinoamericano y del Caribe de Plantas Medicinales y Aromáticas* 11 (3):249-268.

[44] Chanda, S. (2014). Importance of pharmacognostic study of medicinal plants: An overview. *Journal of Pharmacognosy and Phytochemistry* 2 (5).

In: Phytochemicals from Mexican … ISBN: 978-1-53617-397-0
Editors: J. Basilio Heredia et al.

Chapter 2

REGULATION AND SAFETY OF MEDICINAL PLANTS

M. R. Tapia-Rodríguez, L. A. Ortega-Ramírez, R. Pacheco-Ordaz, G. A. González-Aguilar and J. F. Ayala-Zavala*

Centro de Investigación en Alimentación y Desarrollo, A. C. (CIAD, AC), Sonora, México

REGULATION OF MEDICINAL PLANTS IN MEXICO

Within the framework of this chapter is to analyze the actual regulations and safety of medicinal plants in Mexico, examine the scientific approach and legislative acts related to their safe use. The use of medicinal plants as therapeutic agents is less frequent compared to synthetic drugs; however, well-controlled clinical trials and numerous individual case reports have now confirmed that such effects exist [1, 2].

*Corresponding Author: J.F. Ayala-Zavala, Centro de Investigación en Alimentación y Desarrollo, A.C. (CIAD, AC). Carretera Gustavo Enrique Astiazarán Rosas, #46, 83304 Sonora, México. Email: jayala@ciad.mx.

The safety of herbal medicines is guaranteed in some cases by their prolonged use, but this is not always the case. In addition, modern phytotherapy includes the intake of new plants or new uses of old plants; however, unknown risks may occur [3-5].

The ethnopharmacological relevance of the use of medicinal plants in Mexico has been documented since prehispanic times. Nevertheless, the level of use of medicinal plants by health professionals in Mexico remains to be explored [1]. In Mexico, the regulation of products based on medicinal plants is not well established, and these are widely sold as food supplements [6]. Consequently, the lack of standard regulation and their sale as a food supplement has meant that the harmonization of their use has been not possible until recently. It is important to mention that in Mexico, there are various political positions concerning the practice of traditional medicine in different governmental instances.

On the one hand, traditional therapists are outside the conventional health system, even when some transnational laboratories use them to give authenticity and commercialize certain products of natural origin [7]. However, The General Health Law (article 79) only recognizes the healing practice of medical professionals [7]. In 1998, Mexico was in the process of preparing an official permit for the exercise of traditional therapists endorsed by the National Indigenous Institute, the Ministry of Health, and the Mexican Institute of Social Security [8]. Among the problems that the use of medicinal plants face is the lack of respect for indigenous knowledge, limited legal and financial support for the development of traditional medicine [8]. Some states have developed initiatives to reform their health law and regulate the practice of traditional medicine [8].

There is a document called the Mexican Pharmacopoeia in which there is a section (Mexican Herbal Pharmacopoeia) where the properties of different medicinal plants are listed [9]. The purpose of this section is to record existing medicinal species in Mexico that are sustainable from an ethnobotanical and taxonomic point of view. However, there are still not many studies (chemical and clinical) to validate therapeutic use in the framework of biomedicine [10]. For conventional medical treatments exists the idea of considering only those medicines, which are fully identified in their chemical composition and specific mode of action. On the other hand, those who lack studies of this type are mostly neglected even though they have an evident beneficial activity on specific conditions. Thus the pharmacopeia section collects species of medicinal plants potentially usable in the future, intending to encourage their study within the sector and thus filling in the empty spaces [11].

Mexican health authorities have shown interest in the research of medicinal plants, and the General Health Office published the Herbal Pharmacopeia in 2001 and 2013 [12]. Nevertheless, only a small group of medicinal plants has been studied for their pharmacological, phytochemical, and toxicological effects, as well as for their pharmacokinetics. In 2012, 52.6% of homes were covered by a public health service, which means that almost half of the population had no access to a public health institution. Therefore, this population needed to appeal to alternative medicine, including medicinal plants, which are prescribed by traditional healers [1, 13, 14]. The use and acceptance of medicinal plants by the general population in Mexico has been documented carried out a survey in the state of Sonora about the use and acceptance of herbal medicine by physicians. The Mexican Ministry of Health (SSA) issued a document called "Good pharmacopoeial practices" derived from the institutional participation of the Mexican Pharmacopoeia (FEUM) in 2012. Which mentioned that a pharmacopeia's core mission is to protect public health by creating and making available public standards to help ensure the quality of medicines. The pharmacopeia standards support the regulatory authorities in controlling the quality of pharmaceutical substances, their finished products, and related materials, to offer an independent judgment regarding quality and safety to guard the public health [15].

The European Union has implemented policies for the use of medicinal plants on the treatment of different pathologies. The document 2004/24/EC refers to traditional herbal medicines for human use. This directive should be the basis of regulation for the future use of herbal medicines in Europe [16]. Once implemented in the EU Member States, this new directive should be helpful to smooth marketing authorizations of herbal substances and preparations as traditional medicinal products [16, 6]. The mentioned directive establishes that a medicinal herbal product is a product that exclusively contains active ingredients, one or more herbal substances, one or more preparations, or one or more herbals substances in combination with one or more herbal preparations.

But nevertheless, the first real novelty in amending a previous Directive, the 2001/EC/83, by establishing that herbal medicines will need authorization like any medicine and they will be released only after the approval of the full application of the national regulatory authorities in European countries.

The objective of harmonization of the European market for medicinal plants, proposed by the new legislation, introduces important regulations. Without, however, beyond the market, the adoption of this new regulation in its set is an opportunity that will help to contribute to a safer use of herbal substances.

In countries like the United States, India, and the European Union, some authorities regulate the use of medicinal plants for therapeutic purposes. In the United States of America, the USFDA and Center for Drug Evaluation and Research are responsible for Botanical drug definition and Regulation of herbal product [17]. On the other side, the Center for Biologics Evaluation and Research is responsible for the marketing procedures of botanical drugs, regulation of Allergenic extracts and vaccines that contain botanical ingredients (Regulation: 201 (g) (1) (B), Federal Food, Drug, and Cosmetic Act; Dietary Supplement Health and Education Act of 1994; 21 CFR 10.20, 10.30, 312, 314, 321, 324, 330, 331-358) [18].

In India, the Department of Ayurveda, Yoga & Naturopathy, Unani, Siddha and Homoeopathy (AYUSH) is responsible for the Production and marketing of ASU drugs (Ayurveda, Siddha and Unani) and their good manufacturing practices (regulation: Drugs & Cosmetics Act, 1940 Drugs & Cosmetics Rules, 1945 and Schedule T, Drugs & Cosmetics Act, 1940) [19, 20].

In the European Union, the European Medicines Agency, through the Committee on Herbal Medicinal Products, is responsible for the

Establishment of the Committee on Herbal Medicinal Products (HMPC) and the regulation of herbal medicine. Under the regulation 2004/24/EC (Traditional Herbal Medicinal Products Directive) and Regulation (EC) No 726/2004; articles 16a to 16i of Directive 2001/83/EC [21, 22].

Within the Pharmacopoeia of the United Mexican States there are some laws that regulate the use of medicinal herbs, such as the Official Mexican Standard NOM-072-SSA1-2012 that regulates the labeling of medicines and herbal remedies [23]; the PROY-NOM-248-SSA1-2006 that regulates good manufacturing practices for establishments dedicated to the manufacture of herbal remedies [24]. Besides, there is an "agreement" listing the prohibited or permitted plants for teas, infusions, and edible vegetable oils [12]. However, only a few plants are mentioned in this document; therefore, it is necessary to include more information to have a broader knowledge of their possible applications. Although the herbal pharmacopeia has come to fill the gap regarding the official literature on medicinal plants in our country, it is clear that there is still much work to be done. The interest in these types of products must be reflected in a significant presence in the official code. In Europe, this is true; The European Pharmacopoeia has a significant number of monographs on medicinal plants, many of them from America, and far exceeds that of Mexico records [6]. In Mexico, official monographs do not reach 60 records, even when the official number of medicinal plants used in the country is 6000.

Toxicology of Mexican Medicinal Plants

The use of Mexican medicinal plants has increased as natural extracts or food supplements due to its bioactive properties and general acceptance. However, their toxicity and secondary effects after its consumption remain to be evaluated. Besides, many of these medicinal plants need scientific validation for its use in humans to treat different diseases or health care [25]. In most cases, the doses for the secondary toxic effect by intake of

Table 1. Toxicological evidence of medicinal plants from Mexico

Plant	Majority compound	Pharmaceutical use	Doses of toxicity	Reference
Agrimonia eupatoria L.	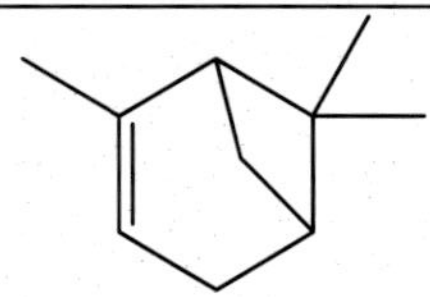 α-Pinene	Analgesic Anti-carcinogenic Antioxidant Immunomodulation	LD50: 1190mg/kg	Al-Snafi [26]
Allium sativum L.	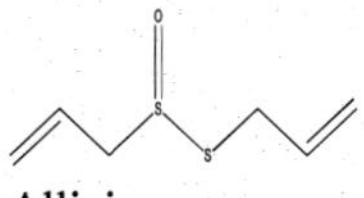 Allicin	Anti-carcinogenic Antioxidant Antimicrobial	LD50: 874mg/kg	Bender-Bojalil and Bárcenas-Pozos [27]
Cynara scolymus L	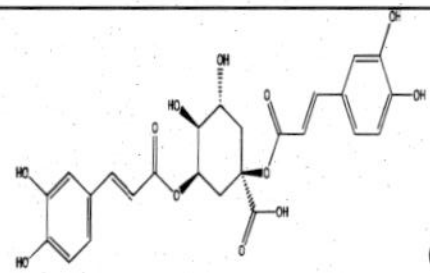 Cynarin	Antioxidant Antibacterial Prebiotic	LD50: 5000mg/kg	Wang et al. [28]
Aloe vera (L.) Burm	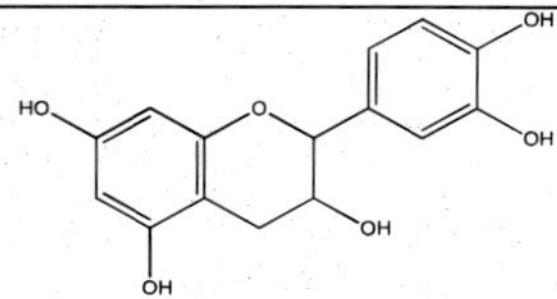 Catechin	Anti-inflammatory Anti-carcinogenic Antioxidant Therapeutic properties	LD50: 10000mg/kg.	López et al. [29]

Plant	Majority compound	Pharmaceutical use	Doses of toxicity	Reference
Vaccinium myrtillus L.	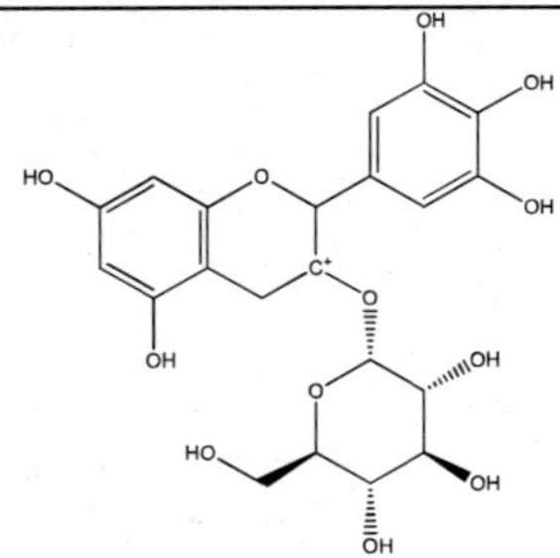 Myrtillin	Antioxidant Antimicrobial	LD50: 5000mg/kg	Lätti, Riihinen, and Kainulainen [30]
Calendula officinalis L.	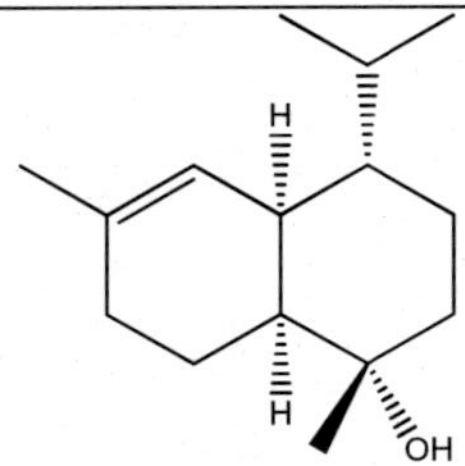 T-Muurolol	Antifungal Antioxidant Anti-carcinogenic	LD50: 2830mg/kg	Okoh et al. [31]
Cinnamomum verum	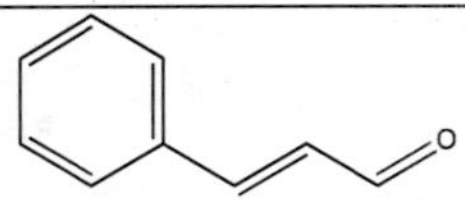 Cinnamaldehyde	Antioxidant Antimicrobial Cardio-protector	LD50: 1850mg/kg	Mazimba et al. [32]

Table 1. (Continued)

Plant	Majority compound	Pharmaceutical use	Doses of toxicity	Reference
Elettaria cardamomum L Maton	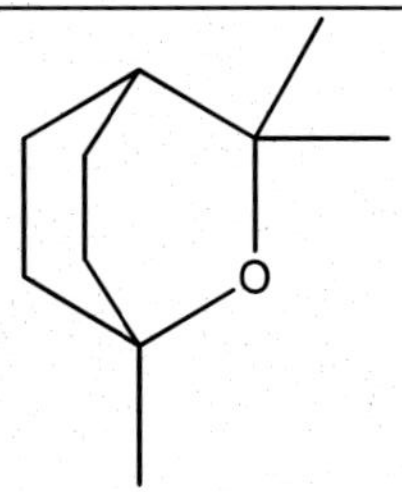 1,8-Cineole	Antioxidant Antifungal Blood pressure lowering properties	LD50: 2480mg/kg	Ashokkumar et al. [33]
Turnera diffusa	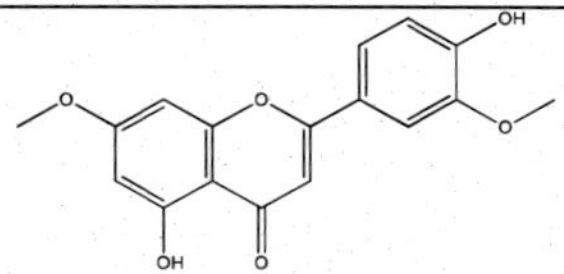 Velutin	Aphrodisiac potential Antioxidant	LD50: 4000mg/kg	Zhao et al. [34]
Echinacea angustifolia	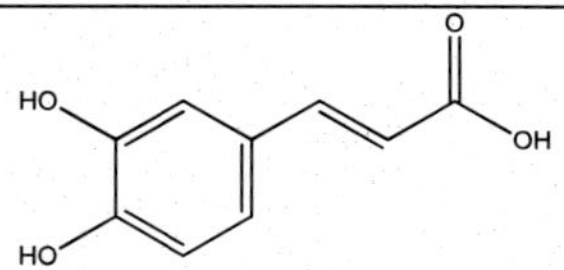 Caffeic acid	Anti-carcinogenic Antioxidant	LD50: 2980mg/kg	Barnes et al. [35]

Plant	Majority compound	Pharmaceutical use	Doses of toxicity	Reference
Melissa officinalis	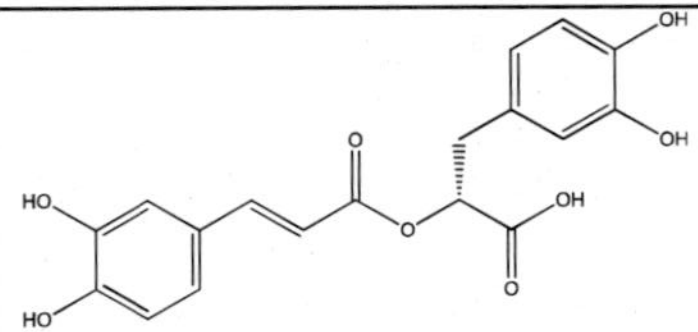 Rosmarinic acid	Anti-inflammatory immune-modulatory activities	LD50: 5000mg/kg	Mimica-Dukic et al. [36]
Ginkgo biloba	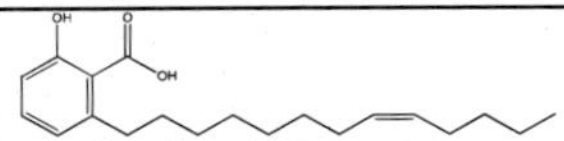 Ginkgolic acid	Anti-depressant Neuroprotective properties	LD50: 1000mg/kg	Polich and Gloria [37]
Mentha x piperita L.	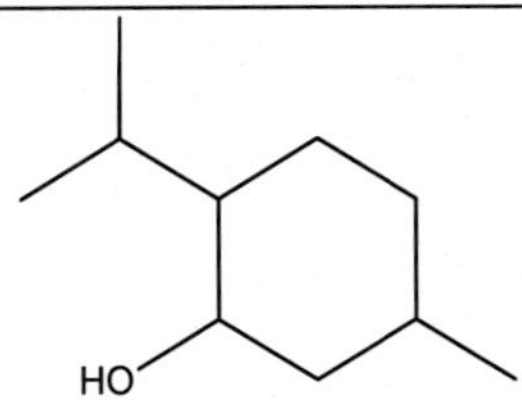 Menthol	Antioxidant Antimicrobial	LD50: 940mg/kg	Yang et al. [38]
Citrus aurantium L.	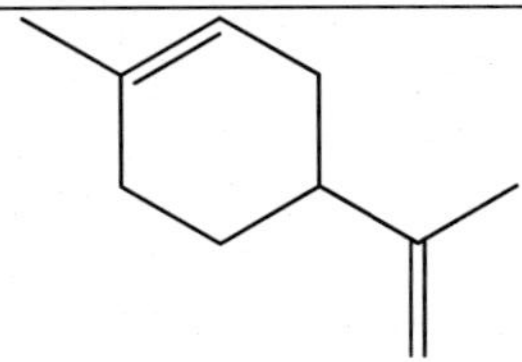 Limonene	Antioxidant *Antiviral* Anticancer properties	LD50: 4400mg/kg	Suntar et al. [39]

Table 1. (Continued)

Plant	Majority compound	Pharmaceutical use	Doses of toxicity	Reference
Origanum vulgare L.	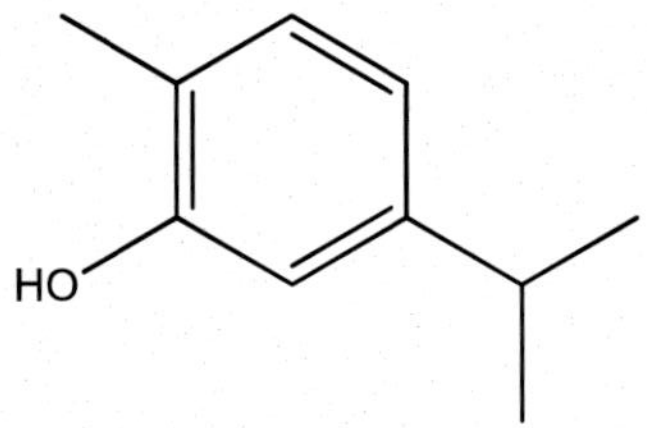 Carvacrol	Antimicrobial Anti-inflammatory, Antiseptic Analgesic and sedative effects	LD50: 810mg/kg	Fikry, Khalil, and Salama [40]

LD50: lethal dose required to kill 50% of the test population.

medicinal plants is not indicated before its consumption. Usually, the consumers of Mexican medicinal plants think that incrementing the doses of the nutraceutical product will increase the benefit of the treatment.

Regrettably, the risk of toxicity and several complications arise if the recommended doses are exceeded.

Table 1 lists some examples of the Mexican medicinal plants and signs of toxicity.

REFERENCES

[1] Alonso-Castro, A. J., Domínguez, F., Maldonado-Miranda, J. J., Castillo-Pérez, L. J., Carranza-Álvarez, C., Solano, E., Isiordia-Espinoza, M. A., del Carmen Juárez-Vázquez, M., Zapata-Morales, J. R., Argueta-Fuertes, M. A., Ruiz-Padilla, A. J., Solorio-Alvarado, C. R., Rangel-Velázquez, J. E., Ortiz-Andrade, R., González-Sánchez, I., Cruz-Jiménez, G. and Orozco-Castellanos, L. M. (2017). Use of medicinal plants by health professionals in Mexico. *Journal of Ethnopharmacology,* 198:81 - 86. doi: https://doi.org/10.1016/j.jep.2016.12.038.

[2] Calixto, J. B. (2000). Efficacy, safety, quality control, marketing and regulatory guidelines for herbal medicines (phytotherapeutic agents). *Brazilian Journal of Medical and Biological Research,* 33:179 - 189.

[3] Calapai, G. J. D. S. (2008). *European Legislation on Herbal Medicines,* 31 (5):428 - 431. doi: 10.2165/00002018-200831050-00009.

[4] Cos, P., Vlietinck, A. J., Berghe, D. V. and Maes, L. (2006). Anti-infective potential of natural products: How to develop a stronger in vitro 'proof-of-concept'. *Journal of Ethnopharmacology,* 106 (3):290 - 302. doi: https://doi.org/10.1016/j.jep.2006.04.003.

[5] Ernst, E. (1998). Harmless Herbs? A Review of the Recent Literature. *The American Journal of Medicine,* 104 (2):170 - 178. doi: https://doi.org/10.1016/S0002-9343(97)00397-5.

[6] Sahoo, N., Manchikanti, P. and Dey, S. (2010). Herbal drugs: Standards and regulation. *Fitoterapia,* 81 (6):462 - 471. doi: https://doi.org/10.1016/j.fitote.2010.02.001.

[7] Nigenda, G., Mora-Flores, G., Aldama-López, S. and Orozco-Núñez, E. (2001). La práctica de la medicina tradicional en América Latina y el Caribe: el dilema entre regulación y tolerancia [The practice of traditional medicine in Latin America and the Caribbean: a dilemma between regulation and tolerance]. *Salud Pública de México,* 43: 41 - 51.

[8] Schifter Aceves, L. (2014). Las Farmacopeas Mexicanas en la construcción de la identidad nacional [Mexican Pharmacopoeias in the Construction of National Identity]. *Revista Mexicana de Ciencias Farmacéuticas,* 45 (2):43 - 54.

[9] Quinlan, M. B. and Quinlan, R. J. (2007). Modernization and Medicinal Plant Knowledge in a Caribbean Horticultural Village. *Medical Anthropology Quarterly,* 21 (2):169 - 192. doi: 10.1525/maq.2007.21.2.169.

[10] Gutiérrez-Grijalva, E. P., Contreras-Angulo, L. A., Emus-Medina, A., Vázquez-Olivo, G. and Heredia, J. B. 2019. Anticancer Properties of Medicinal Plants Listed in the Herbal Pharmacopoeia of The United Mexican States. In *Medicinal Plants: Chemistry, Pharmacology, and Therapeutic Applications*, edited by Mallappa Kumara Swamy, Jayanta Kumar Patra and Gudepalya Renukaiah Rudramurthy. CRC Press.

[11] Carmona, R. H. and Liponis, M. 2017. *Integrative Preventive Medicine*: Oxford University Press.

[12] Secretaría de Salud. 2013. *Farmacopea Herbolaria de los Estados Unidos Mexicanos* [*Herbal Pharmacopoeia of the United Mexican States*]. 2nd ed. D.F., México: Secretaría de Salud.

[13] Ortega-Ramirez, L. A., Rodriguez-Garcia, I., Leyva, J. M., Cruz-Valenzuela, M. R., Silva-Espinoza, B. A., Gonzalez-Aguilar, G. A., Siddiqui, M. W. and Ayala-Zavala, J. F. (2014). Potential of Medicinal Plants as Antimicrobial and Antioxidant Agents in Food

Industry: A Hypothesis. *Journal of Food Science,* 79 (2):R129 - R137. doi: 10.1111/1750-3841.12341.

[14] Pérez-Valladares, C. X., Velázquez, A., Moreno-Calles, A. I., Mas, J.-F., Torres-García, I., Casas, A., Rangel-Landa, S., Blancas, J., Vallejo, M. and Téllez-Valdés, O. (2019). An expert knowledge approach for mapping vegetation cover based upon free access cartographic data: the Tehuacan-Cuicatlan Valley, Central Mexico. *Biodiversity and Conservation,* 28 (6):1361 - 1388. doi: 10.1007/s10531-019-01723-w.

[15] World Health Organization. 2016. Good Pharmacopoeial Practices. *WHO Technical Report Series*, No. 996, 2016, Annex 1.

[16] Knoess, W. and Wiesner, J. (2019). The Globalization of Traditional Medicines: Perspectives Related to the European Union Regulatory Environment. *Engineering,* 5 (1):22 - 31. doi: https://doi.org/10.1016/j.eng.2018.11.012.

[17] Ross, S. (2000). Functional Foods: the Food and Drug Administration perspective. *The American Journal of Clinical Nutrition,* 71 (6):1735S - 1738S. doi: 10.1093/ajcn/71.6.1735S %J The American Journal of Clinical Nutrition.

[18] Shankar, G., Shores, E., Wagner, C. and Mire-Sluis, A. (2006). Scientific and regulatory considerations on the immunogenicity of biologics. *Trends in Biotechnology,* 24 (6):274 - 280. doi: 10.1016/j.tibtech.2006.04.001.

[19] Bharati, K. and Ganguly, N. K. 2014. Status of Biotherapeutics Development in India. In *Biotechnology*, edited by K. K. Bhutani and J. N. Govil. Studium Press LLC, USA.

[20] Yoga, C. 2014. Ministry of Ayurveda, *Yoga and Naturopathy, Unani, Siddha and Homoeopathy*. New Delhi.

[21] Steinhoff, B. (2019). Review: Quality of herbal medicinal products: State of the art of purity assessment. *Phytomedicine,* 60:153003. doi: https://doi.org/10.1016/j.phymed.2019.153003.

[22] Vlietinck, A., Pieters, L. and Apers, S. (2009). Legal Requirements for the Quality of Herbal Substances and Herbal Preparations for the

Manufacturing of Herbal Medicinal Products in the European Union. *Planta Medica,* 75 (07):683 - 688. doi: 10.1055/s-0029-1185307.

[23] Norma Oficial Mexicana. (2012). *NORMA Oficial Mexicana NOM-072-SSA1-2012, Etiquetado de medicamentos y de remedios herbolarios* [*Official Mexican Standard NOM-072-SSA1-2012, Labeling of medicines and herbal remedies*].

[24] Norma Oficial Mexicana. (2012). *NORMA Oficial Mexicana NOM-248-SSA1-2011, Buenas Prácticas de fabricación para establecimientos dedicados a la fabricación de remedios herbolarios* [*Official Mexican Standard NOM-248-SSA1-2011, Good Manufacturing Practices for establishments dedicated to the manufacture of herbal remedies*].

[25] Alonso-Castro, A. J., Domínguez, F., Ruiz-Padilla, A. J., Campos-Xolalpa, N., Zapata-Morales, J. R., Carranza-Alvarez, C. and Maldonado-Miranda, J. J. (2017). Medicinal Plants from North and Central America and the Caribbean Considered Toxic for Humans: The Other Side of the Coin. *Evidence-Based Complementary and Alternative Medicine,* 2017:28. doi: 10.1155/2017/9439868.

[26] Al-Snafi, A. E. (2015). The pharmacological and therapeutic importance of Agrimonia eupatoria-A review. *Asian Journal of Pharmaceutical Science and Technology,* 5 (2):112 - 117.

[27] Bender-Bojalil, D. and Bárcenas-Pozos, M. (2013). El ajo y sus aplicaciones en la conservación de alimentos [Garlic and its applications in food preservation]. *Revista Temas Selectos Ingeniera en Alimentos,* 7 (1):25 - 36.

[28] Wang, M., Simon, J. E., Aviles, I. F., He, K., Zheng, Q.-Y. and Tadmor, Y. (2003). Analysis of antioxidative phenolic compounds in artichoke (*Cynara scolymus* L.). *Journal of agricultural and Food Chemistry,* 51 (3):601 - 608.

[29] López, A., de Tangil, M., Vega-Orellana, O., Ramírez, A. and Rico, M. (2013). Phenolic constituents, antioxidant and preliminary antimycoplasmic activities of leaf skin and flowers of *Aloe vera* (L.) Burm. f.(syn. *A. barbadensis* Mill.) from the Canary Islands (Spain). *Molecules,* 18 (5):4942 - 4954.

[30] Lätti, A. K., Riihinen, K. R. and Kainulainen, P. S. (2007). Analysis of anthocyanin variation in wild populations of bilberry (*Vaccinium myrtillus* L.) in Finland. *Journal of Agricultural and Food Chemistry,* 56 (1):190 - 196.

[31] Okoh, O., Sadimenko, A., Asekun, O. and Afolayan, A. (2008). The effects of drying on the chemical components of essential oils of Calendula officinalis L. *African Journal of Biotechnology,* 7 (10).

[32] Mazimba, O., Wale, K., Tebogo, E., Tebogo, E. and Kwape Shetonde, O. (2015). Cinnamomum verum: Ethylacetate and methanol extracts antioxidant and antimicrobial activity. *Journal of Medicinal Plants Studies,* 3 (3):28 - 32.

[33] Ashokkumar, K., Murugan, M., Dhanya, M., Raj, S. and Kamaraj, D. (2019). Phytochemical variations among four distinct varieties of Indian cardamom *Elettaria cardamomum* (L.) Maton. *Natural Product Research*, 1 - 4.

[34] Zhao, J., Pawar, R. S., Ali, Z. and Khan, I. A. (2007). Phytochemical investigation of Turnera diffusa. *Journal of natural products,* 70 (2):289 - 292.

[35] Barnes, J., Anderson, L. A., Gibbons, S. and Phillipson, J. D. (2005). *Echinacea* species (*Echinacea angustifolia* (DC.) Hell., *Echinacea pallida* (Nutt.) Nutt., *Echinacea purpurea* (L.) Moench): a review of their chemistry, pharmacology and clinical properties. *Journal of Pharmacy and Pharmacology,* 57 (8):929 - 954.

[36] Mimica-Dukic, N., Bozin, B., Sokovic, M. and Simin, N. (2004). Antimicrobial and antioxidant activities of *Melissa officinalis* L.(Lamiaceae) essential oil. *Journal of Agricultural and Food Chemistry,* 52 (9):2485 - 2489.

[37] Polich, J. and Gloria, R. (2001). Cognitive effects of a *Ginkgo biloba*/vinpocetine compound in normal adults: systematic assessment of perception, attention and memory. *Human Psychopharmacology: Clinical and Experimental,* 16 (5):409 - 416.

[38] Yang, S.-K., Yap, P. S.-X., Krishnan, T., Yusoff, K., Kok-Gan, C., Yap, W.-S., Kok-Song, L. and Swee-Hua, E. L. (2018). Mode of Action: Synergistic Interaction of Peppermint (*Mentha x piperita* L.

Carl) Essential Oil and Meropenem Against Plasmid-Mediated Resistant E. col Shun-Kai Yang, Polly Soo-Xi Yap, Thiba Krishnan, Khatijah Yusoff, Kok-Gan Chan, Wai-Sum Yap, Kok-Song Lai and Swee-Hua Erin Lim. *Records of Natural Products,* 12 (6):582.

[39] Suntar, I., Khan, H., Patel, S., Celano, R. and Rastrelli, L. (2018). An overview on *Citrus aurantium* L.: its functions as food ingredient and therapeutic agent. *Oxidative Medicine and Cellular Longevity,* 2018.

[40] Fikry, S., Khalil, N. and Salama, O. (2019). Chemical profiling, biostatic and biocidal dynamics of *Origanum vulgare* L. essential oil. *AMB Express,* 9 (1):41.

In: Phytochemicals from Mexican … ISBN: 978-1-53617-397-0
Editors: J. Basilio Heredia et al.

Chapter 3

OBESITY

Sara Avilés-Gaxiola,
Luis Angel Cabanillas-Bojórquez,
Gabriela Vázquez-Olivo,
Marilyn Shomara Criollo-Mendoza
and J. Basilio Heredia*
Centro de Investigación en Alimentación y Desarrollo,
A. C. Coordinación Culiacán, Culiacán, Sinaloa CP, México

ABSTRACT

Obesity is a multifactorial and complex disease. It is defined as a health condition in which people have an enormous increase in adipose tissue. Mexico is among the nations with the highest obesity rates among children and adults, being one of the countries spending the most on obesity and obesity-related illness with a total of 7800 million dollars spent each year. It is expected that the global cost of obesity will reach 1.2 trillion dollars a year in 2025. Sustained prevention efforts started around a decade ago, promoting the joint work of governments, international organizations, the private sector, and civil society. The

* Corresponding Author's Email: jbheredia@ciad.mx.

consumption of natural foods has been suggested, for example, the use of native plants. Until recently, the responsible medical constituents of Mexican plants, and their mechanism of action were unknown. In this chapter, the anti-obesity potential of Mexican plant phytochemicals is discussed.

Keywords: obesity, phytochemicals, Mexican plants, mechanism of action

INTRODUCTION

Hippocrates first introduced obesity in ancient Greek. He defined it as a condition affecting health, and later it was established that "sudden death is more common in those who are naturally fat than in the lean" [1]. For a long time, obesity was thought of as abnormal behavior. Nevertheless, from 1998, different worldwide organizations declared obesity as a disease, since it requires treatments and prevention efforts [2]. Nowadays, obesity is defined as a health condition in which people have an enormous increase of adipose tissue, mainly promoted by excessive energy intake and deficient energy expenditure [3]. Body Mass Index (BMI) is the parameter by which obesity is formally detected. BMI is defined as a ratio between the weight to the squared height (kg/m^2) of a subject [4]. According to the BMI, a person can be classified into one of five categories: underweight (<18.5 kg/m^2), normal weight (18.5-25.0 kg/m^2), overweight (25-29.9.0 kg/m^2), obese (30.0-40.0 kg/m^2) and extreme obese (>40.0 kg/m^2) [5].

Obesity etiology is very complex, and it includes different components such as environmental, psychological, genetic, physiologic, economic, and social, which implication differs in each particular case [6]. As for the physiological factors, obesity has been related to energy metabolism, which regulates body energy stores, is responsible for how much energy is consumed, and how much is expended. Several studies have shown that energy metabolism varies among people, making some more prone to develop obesity, which has also been associated with gender and ethnic groups. Obesity has also been linked to hypothyroidism, since thyroid hormones regulate basal metabolism, decrease thermogenesis, and are involved in glucose and lipid metabolism, as well as fat oxidation [7]. Also, a lot of scientific evidence has strongly suggested a genetic component to be implicated in obese development. From 70 to 90% of monozygotic obese twins share this health problem, while the adoptive people´s body max index is related to biological and not to adoptive parents eating patterns [1]. The genetic component of obesity is associated with to the leptin/melanocortin pathway, which is critical in the whole body-energy homeostasis regulation, being involved in increased appetite and diminished satiety [8].

When it comes to determining when and how this disease begins, the psychological factor becomes determinant. Disorder eating behaviors related to an attempt to overcome negative feelings or experiences are in around 10% of the cases, which means that people eat in response to emotions, lacking from adaptive emotion regulation strategies and skills [9]. One of the factors most recently associated with obesity is gut bacteria. It was introduced around a decade ago when it was found that two phyla (*Firmicutes and Bacteroidetes*) differ between lean and obese population. These bacteria interfere in energy metabolism, and carbohydrate and lipid digestion, as well as, influence the expression of host genes that regulate both energy expenditure and storage [10]. Also, it has been reported that the entire embryo-fetal and perinatal periods are critical for obesity predisposition since maternal obesity brings complications for both the mother and the fetuses [11]. While the mother may suffer from preeclampsia, gestational diabetes mellitus, and non-planned cesarean delivery, the babies become more prone to develop diseases later in life, including hypertension, heart disease, non-insulin dependent diabetes and obesity itself. As for the latter, the mother excess weight impacts the uterus environment and therefore the fetal development, which promotes for babies born to obese women to accumulate adiposity easier than those coming from normal weight woman [12]. In recent years, the epigenetic component of obesity has been studied and it has been confirmed that particular epigenetic modifications are involved in the development of obesity, where obese people have a pattern of epigenetic marks differing from those of nonobese people, which in turn are inherited to their offspring [13].

Obesity increase over time has been such that since 1975, the cases have tripled, and today more people die due to problems associated with overweight, in comparison to malnutrition. The most recent worldwide data, provided by the Organization for Economic Co-operation and Development (2017), show that 41 million children under the age of 5340 million children and adolescents aged 5-19 and 13% of adults over 19 are obese. The countries with the highest obesity rates among adults, according to the population, are the United States of America (USA) (38.2%), Mexico (32.4%), New Zealand (30.7%), Hungary (27.9%) and Australia (26.9%). As for childhood obesity, the countries with the highest rates are Nauru (36.3%), USA (22.7%), Chile and New Zealand (18.7%), Saudi Arabia (18.5%), and Mexico (17.5%). This pandemic has promoted an increase in medical spending, mainly due to prescription drugs and hospitalization. Nowadays, obesity requires approximately 21% of developed countries' healthcare expenditure. For instance, in the USA, the cost of healthcare is of 12,000 US dollars for a healthy weight 10-year-old and of 19,630 US dollars for an obese one. These costs directly increment as it does the patient's lifetime [14]. The countries spending the more in obesity and obesity-related illness are US, China, Germany, Russia, France, Brazil, UK, Canada, Italy, and Mexico, and it is estimated that the global cost of obesity will reach 1.2 trillion US dollars a year in 2025 [15, 16].

In Mexico, obesity has a profound impact on the health care system, with 7,800 US million dollars spent each year. The main factors increasing obesity incidence are high consumption rates of cookies, deep-fried foods, and sugary drinks. As for the latter, Mexico is the primary consumer of carbonated soft drinks in the world, for which it is considered a country with a high obesogenic environment [17]. Sustained prevention efforts started around a decade ago. In this regard, there has been promoted the joint work of governments, international organizations, private sector, and civil society aiming to develop a coordinated approach for obesity monitoring, prevention, and control [18]. Strategies implemented in public health for obesity prevention focus mainly on targeting food habits, physical activity, and the broader socioeconomic environment. The critical targets for obesity prevention, especially in the scholar system, have been to promote healthy choices as the easier ones and to encourage physical activity to reduce the sedentary lifestyle. As for this, the main objective has been to build recreation facilities in communities, more extensive and more school grounds, as well as promote the use of stairs instead of elevators. Another line of action is related to processed foods, where the primary purpose has been to reformulate high demanded food products and to restrict unhealthy food marketing, as well as to raise their prices [19]. Nevertheless, the strategies attempting to change diet and physical activity patterns have not succeeded yet, since only have resulted in around 8% of weight loss, which is gained within one year. Anti-obesity drugs (Table 1) are sometimes recommended as adjuvant therapy for obese or overweight patients, although they only reduce a maximum of 10% body weight [20]. These pills act mainly by four modes of action, which are appetite suppression, inhibition of fat absorption, increased energy consumption, and thermogenesis. Nevertheless, the typical pattern is for Food and Drug Administration to approve them and subsequently remove them from the market due to reported side effects, such as hyperthyroidism, addiction, tachycardia, hypertension, hemorrhagic stroke, among others [21]. As a result, natural products derived mainly from plants, have emerged as an option to counteract obesity. The main constituents of these products are phytochemicals, known for their high potential both in preventing and treating diseases [22]. Phytochemicals are bioactive compounds not classified as essential nutrients. They are usually low-molecular-weight particles that regulate plant growth and development. They are also responsible for plant color, essence, and odor. The main phytochemicals are polyphenols, glucosinolates (indoles and isothiocyanates), phytoestrogens (lignans and isoflavones), and phytates [23]. In Mexico, the indigenous cultures have exploited the very diverse flora, having an extensive pharmacopeia of medicinal plants. This knowledge has been passed through generations, perpetuating their use and making it common to relate an illness with some plant infusion [24]. Nevertheless, until recently, the responsible medical constituents of Mexican plants, and their mechanism of action were unknown. In this chapter, the anti-obesity potential of Mexican plant phytochemicals, is discussed.

Table 1. Pharmacological treatment of obesity

Drug	Mechanism of action	Status	References
Orlistat	Reversible active-site inhibitor of gastrointestinal lipases.	Approved	[25]
Phentermine/ topiramate	Stimulates the hypothalamic release of norepinephrine and reduces hunger sensation. It also causes the release of epinephrine and adrenaline outside of the brain, resulting in the breakdown of stored fat.	Approved	[26]
Lorcaserin	Targets the 5HT2C receptor, altering body weight by regulating satiety.	Approved	[27]
Bupropion/ Naltrexone	Norepinephrine–dopamine reuptake inhibitor, which commonly induces modest weight loss.	Approved	[28]
Liraglutide	Glucagon-like peptide-1 receptor agonist and a neuroprotective agent.	Approved	[29]
Metformin	Reduces blood glucose levels by decreasing the production of glucose in the liver, decreasing intestinal absorption, and increasing insulin sensitivity.	Non-approved	[30]
Pramlintide	Regulates blood glucose by slowing gastric emptying promoting satiety via hypothalamic receptors (different receptors than for GLP-1) and inhibits inappropriate secretion of glucagon.	Non-approved	[31]
Topiramate	Antagonizes kainate/AMPA subtype of the glutamate receptors.	Non-approved	[32]
Zonisamide	Blocks sodium and calcium channel, stabilizing neuronal membranes and suppressing neuronal hyper-synchronization. It also facilitates both dopaminergic and serotonergic neurotransmission.	Non-approved	[33]

MEXICAN MEDICINAL PLANTS WITH ANTI-OBESITY POTENTIAL

Phytochemical Constituents

Mexican flora diversity is reflected in the vast knowledge of herbal remedies for different disorders [34]. The anti-obesity potential of Mexican plants (Table 2) was first tested over enzymes associated with nutrient absorption such as α-amylase, α-glycosidase, trypsin, and mainly lipase. Later, over cellular in vitro models and recently, over both obese animals and humans, finding that some of these plants extracts and isolated compounds show potential on reducing obesity parameters, such as total cholesterol, high-density lipoprotein (HDL), low-density lipoprotein (LDL), triglycerides, food intake, fat content, etc. [35].

In Vitro Studies

Different Mexican plant extracts have been obtained and tested over in vitro models, including enzymes implicated in the digestion process and adipocytes cell lines.

In Vitro Enzymatic Studies

The biological activity observed in enzymatic studies tends to validate some of the traditional claims of *Aloe vera* (aloe) as a functional food. It has been found that 1.23 μmol/min*g of *Aloe vera* dry aqueous extract, inhibits α-glycosidase in 35.46 $\mu M\ min^{-1}\ g^{-1}$, showing potential as adjuvant therapy in the treatment of obesity [89]. Also, its lipase inhibitory activity was compared to that of Orlistat, where it was found that aloe gel inhibited pancreatic lipase in 85.56%, while Orlistat inhibition percentage was of 70.58% [90]. Lipase activity is also inhibited by methanolic extracts of

Table 2. Description of Mexican plants and its common uses

Plant	Common name	Producer regions	Popular uses	Anti-obesity constituents	References
Agrimonia eupatoria L.	Agramon, agrimony, sticklewort and church steeples.	Mexico, USA, Canada, and Western Europe.	Its tea is used as a dietary supplement and as decongestant of the digestive system.	Not reported.	[36]
Allium sativum L.	Garlic	China, India, Bangladesh, Egypt, and South Korea.	Entire cloves are consumed as a dietary supplement for different purposes, including high blood cholesterol and pressure levels stabilization.	Diallyl disulfide and s-**allyl cysteine sulfoxide.**	[37, 38]
Cynara scolymus L.	Artichoke	Italy, Egypt, Spain, Argentina, and Peru.	Flower tea is used for its potential in lowering cholesterol levels.	Fiber, chlorogenic acid, cynarin, 1,5-di-o-caffeoylquinic, luteolin, apigenin, and phytohemagglutinin.	[39, 40]
Aloe vera L. Burm	Burn plant and aloe	Honduras, Mexico, Brazil, Dominican Republic, and India.	Liquefied pulp and the leafy parts are consumed for its glucose-lowering, constipation and laxative effects, as well as ulcerative colitis and inflammatory bowel disease dissipation.	Aloesin.	[41]
Vaccinium myrtillus L.	Bilberry, wimberry, whinberry, blueberry, and whortleberry.	USA, Canada, Peru, Mexico, and Spain.	Fruit and leaves infusions are consumed for its hypoglycemic effect.	Anthocyanins, apigenin-7-glucuronide, vitexin-2″-rhamnoside, and isoorientin.	[42, 43]
Calendula officinalis L.	Marigold, ruddles, common marigold, and Scotch marigold.	-	As balms and creams to alleviate inflammation.	Hesperidin and epigallocatechin gallate.	[44]

Plant	Common name	Producer regions	Popular uses	Anti-obesity constituents	References
Cinnamomum verum J. Presi	True cinnamon and Ceylon cinnamon.	Sri Lanka, Indonesia, China, and Viet Nam.	Cinnamon sticks infusion is consumed to speed metabolism and to reduce abdominal body fat.	Alkaloids, flavonoids, tannins, phenols, saponins, and cinnamaldehyde.	[45]
Aesculus hippocastanum	Horse chestnut, false chestnut, crazy chestnut, chestnut piloga, and chestnut edge.	Northern Greece and Albania.	Leaves and seeds infusions are used as anti-inflammatories.	Barringtogenol C, barrigenol R1 and escina.	[46, 47]
Turnera diffusa Wild	Damiana, grass of the shepherdess, deer grass, break shirt male, and Mexican tea.	South America, USA, and Mexico.	Leaf infusion and essential oil are used as a mild laxative.	Acacetin, apigenin, gonzalitosin, laricitrin, vitexin, luteolin-8 propenoic acid, orientin, orientin-300- ketone, pinocembrin, and syringetin.	[48]
Echinacea pallida	Pale purple coneflower and narrow-leaved echinacea	Mexico, USA, and Canada.	All its anatomical parts are used to make teas and capsules with anti-inflammatory properties.	Caffeic acid derivates.	[49]
Ginkgo biloba	Tree of the forty shields and walnut of Japan.	China, Korea, USA, France, and Spain.	Its leaves are used to improve digestion.	Bilobalide, ginkgolides A, B, C, flavonol glycosides, and ginkgolic acids.	[50-52]
Psidium guajava L.	Guava.	India, China, Thailand, Pakistan, and Mexico.	Leaf infusion and pulp are both consumed to treat diarrhea, gastroenteritis, stomachache, and as an anti-inflammatory agent.	Quercetin, guajaverin, gallic acid, syringic acid, o-coumaric acid, stilbenoids such as resveratrol and tannins such as epigallocatechin.	[53-56]

Table 2. (Continued)

Plant	Common name	Producer regions	Popular uses	Anti-obesity constituents	References
Hypericum perforatum L.	St. John's Wort, Klamath weed, Tripton weed, Enola weed, and goat weed.	Europe, Western Asia, North Africa, North America, and Australia.	Capsules and infusions are used to treat gastritis and water retention.	Hypericin, hyperforin, naringenin, quercetin glucosides, and hydroxycinnamic acids.	[57-61]
Hibiscus sabdariffa L.	Roselle and rosella.	China, India, Sudan and Mexico.	Red calyces' infusion is used to treat hypertension, gastrointestinal disorders, and hypercholesterolemia.	Ellagic acid, rosmarinic acid, cyaniding, pelargonidin, malvidin, peonidin cyaniding-3-sambubioside, delphinidin-3-sambubioside, chlorogenic acid, tetra-O-methyljeediflavanone, quercetin-3-sambubioside gallic acid, catechin, epicatechin, and delphinidin 3-O-sambubioside.	[62-66]
Zingiber officinale Roscoe	Ginger.	India, China, Nepal, Indonesia, and Nigeria.	Rhizome infusion is used to improve digestion and to treat nausea and vomit in a pregnant woman.	Gingerols and shogaols.	[67, 68]
Humulus lupulus L.	Hop.	USA, Ethiopia, Germany, China, and Czech Republic.	Capsules and infusion are used to treat gastric disorders.	Xanthohumol, iso-alpha acids, procyanidins, astragalin and isoquercitrin.	[69-72]

Plant	Common name	Producer regions	Popular uses	Anti-obesity constituents	References
Melissa officinalis	Mint, lemongrass, lemon balm, lemon leaf and Melissa.	Southeast Europe, Asia Minor, South Africa, Mexico, and Cuba	Its leaves are used as infusions or as condiments to relieve poor digestion, feeling of emptiness, gastrointestinal spasms, diarrhea, gastritis, and abdominal swelling	Rosmarinic acid, apigenin, quercetin, and luteolin.	[73, 74]
Citrus aurantium L.	Seville orange, sour orange, and bitter orange.	Brazil, China, USA, and Mexico.	Citrus peel is used as a remedy for indigestion as well as an anti-inflammatory.	Nobiletin, tangeretin, hesperidin, naringin and p-synephrine.	[75-77]
Rheum palmatum L	Turkish rhubarb	Turkey, China, Tibet, Mongolia, and Siberia.	Root and leaf infusions are used to treat constipation, gastric bleeding, and ulcers.	Anthraquinone (chrysophanol, physcion, emodin and aloe-emodin), dianthrones derivatives (sennosides A, B, C, and D), catechin, emodin, and ursolic acid.	[78-82]
Rosmarinus officinalis L.	Rosemary and alecrim	South Africa, Mediterranean countries, Northern Africa, England, and Mexico.	Leaves are used to treat inflammation problems.	Carnosol and o-diphenolic diterpene carnosic acid.	[83-85]
Trigonella foenum-graecum L.	Fenugreek	Pakistan, India, China, USA, and Mexico.	Seeds are used to treat inflammation, stomach irritation, constipation, and diarrhea.	Naringenin, diosgenin, and alkaloids.	[86-88]

Hibiscus sabdariffa flower with an IC_{50} of 35.8 μg/ml [91]. While α-amylase is also inhibited by ethanolic extracts of *Borago officinalis* L. with an IC_{50} of 31.61 μg/ml (Marrelli et al. [92]. Another important target for obesity treatment is tyrosine phosphatase 1B, which is inhibited by *Hypericum perforatum* (*H. perforatum*) ethanolic extract [93].

The inhibitory effects of aloe, *B. officinalis*, *H. sabdariffa,* and *H. perforatum* were not attributed to any particular compound. Nevertheless, the enzymatic inhibitory effect of other Mexican plants has been attributed to specific constituents. For example, epigallocatechin gallate and resveratrol found in *Psidium guajava* L. (guava) leaves, are responsible for its lipase inhibitory effects [54]. In another study, guajaverin, a flavonol also found in guava leaves, showed an inhibitory effect over α-glycosidase activity [94]. Carnosic acid and carnosol also have anti-obesogenic potential. Patented *Rosmarinus officinalis* (rosemary) rich extracts inhibit pancreatic lipase activity with an IC_{50} of 12 and 4.4 mg/ml, respectively [84, 95-97]. Similarly, Zheng et al. [82] found that catechin, emodin, and ursolic acid, highly found in *Rheum palmatum* (*R. palmatum*) methanolic extract, inhibited lipase by 50% at a concentration of 200 μg/ml. A higher lipase inhibition (70.11%) was achieved by *Trigonella foenum-graecum* L. (*T. foenum-graecum*), where Fernando et al. [98], reported that water extract was more effective than methanolic, ethanolic and hexane fractions. This effect was attributed to alkaloids and saponins [99, 100].

Thanks to in vitro studies, there is already a commercial anti-obesity dietary supplement called Boldocynara® which is based in a mix of medicinal plants, being *Cynara scolymus* (artichoke), the main component (40%). The first test to determine Boldocynara® efficacy (100 μg/ml) was over enzymatic models. Its effects were compared with those of each plant extract alone. Both artichoke and Boldocynara® inhibited α-glucosidase and pancreatic lipase in the same proportion (20 and 18%, respectively). Nevertheless, angiotensin-converting enzyme inhibitory activity was 3-fold higher when using artichoke alone (31%) [101]. By this investigation, it is concluded that Boldocynara® is an effective drug for metabolic syndrome-related enzymes mild modulation. Nevertheless, for a particular enzyme modulation, better effects are found when consuming a single plant, like artichoke.

Cell Lines Studies

Also, the effects of Mexican plants have been evaluated over obesity complications. For example, nonalcoholic fatty liver disease, which is an inflammation caused by an accumulation of fat in the liver, and it is a very common outcome of obesity. This condition may be induced in HepG2 cells using oleic acid, and it has become a model to study the potential of medicinal plant extracts on treating obesity complications. One of these plants is *Agrimonia eupatoria* L. (*A. eupatoria*), from which different extracts have been obtained (ethanol, n-hexane, dichloromethane, ethyl acetate, n-butanol, and H_2O). The highest lipid accumulation inhibitory effect was the one obtained with the ethanolic extract. This extract suppressed triglyceride accumulation and inhibited the expression of peroxisome proliferator-activated receptor γ (*PPAR-γ*) in a dose-dependent manner. PPAR-γ modulates lipid and glucose metabolism and the overall energy homeostasis, mitigating obesity, and inflammation [102].

Given the above, *A. eupatoria* ethanolic extract was also tested over 3T3-L1 preadipocytes in a proportion of 2.5%. This treatment **r**esulted in a significant decrease in mRNA levels of Monocyte Chemoattractant Protein-1 (***MCP-1***), which attracts proinflammatories macrophages to the adipose tissue. These macrophages produce and secrete cytokines increasing, even more, the proinflammatory preadipocyte state, and it has become a target when treating obesity. Also, ***A.*** *eupatoria ethanolic* extract significant increased interleukin 6 (***IL-6***) transcription levels. Human IL-6 is a single polypeptide chain of 185 amino acids that forms a bundle of four α-helices. It is regarded as the major inflammatory mediator, improving energy metabolism and glucose homeostasis [103]. Preadipocytes 3T3-L1 have also been used as a model to determine the effect of *Cinnamomum verum* J. Presi (cinnamon), both supercritical fluids and methanol extracts. Both extracts reduced the mRNA expression of the transcription factor *PPAR-γ*, the sterol regulatory-element-binding protein-1c (*SREBP1c*), and the CCAAT enhancer-binding-protein α (*C/EBPα*) in a concentration-dependent manner, decreasing its differentiation rates. Also, the extract obtained by supercritical fluid down-regulated acyl-CoA synthetase-1 (*ASC1*), fatty acid synthase (*FASN*), fatty acid transport-1 (*FATP1*), fatty acid-binding protein 4 (*FABP4*), and perilipin [104].

H. perforatum flower tops Indena® commercial extract has also shown in vivo anti-obesity effect. It diminished inflammation to a low-grade by promoting the secretion of anti-inflammatory cytokines [105]. Further, *H. perforatum* inhibits adipogenesis in mature fat cells [106]. Currently, there is a product called OB-X®, based on Mexican plants. It is a mixture of *Melissa officinalis* (*M. officinalis*), *blackberry,* and artemisia extracts. OB-X® was found to suppress adipogenesis by inhibiting 3T3-LI preadipocytes into adipocytes differentiation, which was suggested to be promoted by a change in lipogenesis genes expression [74]. Hong, Kim, and Yoon [107]. In a latter study, it was found that the *M. officinalis* balm ALS-L 1023, already in clinical trials, inhibited lipid accumulation and the formation of blood tubes also in 3T3-L1 adipocytes. These cells exhibited a reduction in angiogenic factors expressions, such as the vascular endothelial growth factor A (*VEGF-A)* and the fibroblast growth factor 2 (*FGF-2*). Also, an increase in angiogenic inhibitors levels, such as thrombospondins 1 (*TSP-1)* and TIMO metallopeptidase inhibitor 1 and 2 (*TIMP-1* and *TIMP-2*), results that promoted a subsequent *in vivo* study [108].

Besides extracts, single compounds have also been tested. For example, quercetin-3-O-(6′′-feruloyl)-β-D-galactopyranoside, purified from guava leaves, showed a suppressive effect on adipogenesis in 3T3-L1 preadipocytes. This effect was attributed to the inhibition of PPAR-γ and *C/EBPα* expression, via independent pathways of Wnt-β-catenin [109]. *PPAR-γ* and *C/EBPα* expression was also downregulated in 3T3-L1 preadipocytes by both *Zingiber officinale* (*Z. officinale)* and *Humulus lupulus* L. (*H. lupulus*). As for *Z. officinale*, the effect was attributed only to a galanolactone, while xanthohumol was the compound responsible for *H. lupulus* effect. Xanthohumol also inhibited the 3T3-L1 preadipocytes fatty acid-binding protein (aP2) [110, 111]. Also, Alliin (S-allyl cysteine sulfoxide), the main active component of ***Allium sativum*** L. (garlic), was isolated and applied in a dose of 100 µmol/l during 24 h over lipopolysaccharide-(LPS)-stimulated 3T3-L1 adipocytes. It was found that the expression of proinflammatory genes was prevented, mainly extracellular signal-regulated kinases (***ERK1*** and ***ERK2***), ***IL-6***, ***MCP-1,*** and early growth response protein (***ERG-1***), inhibiting differentiation [112]. Another single compound tested over 3T3-L1 adipocytes, is bilobalide, from ***Ginko biloba (G. biloba***). It was found that at a dose of 50 µM, it protected the adipocytes from hypoxia, a key factor responsible for adipose tissue dysfunction in obesity patients. By this, oxidative stress, inflammation, and mitochondria damage were prevented [51].

In the case of *H. sabdariffa,* the adipogenic inhibitory effect was more significant when proving the whole aqueous extract, than when testing isolated anthocyanins alone. Besides, the removal of polysaccharides (due to the extraction process) showed better inhibition of the triglyceride accumulation, indicating that polyphenols may be the main contributors to *H. sabdariffa* bioactivity [65].

The investigations here summarized, laid the foundation to claim that the consumption of different Mexican plant extracts, or their specific compounds, can modulate the metabolic function of adipose tissue, attenuating the inflammatory state and therefore improving patient´s quality life. Nevertheless, further in vivo studies were needed to have a better understanding of the Mexican plant´s mechanism of action in the prevention/treatment of obesity.

In Vivo Studies

Agrimonia eupatoria L.

In Mexico, *A. eupatoria* aerial parts tea is commonly consumed, and it is associated with regulation of digestive disorders and weight loss. The first in vivo study aiming to determine *A. eupatoria* anti-obesity potential was carried out in 1998 [113]. Here, it was found that its aqueous extract, either incorporated into solid food or water, promoted weight loss by 22% in streptozotocin-diabetic mice. Nevertheless, the mechanism of action was not elucidated. Also, *A. eupatoria* extracts prevented liver fat accumulation and adipose tissue in metabolic disturbed rats, as well as normalized their serum lipid profile [114]. The last study was carried out in obese people, where it was found that consuming 200 ml of *A. eupatoria* aerial part tea twice a day for 30 days, reduced LDL cholesterol by 7.99%. Also, it regulated energy balance and inhibited hungrily, which was associated with a 58% reduction of leptin serum concentration. The *A. eupatoria* weight loss potential was associated with its high tannin content, which reduces protein absorption [115]

Allium sativum

In terms of obesity, *Allium sativum* is the most studied Mexican plant, mainly for their historical medicinal properties, which includes gastrointestinal disorders relief and weight loss. The most studied anatomical parts are their bulbs, from where different extracts have been obtained. Methanolic extracts, for example, decreased bodyweight, relative masses of fat tissues and the liver, serum triglycerides levels, and increased fecal lipid content in obese Wistar rats. Upregulation of AMP-activated protein kinase *(AMPK)*, forkhead box protein O1 (*FOXO1*), sirtuin 1 (*SIRT1*), adipose triglyceride lipase (*ATGL*), hormone-sensitive lipase (*HSL*), periplin 1 (*PLIN1*), 1-aminocyclopropane-1-carboxylate oxidase (*ACO*), carnitine palmitoyltransferase I (*CPT1*), thermogenin (*UCP*) and adiponectin *(ADIPOQ)* as well as downregulation of cluster of differentiation 36 (*CD36*), plasminogen activation inhibitor-1 (*PAI-1*), resistin and tumor necrosis factor *α (TNF-α*) in adipose tissue, were detected [116]. The same extract was tested in vivo in combination with turmeric ethanolic extract in a dose of 2.4 grams daily for 12 weeks over obese patients. Total cholesterol was lowered as well as LDL cholesterol, triglycerides, and body max index in 6.7. 8.85, 8.87, and 4.3%, respectively. The study was extended to 3 months, where it was proved that the combination of the extracts was safe and well-tolerated [117]. Not only extracts, but also the hole garlic anti-obesity potential has been evaluated in vivo. Two capsules of 250 mg of a combination of bulb flour and metformin daily for 12 weeks reduced total cholesterol, triglycerides, and LDL-C by 6.13, 6.64, and 5.1% in obese patients, while metformin alone reduced them in 2.63, 3.18 and 2.73%, respectively. The increase was double when using garlic plus metformin, making it possible to migrate to more natural adjuvant therapies [118].

Some researchers have found that the bioactivity of garlic bulbs can be increased by fermentation using either *Saccharomyces cerevisiae* or *Lactobacillus plantarum.* A dose of 400 mg/kg of fermented yeast bulbs, given during 28 days to fat-fed mice, decreased body weight gain, abdominal adipocyte diameter, total cholesterol, triglyceride, and LDL by 47.13, 61.56, 22.48, 21.05 and 20.60%, respectively, as compared to the 20, 41.2, 37.33, 33.76 and 33.1% achieved when consuming non-fermented bulbs [119]. On a more extensive study, 500 mg/kg administered to induced obese male C57BL/6J mice for 8 weeks. It was detected a reduction of body weight, lower epididymal, retroperitoneal, and mesenteric adipose tissue mass by 18, 36, 44, and 63%, respectively. The above was attributed to the downregulation in mRNA and protein expression of *PPARγ*, *C/EBPα*, and lipogenic proteins, including *SREBP-1c*, *FASN*, and stearoyl-CoA desaturase (*SCD-1*) [120]. The anti-obesity effects are more pronounced when the yeast fermentation process is used.

Garlic bulb volatile oil has also been tested, mainly its potential on lowering serum lipid levels in hyperlipidemia male Sprague-Dawley rats, to which a dose of 92.6 mg/kg/body weight was intragastrically administered daily for 60 days. Bodyweight gain, triglycerides, total cholesterol, and LDL were reduced by 32, 12.86, 31.83, and 10.12%, respectively, compared to the control group. The ratio of liver to bodyweight and HDL were increased by 24.9 and 93-75%, respectively. The oil was also effective at improving the lipid profile and alleviating hepatic steatosis [121]. The studies above do not attribute its effect to any particular compound. However, most of garlic's beneficial effects on health have been attributed to diallyl disulfide. To determine the above, the effect of this compound alone was compared to the one of the garlic hole essential oil in obesity-triggered nonalcoholic fatty liver disease in obese C57BL/6J mice. The administration of 50 mg/kg garlic essential oil and 20 mg/kg of diallyl disulfide, diminished total fat in 32.52 and 45.04%, respectively. While liver weight was reduced in equal proportions (60%) and very similar results were obtained in fatty acids, triglycerides, and cholesterol decrease. Both diallyl disulfide alone and the essential oil significantly decreased the release of pro-inflammatory cytokines in the liver. These effect over non-alcoholic fatty liver disease was determined to be mediated through down-regulation of *SREBF1*, acetyl-CoA carboxylase (*ACC*), *FASN*, and 3-hydroxy-3-methylglutaryl-coenzyme A reductase (*HMG-CoA*), as well as stimulation of peroxisome proliferator-activated receptor α (*PPARα*) and *CPT1* [122]. It was determined that the tested dose of garlic essential oil is composed of 40% diallyl sulfide, equivalent to the pure compound tested amount, which demonstrates that diallyl disulfide is the main garlic bioactive component.

Besides garlic bulbs, also the stem has been studied. The administration of stem ethanolic extract in a dose of 100 mg/kg daily for 4 weeks to high-fat diet-induced obese mice significantly decreased bodyweight, triglycerides, and total cholesterol by up to 10, 22.6, and 19.6%, respectively. Also, adiponectin and leptin levels were reversed to those of healthy weight people, by increasing adiponectin by 49.4% and decreasing leptin by 46.5%, respectively. Liver and serum fat accumulation were inhibited by regulating the activity of HMG-CoA and glucose-6-phosphate dehydrogenase (G6PDH) enzymes implicated in cholesterol synthesis and hepatic lipogenesis, respectively [123].

Cynara scolymus L.

Fiber is a compound highly associated with anti-obesogenic effects, and one of the main components of artichoke, to which different effects over metabolic syndrome, have been attributed. Nevertheless, it was suspected that, besides fiber, other *C. scolymus* compounds were also responsible. So, the anti-obesity potential of the flowering head, which does not contain fiber, was studied over genetically obese Zucker rats. Treatment with 500 mg/kg of the extracts reduced glycemia (mg/dL) in around 11.54 and 30.77% after 60 and 360 minutes, respectively, while the a of 1500 mg/kg reduced it in 19.23 and 38.46% at the same time lapses [124]. These effects were associated with phenolic compounds;

nevertheless, they were not identified. As for those fond un *C. scolymus* leaves and heads, they were recovered by methanolic extractions and fount to be mainly chlorogenic acid, cynarin, 1,5-di-o-caffeoylquinic, luteolin and, apigenin. In a dose of 1.5 g/kg/day over albino rats, this extract had hypoglycemic and hypocholesterolemia effects, especially the leave extract. It was found that glucose, total cholesterol, LDL, and triglycerides were diminished by 61.31 and 45.8%, 28.04 and 18.74%, 60.8 and 38.21%, and 39.73 and 27.67%, for leaves and head, respectively [125].

A luteolin rich artichoke leaf extract was tested over obese C57BL/6N mice. The mice diminished their body weight gain, food deficiency ratio, adipose tissue weight, free fatty acids, total triglycerides, and total cholesterol in around 16 and 10, 18, 12.5, 16.67, and 20.7%, respectively after 16 weeks. The effects were attributed to the upregulation of genes such as *CD36*, lipoprotein lipase (*LPL*), *SREBP1*, adrenoreceptor Beta 3 (*ADRB3*), *CPT1,* and peroxisome proliferator-activated receptor-gamma coactivator 1-beta (*PGC-1 beta*), all involved in adipocyte fatty acid intake and lipogenesis. Also, insulin sensitivity was reestablished to normal levels in response to the reduction of hepatic glucogenic enzyme activity, which was, in turn, the effect of plasma interleukin-6, interleukin-1β, and plasminogen activator reduction. The results of this study suggest that 16 weeks of supplementation with artichoke luteolin enriched extract, prevents obesity and metabolic disorders such as dyslipidemia, insulin resistance, and inflammation in mice with high-fat diet-induced obesity [126].

Artichoke has also been tested along with other plants such as *Phaseolus vulgaris.* It was found that in rats, this mixture is preferable over every single extract, as it combines the anorectic effect of *Phaseolus vulgaris* with the hypoglycemic effect of artichoke. The best combination dose was 200 plus 400 mg/kg of *P. vulgaris* and *C. scolymus*, respectively, which diminished food intake and glycemic index (after 240 minutes of each meal) by around 28 and 20%, respectively [127].

Artichoke consumption has been associated by many people to a feeling of satiety, that there are currently artichoke supplements in the market. To determine the real effects, in vivo studies proving *C. scolymus* over obese humans, have been carried out. On average, patients consuming artichoke supplements felt 70% more satiation compared to the placebo patients. Even when total cholesterol, LDL cholesterol, HDL cholesterol, and triglycerides were not significantly changed by artichoke supplementation, these findings are relevant since the control of appetite is one of the most limiting factors in obesity treatment. The satiety effect was attributed to phytohemagglutinin, which is known to increase the release of cholecystokinin, glycoproteins, and glucagon-like peptides, which increase the gastric emptying time [128]. Other common problems in obese people are high glucose levels and hypocholesterolemia. It was found that consuming 200 mg of flowering artichoke buds extract daily for 60 days, reduces fasting blood glucose, A1c-derived average glucose, and glycosylated hemoglobin by 9.6, 3.1, and 2.3%, respectively, compared to the placebo group [129]. As for hypocholesterolemia, a homeopathic artichoke dose of 10 drops three times a day for 15 days, over obese people, increased bile production after just 30 minutes of the administration. Luteolin was signaled as responsible for the observed effect [130].

Aloe vera

Aloe is a Mexican plant very well known for its medicinal properties. It is mainly used for skin disorders, especially the leaf gel, but it has also been tested over metabolic syndrome, including obesity. In C57BL/6J, obese mice treated with 100 mg/kg of leaf gel daily for 8 weeks, blood glucose, hepatic triglycerides, plasma triglycerides and fat cell size were reduced by 48.15, 48.99, 27.31, and 20%, respectively [131]. This result motivated a human study in obese patients. Under the same condition's total cholesterol, triglycerides, LDL-C, glucose, and fructose were reduced by 9.71, 5.13, 5, 5.4, and 15.19%, respectively [132].

It was suspected that the aloe gel compound promoting the anti-obesity effects was aloesin since, in previous studies, it was found that it modulates AMP-activated protein kinase (AMPK), which is a focus of attention and a target when it comes to obesity treatment. This protein is an energy sensor implicated in glucose and lipid metabolism regulation, and it plays an essential role in glucose and fatty acid catabolism as well as in inflammation. The implication of aloesin in obesity was determined by a study where male C57BL/6 obese mice were treated with 10 mg/kg of aloe gel containing 2% of aloesin. This treatment resulted in fat size down-regulation and epididymal fat pad reduction (by around 56%). The results were attributed to a detected increase in AMPK activation in both adipocyte and muscle tissue. Also, obesity-induced inflammatory cytokines (*IL-1β* and *IL-6*) and hypoxia-inducible factor 1-*α* (*HIF1-α*) were markedly decreased, as well as it did macrophage infiltration [133].

Vaccinium myrtillus L.

V. myrtillus, also known as bilberries, have been for a long time, traditionally used for treating metabolic syndrome complications. Nevertheless, its therapeutical action was not defined until later, when anthocyanins were found as its main bioactive component. The first study evaluating the anti-obesity potential of *V. myrtillus* was carried out in 1996. There, hydroalcoholic leaves extracts were orally administered to streptozotocin-diabetic rats (once a day for four consecutive days in a 2% Arabic gum suspension), reducing characteristic obesity parameters, such as plasma triglyceride, which was decreased by 39%. Also, this bilberry extract hypolipidemic effect was compared to that of Ciprofibrate, a well-established hypolipidemic drug at the time. It was found that plasma triglycerides and free fatty acids were reduced by 30 and 60% and by 31.8 and 55.6% by the extract and Ciprofibrate over obese Yoshida rats, respectively [134].

Bilberry is also consumed as a whole, and its flour administration over Wistar rats, two times a day for four weeks, diminished LDL and triglycerides by 5.1 and 49.43%, respectively [135]. Hole bilberry effect was tested for a longer period, where C57BL obese mice were fed in a proportion of 10% of its total diet with *V. myrtillus* flour for three months. It was found that bilberries diminished type 1 pro-inflammatory responsiveness, which is induced by a high-fat diet. Total weight and systolic blood pressure levels were decreased by 31.25 and 13.64%, respectively. Also, pro-inflammatory cytokines, chemokines, leptin, resistin, T helper 1, cytotoxic T cells, and interferon-gamma levels were significantly reduced, while adiponectin was increased by 39.8%. In this study, bilberries were found to reduce the development of systemic inflammation and prevent the progression of chronic hypertension, thus supporting their potential role in alleviating the adverse health effects associated with obesity [136].

Besides anthocyanins, apigenin-7-glucuronide, vitexin-2″-rhamnoside, and isoorientin are known for their lipid-lowering profile, and they are all found in *V. myrtillus* leaf, so a rich ethanolic extract was obtained and given to high-fed-diet male Wistar rats in a 2% of their total diet during 50 days. This treatment normalized lipid metabolism parameters. Also, weight gain and triacylglycerols were reduced by 40 and 19.34%, respectively. This study confirmed the lipid-lowering potential of the compounds mentioned above [137]. The anti-obesity and lipid-lowering properties of *v. myrtillus* leaves were also tested in obese Syrian hamsters. In this study, 35 mg/kg/day of leaf flour was administered for two weeks, using Styfimol (6.2 mg/kg/day) as a positive control drug. The given dose caused a significant reduction in body weight gain, visceral fat mass, serum triacylglycerols, free fatty acids, total cholesterol, and LDL. Besides, the hamsters treated with powder extract had the closest intact value ratio of HDL and LDL compared with positive control animals [138].

Calendula officinalis

The anti-obesogenic effect of Mexican plants has not only been evaluated in terms of weight loss, but also its action on obesity caused health complications, for example, cardiovascular diseases, which are mainly related to the obesity underlying the inflammatory process. *C. officinalis* infusion (*ad libitum* for 12 weeks) was proven against cardiovascular complications developed in obese Male Sprague-Dawley rats. By this treatment, the C-reactive protein level was reduced. This protein plays a direct role in promoting vascular inflammation, vessel damage, and all clinical cardiovascular disease events. Also, this infusion decreased bodyweight, abdominal fat mass, triglycerides, total cholesterol, and LDL in 12.73, 28, 45.05, 27.7, and 48.6%, respectively. The mechanism underlying the effect was triglycerides digestion inhibition and absorption, being hesperidin, and epigallocatechin gallate the components responsible for this effect. These results suggest that *C. officinalis* infusion may be used as an alternative for obesity-related cardiovascular risk treatment [60].

Hypericum perforatum

C-reactive protein levels were also reduced by *H. perforatum* aqueous extract in obese rats in an *at libitum* dose of 1% w/v [60]. Also, the administration of *H. perforatum* dry alcoholic extract (200 mg/kg of body weight for 15 days) to obese rats, fed either with high-fat-diet or fructose-diet, decreased total cholesterol, triglycerides and LDL levels by 23, 23 and 46%, respectively [139]. The anti-obesity activity of *H. perforatum* was also tested in ovariectomized rats. The ethanolic extract promoted an estrogen-like effect against adiposity. The last in vivo study of *H. perforatum* concluded that aqueous extracts from flowers, leaves, and stem (2% solution at libitum during 6 months), prevented the development of hepatic steatosis in obese rats [59].

Psidium guajava L.

Sprague-Dawley rats were also used as a model to determine guava in vivo anti-obesity effect. Guava pure in a dose of 2000 mg/kg significantly reduced body weight by 11% [140]. In another study, guava leaves prevented the glucose intolerance and fat deposition in obese Wistar rats fed with a high carbohydrate-high fat diet [141]. Also, in a dose of 5 mg/kg, guava leaves reduced serum lipid markers in C57BL/6J obese rats [142].

Cinnamomum verum

Another very popular Mexican plant is *C. verum*. Cinnamon is thought to reduce overweight and their derived complications. Investigations results prove that its effect is not directly on obesity, but on complications derived from it. For example, cinnamon extracts in a dose of 200 mg/kg bodyweight for 2 months did not show any significant changes in plasma levels of obesity-related parameters, including lipid profiles, leptin, and lipase enzyme on obese and diabetic Wistar rats. However, liver histology showed a decrease in glycogen accumulation. The antidiabetic action of cinnamon extract was confirmed since treated obese-diabetic rats showed an increase in mRNA expression of leptin (*LEP*), *PPAR-γ,* and *ADIPOQ* [143].

The above results were confirmed when treating obese patients with a dose of 250 mg of dried aqueous extract twice a day for 12 weeks. By this treatment, it was only found a reduction of 10.53% of fasting glucose [144].

In conclusion, cinnamon extracts are not anti-obesity, but they have antidiabetic and insulin-mimetic actions through the regulation of genes related to lipid and glucose metabolism. These effects are attributed to alkaloids, flavonoids, tannins, phenols, and saponins [145].

Regarding cinnamaldehyde alone, the compound given cinnamon its characteristic flavor, it was proven to be an agonist of transient receptor potential ankyrin receptor 1 (TRPA1), achieving a significant reduction of food intake and gastric emptying rates, as well as plasma adiponectin level in obese mice. Additionally, obese mice fed for five weeks with cinnamaldehyde, reduce their body weight gain and lean mass gain by around 13.5 and 73.7 %, respectively. The results were also attributed to the up-regulation of the genes related to fatty acid oxidation acyl-CoA synthetase long-chain family member 4 (*ACSL4*), glucose transporters (*GLUT1*, *GLUT8*, and *GLUT12*), *CPT1A* and insulin receptor precursor (*INSR*) [146].

Hibiscus sabdariffa

Another Mexican plant intensely studied in terms of anti-obesity properties, is *H. sabdariffa.* Many studies have shown that this plant effectively reduces rat´s body weight [147-150, 64]. Furthermore, *H. sabdariffa* (in a dose of 33 mg of total anthocyanins/kg body weight) decreased the accumulation of visceral fat and elevated HDL levels in high fat diet-induced obese mice. The mechanisms of action are associated with the regulation of the transcription factors *PPAR-γ* and *SREBP-1c*, both regulators of the hepatic lipid metabolism [147]. Also, *H. sabdariffa* down-regulated adipogenesis-related genes, pro-inflammatory cytokines, and leptin serum, while it upregulated the thermogenesis-related genes and promoted AMPK activation in C57BL/6J mice [150]. In other in vivo study, it was found that the administration of 100 mg for 10 weeks of *Hibiscus sabdariffa* aqueous extract inhibited gut lipid absorption, which was demonstrated by cholesterol and triglycerides reduction in the liver of obese hamsters [62]. Furthermore, *H. sabdariffa* grounded calyces administrated to obese mice effectively reduced total cholesterol and glucose in plasma by 30 and 11%, respectively [148]. Likewise, the administration of 500 mg/kg of *H. sabdariffa* calyx extract prevented the hypercholesterolemia induced by dietary fructose in Sprague-Dawley rats [151]. Also, the calyxes administered to obese rats reduced the triglyceride storage and adipocyte differentiation [64].

Clinical trials testing *H. sabdariffa,* have also been carried out. There it was found that consuming *H. sabdariffa* in combination with *Lippia citriodora* L. at a dose of 500 mg/day during 60 days, has shown to be safe, and improved anthropometric parameters [152, 153]. Also, it regulates appetite-related peptides. The mechanism of action was associated with AMPK activation [152]. In another clinical trial, *H. sabdariffa* encapsulated powder (100 mg/day) was administrated to individuals with metabolic syndrome. After 30 days, cholesterol and triglyceride levels were reduced by 10% [154]. Also, in a higher dose (450 mg/day during 12 weeks), these capsules reduced obese people´s body weight by 4% [155]. Overall, *H. sabdariffa* has shown promising anti-obesity effects due to both phenolics and fiber content. Their main results are bodyweight reduction, modulation of lipid homeostasis, and glucose uptake, mainly through adipocyte differentiation and free fatty acid synthesis regulation.

Zingiber officinale

Zingiber officinale, known as ginger, has been widely studied for its properties in the prevention and treatment of different physical disorders, including obesity. It has been documented that low rhizome aqueous extract doses of 20 mg/kg, reduce body weight gain, cholesterol, LDL, serum insulin, and leptin levels in obese rats by 10, 11, 21, 13, and 10%, respectively [156]. Another study showed that ginger ethanolic extract supplementation to obese C57BL/6J mice suppressed the body weight gain and the perirenal, retro perirenal and epididymal fat accumulation, as well as it reduced total cholesterol, triglycerides and leptin concentration by 29, 34 and 89%, respectively [157]. Ginger rhizome also reduces body mass index in obese subjects when administered as 1 g per day during 12 weeks [158]. When the dose is raised to two tablets a day, it had a moderate effect on the total cholesterol, triglyceride, and serum glucose, and increased HDL in obese patients [159]. In another study, ginger was supplemented to obese women during 12 weeks; compared to the placebo control, all the patients improve their anthropometric indicators, also triglycerides concentration was reduced, however, there were no statistically significant differences in lipid concentration and serum glucose [160].

Humulus lupulus L.

Hop pomace polyphenols reduced body weight in obese mice after 10 weeks. Also, it significantly reduced HDL concentration. Furthermore, hop pomace polyphenols suppressed the activity of enzymes involved in liver fatty acid synthesis [71]. The main anti-obesity *H. lupulus* phytochemical is xanthohumol. It has been demonstrated that it reduces cholesterol and plasma glucose levels as well as ameliorate insulin and leptin resistance in C57BL/6J mice. The anti-obesity effect of xanthohumol has also been associated with its anti-inflammatory potential [161]. In another study, xanthohumol reduced the visceral fat weight and plasma leptin concentrations in KK-Ay mice. Although plasma lipids levels were not modified, total liver cholesterol was significantly reduced. It was found that xanthohumol downregulated *SREBP1c* and upregulated *PPARα* [162]. In Zucker fa/fa rats, a rat model of genetic obesity, the administration of xanthohumol at a dose of 169 mg/kg of body weight during six weeks, lowered bodyweight [163]. The effect of xanthohumol on body weight reduction might be associated with its ability to regulate hepatic fatty acid metabolism and intestinal fat absorption inhibition [164]. Besides xanthohumol, Iso-alpha acids have been identified in hops. These compounds had demonstrated to reduce body weight gain, expansion of adipose tissue, and insulin resistance in rats [165]. Also, iso-alpha acids have shown to decrease the development of fat mass [166, 167].

Aesculus hippocastanu

A study conducted by Avci et al. [47] determined that *Aesculus hippocastanu seeds* reduced leptin and FT_4 (serum thyroid-stimulating hormone) levels by 31.6 and 36%, respectively in obese Swiss mice. Also, a hypoglycemic activity was observed. The effects were attributed to seed saponins and sapogenins [168].

Turnea diffusa

There is a patented product called YGD based on *T. diffusa*, yerba Mate and Guarana. It was found that YGD diminished adiposity, food intake, and hunger sensation in obese women. However, the effects were not attributed to any particular compound or plant [169]. In a latter study, it was found that YGD reduces food intake through the modulation of gastrointestinal hormones, such as acylated ghrelin and glucagon-like peptide 1 (GLP-1) in obese and overweight women. As for Damiana, the effect was associated with flavonoids like vitexin, which in previous studies promoted a reduction in postprandial blood glucose levels in rats [48, 170, 171].

Echinace pallida

Very little is known about *Echinacea pallida* anti-obesity potential. In a study carried out by [49] over obese patients, there were found no significant differences over cholesterol and triglycerides levels compared to the control group.

Ginko biloba

Hong et al. [172] evaluated the effect of the *Ginko biloba* standardized acetone extract EGb 761®. The extract was given to obese mice in a dose of 50 mg/kg every two days for 40 days, lecithin cholesterol acyltransferase levels were increased, and therefore HDL-C concentration incremented.

Melissa officinalis

The ethanolic leaf extract of *M. officinalis* in a dose of 200 mg/kg/day during six weeks, was evaluated for the treatment of insulin resistance and dyslipidemia in C57BL/6 obese mice. It was found that the extract activates PPAR-γ. It also reduces plasma cholesterol, on-esterified fatty acids, and LDL levels [173]. Also, *Melissa officinalis* essential oil was evaluated in human APOE2 transgenic mice (12.5 μg/d for 2 weeks). It increased the synthesis of bile acids and therefore diminished cholesterol levels. It was also found to decrease plasma triglycerides concentration [174].

Kim et al. [175]_tested *M. officinalis* for 15 weeks over high-fat diet-fed C57BL/6J mice. Bodyweight was reduced without affecting calorie consumption. In addition, serum levels ofmalanine transaminase and aspartate transaminase, which metabolizes liver fat, were changed, improving liver lipid metabolism. These results match with a similar study where *M. officinalis* over female wild-type C57BL/6J ovariectomized mice were tested. There, a decrease in the accumulation of hepatic lipids and inflammatory cells was detected [176].

Park et al. [177] analyzed the patented *M. officinalis* leaves extract ALS-L1023 over obese mice. It decreased adipose tissue mass and adipocytes size, as well as its diminished liver lipid accumulation after 8-week treatment. The effects were due to a modulation in angiogenic factors and their inhibitor's expression. In another study conducted in obese C57BL / 6J mice, it was found that ALS-L1023 not only decreased the adipose tissue mass and the size of the adipocytes but also upregulated MPP and angiogenic factors in adipose tissue [108].

The OB-X *M. officinalis*-based product, introduced in the in vitro section, was also tested over in vivo models. It regulated adipose tissue growth through changes in the expression of genes involved in angiogenesis and the MMP system [178]. In another study performed over obese ob/ob mice, OB-X® promoted a bodyweight gain reduction during a high-fat diet trial without affecting other organs weight, suggesting that OB-X® acts specifically on adipose tissue [179]. In later investigations, OB-X® was tested over humans. It reduced fast-growing abdominal fat, mainly visceral fat by 20.5%, by angiogenesis inhibition. These results are in agreement with what was previously found in animal studies [74].

Rosmarinus officinalis L.

The first reports of the anti-obesogenic effect of rosemary were carried out by Harach et al. [84]. They found that a dose of 200 mg/kg bodyweight for 49 days reduced bodyweight and fat by 64 and 57% in obese male mice, respectively. Higher feces lipid concentration was found, and the effect was attributed to carnosic acid and carnosol, which inhibited pancreatic lipase.

Later it was reported that a higher dose (500 mg of rosemary extract/kg body weight, daily for 80 days) reduced body weight by 69% in obese male C57BL/6J mice (Ibarra et al., 2011). A lower effect was obtained when feeding female obese Zucker rats in a ratio of 0.5 w/w for 64 days with carnosic acid and carnosol alone, which body weight was diminished only by 15%, suggesting that the effect is more significant when consuming whole rosemary (Romo Vaquero et al. [97]. Wang et al. [180] reported that carnosic acid and carnosol activated glutathione metabolism in obese mice. These compounds were also found to diminish intestinal lipid absorption [84, 95, 181].

Citrus aurantium

Colker et al. [182] reported that a commercial mixture of 975 mg of *Citrus aurantium* and 900 mg of St. John's Wort extracts, reduced body weight and body fat in 1.4 kg and 2.9%, respectively in obese patients after a six weeks treatment. This effect was attributed to the synergism between the compounds within the mixture. As for *C. aurantium,* the main components were alkaloids [183]. Similarly, Jones [184] reported that *C. aurantium,* along with *Panax Ginseng*, *Ginkgo Biloba,* and *Paullinia cupana* (50 mg/ kg of body weight for two weeks) reduced obese women body weight by 5% after a two-week study.

C. aurantium obesogenic potential was found to be due to an increase in the resting metabolic rate [185, 186], which in turn leads to weight loss. The most recent study evaluating *C. aurantium* anti-obesity potential was carried out over obese male mice. *C. aurantium* ethanolic peel extract (100 mg/kg of body weight for 8 weeks) significantly reduced body and liver weight by 3 and 64%, respectively. The effect was mainly attributed to nobiletin, conforming to the extract in 27% (Han et al. [77]. The same peel extract was reported to decrease triglycerides and cholesterol plasma levels by 3 and 6%, respectively [187, 188].

Rheum palmatum L.

Very little is known about the anti-obesity potential of *Rheum palmatum* L. nevertheless, there are drugs based on it. The polyherbal drugs Gyeongshingangjeehwan 18 (GGEx18) and Gangjihwan, both composed by *Rheum palmatum* L. and other plants, decreased body weight by 21% in obese male C57BL/6J mice at a dose of 500 mg/kg/day during a 13-week study. Fat accumulation was also diminished (Lim et al. [189] Jang et al. [78].

Trigonella foenum-graecum L.

Few reports are testing the effect of *Trigonella foenum-graecum* L over metabolic syndrome disorders. However, Chennappaneskar and Lathadevi [88] reported a reduction of body weight and cholesterol by 5 and 25%, respectively, on male Wistar rats fed with 2 ml of *T. foenum-graecum* emulsion for 12 weeks.

In Figure 1, there are the chemical structures of Mexican medicinal plant compounds with anti-obesity potential. For instance, the stability of the here mentioned anti-obesity compounds and extracts throughout the digestive tract is needed to be studied. As well as their impact over colonic microbiota, as this parameter has been strongly related to the onset of obesity and metabolic syndrome diseases [94, 54].

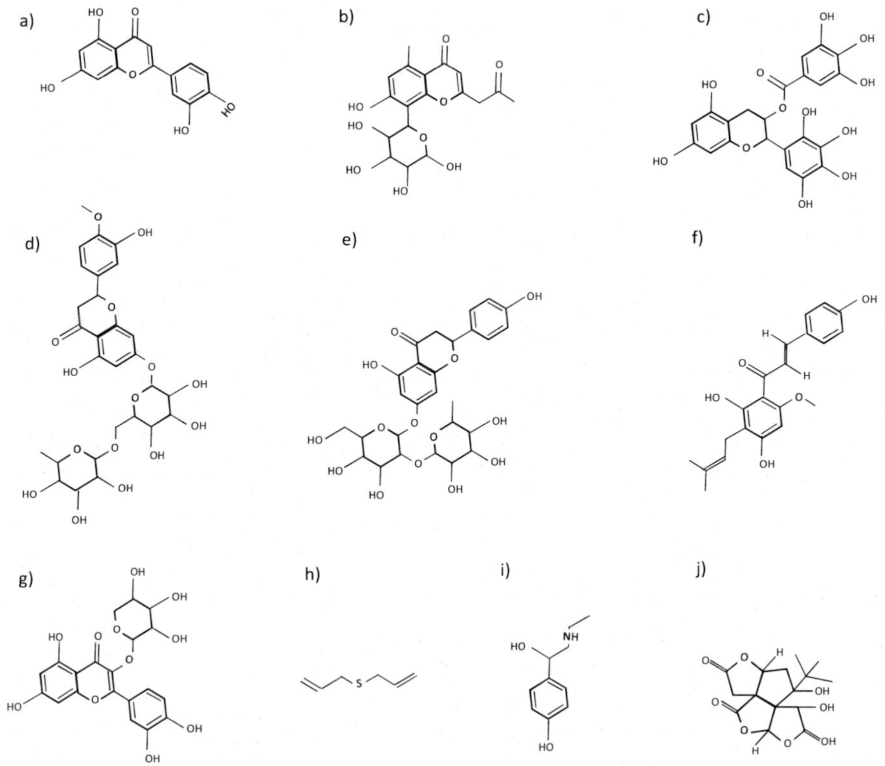

Figure 1. Chemical structure of Mexican medicinal plant compounds with anti-obesity potential. a) luteolin, b) aloesin, c) epigallocatechin gallate, d) hesperidin, e) naringin, f) xanthohumol, g) guajaverin, h) diallyl sulfide, i) p-synephrine, and j) bilobalide.

CONCLUSION

Obesity has become a major public health problem. Despite being a multifactorial disease, the main cause is a psychological and emotional disorder that encourages excessive calorie intake, so available drugs focus mainly on accelerating the energy expenditure rate. However, negative effects have been associated to their consumption, so the use of phytochemicals as an alternative adjuvant therapy has been suggested. For example, those found in Mexican plants. Some of them are known for promoting weight loss and improving obesity parameters by several mechanisms such as suppression of anabolic pathways and promotion of catabolic pathways in adipose tissue, so the consumption of these plants should be promoted along with healthy eating and exercise.

Despite these phytochemicals proved anti-obesity effects, there are still plenty of issues that remain unsolved, so further investigation needs to be done. For example, there is still a lack of pre-clinical and clinical studies. Furthermore, it is necessary to find specific anti-obesity molecules within plants extracts as well as their exact mechanism of action for drug development. Also, the stability of the here mentioned anti-obesity compounds and extracts throughout the digestive tract is needed to be studied, as well as their impact over colonic microbiota, as this parameter has been strongly related to the onset of obesity and metabolic syndrome diseases.

REFERENCES

[1] Xia, Q., and Grant, S. F. (2013). The genetics of human obesity. *Annals of the New York Academy of Sciences* 1281 (1):178.

[2] Mozaffarian, D., Benjamin, E. J., Go, A. S., Arnett, D. K., Blaha, M. J., Cushman, M., De Ferranti, S., Després, J.-P., Fullerton, H. J., and Howard, V. J. (2015). Executive summary: heart disease and stroke statistics—2015 update: a report from the American Heart Association. *Circulation* 131 (4):434-441.

[3] Endo, Y., Yokote, K., and Nakayama, T. (2017). The obesity-related pathology and Th17 cells. *Cellular and Molecular Life Sciences* 74 (7):1231-1245.

[4] De Lorenzo, A., Soldati, L., Sarlo, F., Calvani, M., Di Lorenzo, N., and Di Renzo, L. (2016). New obesity classification criteria as a tool for bariatric surgery indication. *World Journal of Gastroenterology* 22 (2):681.

[5] Nuttall, F. Q. (2015). Body mass index: obesity, BMI, and health: a critical review. *Nutrition Today* 50 (3):117.

[6] Wright, S. M., and Aronne, L. J. (2012). Causes of obesity. *Abdominal Radiology* 37 (5):730-732.

[7] Sanyal, D., and Raychaudhuri, M. (2016). Hypothyroidism and obesity: An intriguing link. *Indian Journal of Endocrinology and Metabolism* 20 (4):554.

[8] Herrera, B. M., and Lindgren, C. M. (2010). The genetics of obesity. *Current Diabetes Reports* 10 (6):498-505.

[9] Gianini, L. M., White, M. A., and Masheb, R. M. (2013). Eating pathology, emotion regulation, and emotional overeating in obese adults with binge eating disorder. *Eating Behaviors* 14 (3):309-313.

[10] Davis, C. D. (2016). The gut microbiome and its role in obesity. *Nutrition Today* 51 (4):167.

[11] Lopomo, A., Burgio, E., and Migliore, L. 2016. Epigenetics of obesity. In *Progress in molecular biology and translational science*, 151-184. Elsevier.

[12] Van Dijk, S. J., Tellam, R. L., Morrison, J. L., Muhlhausler, B. S., and Molloy, P. L. (2015). Recent developments on the role of epigenetics in obesity and metabolic disease. *Clinical Epigenetics* 7 (1):66.

[13] Martínez, J. A., Milagro, F. I., Claycombe, K. J., and Schalinske, K.L. (2014). Epigenetics in adipose tissue, obesity, weight loss, and diabetes. *Advances in Nutrition* 5 (1):71-81.

[14] Hruby, A., and Hu, F. B. (2015). The epidemiology of obesity: a big picture. *Pharmacoeconomics* 33 (7):673-689.

[15] Pineda, E., Sanchez-Romero, L. M., Brown, M., Jaccard, A., Jewell, J., Galea, G., Webber, L., and Breda, J. (2018). Forecasting future trends in obesity across Europe: the value of improving surveillance. *Obesity facts* 11 (5):360-371.

[16] Popkin, B., and Reardon, T. (2018). Obesity and the food system transformation in Latin America. *Obesity reviews* 19 (8):1028-1064.

[17] DiBonaventura, M. D., Sadosky, A., Concialdi, K., Hopps, M., Kudel, I., Parsons, B., Cappelleri, J. C., Hlavacek, P., Alexander, A. H., and Stacey, B. R. (2017). The prevalence of probable

neuropathic pain in the US: results from a multimodal general-population health survey. *Journal of Pain Research* 10:2525.

[18] Swinburn, B. A., Sacks, G., Hall, K. D., McPherson, K., Finegood, D. T., Moodie, M. L., and Gortmaker, S. L. (2011). The global obesity pandemic: shaped by global drivers and local environments. *The Lancet* 378 (9793):804-814.

[19] Chan, R. S., and Woo, J. (2010). Prevention of overweight and obesity: how effective is the current public health approach. *International Journal of Environmental Research and Public Health* 7 (3):765-783.

[20] Igel, L. I., Kumar, R. B., Saunders, K. H., and Aronne, L. J. (2017). Practical use of pharmacotherapy for obesity. *Gastroenterology* 152 (7):1765-1779.

[21] Cheung, B. M. Y., Cheung, T. T., and Samaranayake, N. R. (2013). Safety of antiobesity drugs. *Therapeutic Advances in Drug Safety* 4 (4):171-181.

[22] González-Castejón, M., and Rodriguez-Casado, A. (2011). Dietary phytochemicals and their potential effects on obesity: a review. *Pharmacological research* 64 (5):438-455.

[23] Soyoung, K., Jin, Y., Choi, Y., and Park, T. (2011). Resveratrol exerts anti-obesity effects via mechanisms involving down-regulation of adipogenic and inflammatory processes in mice. *Biochemical Pharmacology* 81 (11):1343-1351.

[24] López-Rubalcava, C., and Estrada-Camarena, E. (2016). Mexican medicinal plants with anxiolytic or antidepressant activity: Focus on preclinical research. *Journal of Ethnopharmacology* 186:377-391.

[25] Sumithran, P., and Proietto, J. (2014). Benefit-risk assessment of orlistat in the treatment of obesity. *Drug Safety* 37 (8):597-608.

[26] Jordan, J., Astrup, A., Engeli, S., Narkiewicz, K., Day, W. W., and Finer, N. (2014). Cardiovascular effects of phentermine and topiramate: a new drug combination for the treatment of obesity. *Journal of Hypertension* 32 (6):1178.

[27] Gustafson, A., King, C., and Rey, J. A. (2013). Lorcaserin (Belviq): A selective serotonin 5-HT2C agonist in the treatment of obesity. *Pharmacy and Therapeutics* 38 (9):525.

[28] Hendrick, V., Suri, R., Gitlin, M., and Ortiz-Portillo, E. (2017). Bupropion Use during Pregnancy: A Systematic Review. *The Primary Care Companion for CNS Disorders* 19 (5).

[29] le Roux, C. W., Astrup, A., Fujioka, K., Greenway, F., Lau, D. C., Van Gaal, L., Ortiz, R. V., Wilding, J. P., Skjøth, T. V., and Manning, L. S. (2017). 3 years of liraglutide versus placebo for type 2 diabetes risk reduction and weight management in individuals with prediabetes: a randomised, double-blind trial. *The Lancet* 389 (10077):1399-1409.

[30] Nasri, H., and Rafieian-Kopaei, M. (2014). Metformin: current knowledge. *Journal of Research in Medical Sciences: the Official Journal of Isfahan University of Medical Sciences* 19 (7):658.

[31] Dunican, K. C., Adams, N. M., and Desilets, A. R. (2010). The role of pramlintide for weight loss. *Annals of Pharmacotherapy* 44 (3):538-545.

[32] Minton, G. C., Miller, A. D., Bookstaver, P. B., and Love, B. L. (2011). Topiramate: Safety and Efficacy of its Use in the Prevention and Treatment of Migraine. *Journal of Central Nervous System Disease* 3:JCNSD. S4365.

[33] Grover, N. D., Limaye, R. P., Gokhale, D. V., and Patil, T. R. (2013). Zonisamide: a review of the clinical and experimental evidence for its use in Parkinson's disease. *Indian Journal of Pharmacology* 45 (6):547.

[34] Jacobo-Herrera, N. J., Jacobo-Herrera, F. E., Zentella-Dehesa, A., Andrade-Cetto, A., Heinrich, M., and Pérez-Plasencia, C. (2016). Medicinal plants used in Mexican traditional medicine for the treatment of colorectal cancer. *Journal of Ethnopharmacology* 179:391-402.

[35] Alonso-Castro, A. J., Villarreal, M. L., Salazar-Olivo, L. A., Gomez-Sanchez, M., Dominguez, F., and Garcia-Carranca, A. (2011). Mexican medicinal plants used for cancer treatment:

pharmacological, phytochemical and ethnobotanical studies. *Journal of Ethnopharmacology* 133 (3):945-972.

[36] Granica, S., Kluge, H., Horn, G., Matkowski, A., and Kiss, A. K. (2015). The phytochemical investigation of *Agrimonia eupatoria* L. and *Agrimonia procera* Wallr. as valid sources of Agrimoniae herba—the pharmacopoeial plant material. *Journal of Pharmaceutical and Biomedical Analysis* 114:272-279.

[37] Farag, M., Ali, S., Hodaya, R., El-Seedi, H., Sultani, H., Laub, A., Eissa, T., Abou-Zaid, F., and Wessjohann, L. (2017). Phytochemical profiles and antimicrobial activities of *Allium cepa* red cv. and A. sativum subjected to different drying methods: a comparative MS-based metabolomics. *Molecules* 22 (5):761.

[38] Fratianni, F., Ombra, M. N., Cozzolino, A., Riccardi, R., Spigno, P., Tremonte, P., Coppola, R., and Nazzaro, F. (2016). Phenolic constituents, antioxidant, antimicrobial and anti-proliferative activities of different endemic Italian varieties of garlic (*Allium sativum* L.). *Journal of Functional Foods* 21:240-248.

[39] Claus, T., Maruyama, S. A., Palombini, S. V., Montanher, P. F., Bonafé, E. G., Junior, O.d. O. S., Matsushita, M., and Visentainer, J. V. (2015). Chemical characterization and use of artichoke parts for protection from oxidative stress in canola oil. *LWT-Food Science and Technology* 61 (2):346-351.

[40] Salem, M. B., Affes, H., Ksouda, K., Dhouibi, R., Sahnoun, Z., Hammami, S., and Zeghal, K. M. (2015). Pharmacological studies of artichoke leaf extract and their health benefits. *Plant Foods for Human Nutrition* 70 (4):441-453.

[41] Rahman, S., Carter, P., and Bhattarai, N. (2017). *Aloe vera* for tissue engineering applications. *Journal of Functional Biomaterials* 8 (1):6.

[42] Müller, D., Schantz, M., and Richling, E. (2012). High performance liquid chromatography analysis of anthocyanins in bilberries (*Vaccinium myrtillus* L.), blueberries (*Vaccinium corymbosum* L.), and corresponding juices. *Journal of Food Science* 77 (4):C340-C345.

[43] Wang, Wu, J., Wang, H. X., Li, S. S., Zheng, X. C., Du, H., Xu, Y. J., and Wang, L. S. (2015). Composition of phenolic compounds and antioxidant activity in the leaves of blueberry cultivars. *Journal of Functional Foods* 16:295-304.

[44] Arora, D., Rani, A., and Sharma, A. (2013). A review on phytochemistry and ethnopharmacological aspects of genus Calendula. *Pharmacognosy Reviews* 7 (14):179.

[45] Nabavi, S., Di Lorenzo, A., Izadi, M., Sobarzo-Sánchez, E., and Daglia, M. (2015). Antibacterial effects of cinnamon: From farm to food, cosmetic and pharmaceutical industries. *Nutrients* 7 (9):7729-7748.

[46] Zhang, X., Zhang, S., Yang, Y., Wang, D., and Gao, H. (2019). Natural barrigenol-like triterpenoids: A comprehensive review of their contributions to medicinal chemistry. *Phytochemistry* 161:41-74. doi: 10.1016/j.phytochem.2019.01.017.

[47] Avci, G., Kucukkurt, I., Kupeli Akkol, E., and Yesilada, E. (2010). Effects of escin mixture from the seeds of *Aesculus hippocastanum* on obesity in mice fed a high fat diet. *Pharmacological Biology* 48 (3):247-52. doi: 10.3109/13880200903085466.

[48] Celestino, M. M., Gomes, A. C., Botelho, P. B., Gambero, A., Mesquita, L. M., Vilegas, W., Ribeiro, M. L., and Mota, J. F. (2017). South American herbal extracts reduce food intake through modulation of gastrointestinal hormones in overweight and obese women. *Journal of Functional Foods* 35:555-563. doi: https://doi.org/10.1016/j.jff.2017.06.015.

[49] Kovitvadhi, A., Gai, F., Dabbou, S., Ferrocino, I., Rotolo, L., Falzone, M., Vignolini, C., Gennero, M. S., Bergagna, S., Dezzutto, D., Barbero, R., Nebbia, P., Rosati, S., Cocolin, L., Zoccarato, I., and Gasco, L. (2016). Rabbit dietary supplementation with pale purple coneflower. 2. Effects on the performances, bacterial community, blood parameters and immunity of growing rabbits. *Animal* 10 (7):1110-7. doi: 10.1017/s1751731115002980.

[50] Mohanta, T. K., Tamboli, Y., and Zubaidha, P. K. (2014). Phytochemical and medicinal importance of *Ginkgo biloba* L.

Natural Products Research 28 (10):746-52. doi: 10.1080/14786419.2013.879303.

[51] Priyanka, A., Nisha, V. M., Anusree, S. S., and Raghu, K. G. (2014). Bilobalide attenuates hypoxia induced oxidative stress, inflammation, and mitochondrial dysfunctions in 3T3-L1 adipocytes via its antioxidant potential. *Free Radical Research* 48 (10):1206-17. doi: 10.3109/10715762.2014.945442.

[52] Wang, M., Carrell, E. J., Chittiboyina, A. G., Avula, B., Wang, Y. H., Zhao, J., Parcher, J. F., and Khan, I. A. (2016). Concurrent supercritical fluid chromatographic analysis of terpene lactones and ginkgolic acids in Ginkgo biloba extracts and dietary supplements. *Analytical and Bioanalytical Chemistry* 408 (17):4649-60. doi: 10.1007/s00216-016-9544-6.

[53] Lim, T. K. 2012. *Psidium guajava*. In *Edible Medicinal and Non Medicinal Plants*, 684-727. Springer.

[54] Simao, A. A., Marques, T. R., Marcussi, S., and Correa, A. D. (2017). Aqueous extract of *Psidium guajava* leaves: phenolic compounds and inhibitory potential on digestive enzymes. *Anais da Academia Brasileira de Ciências* 89 (3 Suppl):2155-2165. doi: 10.1590/0001-3765201720160067.

[55] Castro-Barquero, S., Lamuela-Raventós, M. R., Doménech, M., and Estruch, R. (2018). Relationship between Mediterranean Dietary Polyphenol Intake and Obesity. *Nutrients* 10 (10). doi: 10.3390/nu10101523.

[56] Gutiérrez, R. M. P., Mitchell, S., and Solis, R. V. (2008). *Psidium guajava*: a review of its traditional uses, phytochemistry and pharmacology. *Journal of Ethnopharmacology* 117 (1):1-27. doi: 10.1016/j.jep.2008.01.025.

[57] CABI. 2019. *Hypericum perforatum* (St John's wort). accessed June 1, 2019. https://www.cabi.org/isc/datasheet/28268.

[58] Agyare, C., Boakye, Y. D., Bekoe, E. O., Hensel, A., Dapaah, S. O., and Appiah, T. (2016). African medicinal plants with wound healing properties. *Journal of Ethnopharmacology* 177:85-100. doi: 10.1016/j.jep.2015.11.008.

[59] Pérez-Ramírez, I. F., Gallegos-Corona, M. A., González-Dávalos, M. L., Mora, O., Rocha-Guzmán, N. E., and Reynoso-Camacho, R. (2019). Mechanisms Associated with the Effect of *Hypericum perforatum* and *Smilax cordifolia* Aqueous Extracts on Hepatic Steatosis in Obese Rats: A Lipidomic Approach. *European Journal of Lipid Science and Technology* 121 (2):1800403. doi: 10.1002/ejlt.201800403.

[60] Hernández-Saavedra, D., Pérez-Ramírez, I. F., Ramos-Gómez, M., Mendoza-Díaz, S., Loarca-Pina, G., and Reynoso-Camacho, R. (2016). Phytochemical characterization and effect of *Calendula officinalis, Hypericum perforatum,* and *Salvia officinalis* infusions on obesity-associated cardiovascular risk. *Medicinal Chemistry Research* 25 (1):163-172. doi: 10.1007/s00044-015-1454-1.

[61] Klemow, K. M., Bartlow, A., Crawford, J., Kocher, N., Shah, J., and Ritsick, M. 2011. Medical Attributes of St. John's Wort (Hypericum perforatum). In *Herbal Medicine: Biomolecular and Clinical Aspects.* Edited by Iris F. F. Benzie and S. Wachtel-Galor. Florida: CRC Press/Taylor & Francis.

[62] Huang, T. W., Chang, C. L., Kao, E. S., and Lin, J. H. (2015). Effect of Hibiscus sabdariffa extract on high fat diet-induced obesity and liver damage in hamsters. *Food & Nutrition Research* 59:29018. doi: 10.3402/fnr.v59.29018.

[63] Gamboa-Gómez, C., Pérez-Ramírez, I. F., González-Gallardo, A., Gallegos-Corona, M. A., Ibarra-Alvarado, C., and Reynoso-Camacho, R. (2016). Effect of *Citrus paradisi* and *Ocimum sanctum* Infusions on Blood Pressure Regulation and Its Association with Renal Alterations in Obese Rats. *Journal of Food Biochemistry* 40 (3):345-357. doi: 10.1111/jfbc.12216.

[64] Morales-Luna, E., Perez-Ramirez, I. F., Salgado, L. M., Castano-Tostado, E., Gomez-Aldapa, C. A., and Reynoso-Camacho, R. (2019). The main beneficial effect of roselle (*Hibiscus sabdariffa*) on obesity is not only related to its anthocyanin content. *Journal of the Science of Food and Agriculture* 99 (2):596-605. doi: 10.1002/jsfa.9220.

[65] Herranz-López, M., Fernández-Arroyo, S., Pérez-Sanchez, A., Barrajón-Catalán, E., Beltrán-Debón, R., Menéndez, J. A., Alonso-Villaverde, C., Segura-Carretero, A., Joven, J., and Micol, V. (2012). Synergism of plant-derived polyphenols in adipogenesis: Perspectives and implications. *Phytomedicine* 19 (3):253-261. doi: 10.1016/j.phymed.2011.12.001.

[66] Lim, T. K. 2014. "Hibiscus Sabdariffa." In *Edible Medicinal and Non Medicinal Plants: Volume 8, Flowers*, edited by T. K. Lim, 359. London: Springer.

[67] Saravanan, G., Ponmurugan, P., Deepa, M. A., and Senthilkumar, B. (2014). Anti-obesity action of gingerol: effect on lipid profile, insulin, leptin, amylase and lipase in male obese rats induced by a high-fat diet. *Journal of the Science of Food and Agriculture* 94 (14):2972-2977. doi: 10.1002/jsfa.6642.

[68] Nigam, N., George, J., and Shukla, Y. 2009. Ginger (6-gingerol). In *Molecular Targets and Therapeutic Uses of Spices Modern Uses for Ancient Medicine*, edited by Bharat B. Aggarwal and Ajaikumar B. Kunnumakkara, 225-228. Singapore: World Scientific.

[69] IHGC. 2016. *International Hop Growers' Convention*. accessed June 18, 2019. https://www.usahops.org/img/blog_pdf/34.pdf.

[70] Liu, M., Hansen, E. P., Wang, G., Qiu, L., Dong, J., Yin, H., Qian, Z., Yang, M., and Miao, J. (2015). Pharmacological Profile of Xanthohumol, a Prenylated Flavonoid from Hops (*Humulus lupulus*). *Molecules* 20 (1). doi: 10.3390/molecules20010754.

[71] Yui, K., Uematsu, H., Muroi, K., Ishii, K., Baba, M., and Osada, K. (2013). Effect of dietary polyphenols from hop (*Humulus lupulus* L.) pomace on adipose tissue mass, fasting blood glucose, hemoglobin A1c, and plasma monocyte chemotactic protein-1 levels in OLETF rats. *Journal of Oleo Science* 62 (5):283-92. doi: 10.5650/jos.62.283.

[72] Zanoli, P., and Zavatti, M. (2008). Pharmacognostic and pharmacological profile of *Humulus lupulus* L. *Journal of Ethnopharmacology* 116 (3):383-396. doi: 10.1016/j.jep.2008.01.011.

[73] Habs, M., Binder, K., Krauss, S., Muller, K., Ernst, B., Valentini, L., and Koller, M. (2017). A Balanced Risk-Benefit Analysis to Determine Human Risks Associated with Pyrrolizidine Alkaloids (PA)-The Case of Tea and Herbal Infusions. *Nutrients* 9 (7). doi: 10.3390/nu9070717.

[74] Kang, J.-H., Jeong, I. S., and Kim, M.-Y. (2018). Antiangiogenic Herbal Composition Ob-X Reduces Abdominal Visceral Fat in Humans: A Randomized, Double-Blind, Placebo-Controlled Study. *Evidence-Based Complementary and Alternative Medicine* 2018:11. doi: 10.1155/2018/4381205.

[75] Choi, B. K., Kim, T. W., Lee, D. R., Jung, W. H., Lim, J. H., Jung, J. Y., Yang, S. H., and Suh, J. W. (2015). A polymethoxy flavonoids-rich *Citrus aurantium* extract ameliorates ethanol-induced liver injury through modulation of AMPK and Nrf2-related signals in a binge drinking mouse model. *Phytotherapy Research* 29 (10):1577-1584. doi: 10.1002/ptr.5415.

[76] Farrington, R., Musgrave, I. F., and Byard, R. W. (2019). Evidence for the efficacy and safety of herbal weight loss preparations. *Journal of Integrative Medicine-Jim* 17 (2):87-92. doi: 10.1016/j.joim.2019.01.009.

[77] Han, H. Y., Lee, S. K., Choi, B. K., Lee, D. R., Lee, H. J., and Kim, T. W. (2019). Preventive Effect of *Citrus aurantium* Peel Extract on High-Fat Diet-Induced Non-alcoholic Fatty Liver in Mice. *Biological & Pharmaceutical Bulletin* 42 (2):255-260.

[78] Jang, J., Jung, Y., Chae, S., Cho, S. H., Yoon, M., Yang, H., Shin, S. S., and Yoon, Y. (2018). Gangjihwan, a polyherbal composition, inhibits fat accumulation through the modulation of lipogenic transcription factors SREBP1C, PPARγ and C/EBPα. *Journal of Ethnopharmacology* 210:10-22. doi: https://doi.org/10.1016/j.jep.2017.08.024.

[79] Kon, R., Ikarashi, N., Nagoya, C., Takayama, T., Kusunoki, Y., Ishii, M., Ueda, H., Ochiai, W., Machida, Y., Sugita, K., and Sugiyama, K. (2014). Rheinanthrone, a metabolite of sennoside A, triggers macrophage activation to decrease aquaporin-3 expression in

the colon, causing the laxative effect of rhubarb extract. *Journal of Ethnopharmacology* 152 (1):190-200. doi: https://doi.org/10.1016/j.jep.2013.12.055.

[80] Saleem, A., Walshe-Roussel, B., Harris, C., Asim, M., Tamayo, C., Sit, S., and Arnason, J.T. (2009). Characterisation of Phenolics in Flor-Essence (R)-a Compound Herbal product and its Contributing Herbs. *Phytochemical Analysis* 20 (5):395-401. doi: 10.1002/pca.1139.

[81] Xue, J., Ding, W., and Liu, Y. (2010). Anti-diabetic effects of emodin involved in the activation of PPARγ on high-fat diet-fed and low dose of streptozotocin-induced diabetic mice. *Fitoterapia* 81 (3):173-177. doi: https://doi.org/10.1016/j.fitote.2009.08.020.

[82] Zheng, C.-D., Duan, Y.-Q., Gao, J.-M., and Ruan, Z.-G. (2010). Screening for Anti-lipase Properties of 37 Traditional Chinese Medicinal Herbs. *Journal of the Chinese Medical Association* 73 (6):319-324. doi: https://doi.org/10.1016/S1726-4901(10)70068-X.

[83] Borges, R. S., Ortiz, B. L. S., Pereira, A. C. M., Keita, H., and Carvalho, J. C. T. (2019). Rosmarinus officinalis essential oil: A review of its phytochemistry, anti-inflammatory activity, and mechanisms of action involved. *Journal of Ethnopharmacology* 229:29-45. doi: https://doi.org/10.1016/j.jep.2018.09.038.

[84] Harach, T., Aprikian, O., Monnard, I., Moulin, J., Membrez, M., Beolor, J. C., Raab, T., Mace, K., and Darimont, C. (2010). Rosemary (*Rosmarinus officinalis* L.) Leaf Extract Limits Weight Gain and Liver Steatosis in Mice Fed a High-Fat Diet. *Planta Medica* 76 (6):566-571. doi: 10.1055/s-0029-1240612.

[85] Peng, Y., Yuan, J., Liu, F., and Ye, J. (2005). Determination of active components in rosemary by capillary electrophoresis with electrochemical detection. *Journal of Pharmaceutical and Biomedical Analysis* 39 (3):431-437. doi: https://doi.org/10.1016/j.jpba.2005.03.033.

[86] Belguith-Hadriche, O., Bouaziz, M., Jamoussi, K., El Feki, A., Sayadi, S., and Makni-Ayedi, F. (2010). Lipid-Lowering and Antioxidant Effects of an Ethyl Acetate Extract of Fenugreek Seeds

in High-Cholesterol-Fed Rats. *Journal of Agricultural and Food Chemistry* 58 (4):2116-2122. doi: 10.1021/jf903186w.

[87] Chaudhary, S., Chaudhary, P. S., Chikara, S. K., Sharma, M. C., and Iriti, M. (2018). Review on Fenugreek (*Trigonella foenum-graecum* L.) and its Important Secondary Metabolite Diosgenin. *Notulae Botanicae Horti Agrobotanici Cluj-Napoca* 46 (1):22-31. doi: 10.15835/nbha46110996.

[88] Chennappaneskar, H., and Lathadevi, G. V. (2018). Hypolipidemic Effects of Fenugreek and Atorvastatin-Comparative Study on High Fat Fed Dyslipidemic Rats. *Journal of Clinical and Diagnostic Research* 12 (9):CC1-CC4. doi: 10.7860/jcdr/2018/35372.11997.

[89] Simão, A. A., Corrêa, A. D., and Chagas, P. M. B. (2012). Inhibition of digestive enzymes by medicinal plant aqueous extracts used to aid the treatment of obesity. *Journal of Medicinal Plants Research* 6 (47):5826-5830.

[90] Taukoorah, U., and Mahomoodally, M. F. (2016). Crude Aloe vera gel shows antioxidant propensities and inhibits pancreatic lipase and glucose movement in vitro. *Advances in Pharmacological Sciences* 2016.

[91] Buchholz, T., and Melzig, M. F. (2016). Medicinal Plants Traditionally Used for Treatment of Obesity and Diabetes Mellitus – Screening for Pancreatic Lipase and α-Amylase Inhibition. *Phytotherapy Research* 30 (2):260-266. doi: 10.1002/ptr.5525.

[92] Marrelli, M., Loizzo, M. R., Nicoletti, M., Menichini, F., and Conforti, F. (2014). In vitro investigation of the potential health benefits of wild Mediterranean dietary plants as anti-obesity agents with alpha-amylase and pancreatic lipase inhibitory activities. *Journal of the Science of Food and Agriculture* 94 (11):2217-2224. doi: 10.1002/jsfa.6544.

[93] Tian, J. Y., Tao, R. Y., Zhang, X. L., Liu, Q., He, Y. B., Su, Y. L., Ji, T. F., and Ye, F. (2015). Effect of Hypericum perforatum L. extract on insulin resistance and lipid metabolic disorder in high-fat-diet induced obese mice. *Phytotherapy Research* 29 (1):86-92. doi: 10.1002/ptr.5230.

[94] Müller, U., Stübl, F., Schwarzinger, B., Sandner, G., Iken, M., Himmelsbach, M., Schwarzinger, C., Ollinger, N., Stadlbauer, V., Höglinger, O., Kühne, T., Lanzerstorfer, P., and Weghuber, J. (2018). In Vitro and In Vivo Inhibition of Intestinal Glucose Transport by Guava (*Psidium Guajava*) Extracts. *Molecular Nutrition & Food Research* 62 (11):1701012. doi: 10.1002/mnfr. 201701012.

[95] Ibarra, A., Cases, J., Roller, M., Chiralt-Boix, A., Coussaert, A., and Ripoll, C. (2011). Carnosic acid-rich rosemary (Rosmarinus officinalis L.) leaf extract limits weight gain and improves cholesterol levels and glycaemia in mice on a high-fat diet. *British Journal of Nutrition* 106 (8):1182-1189. doi: 10.1017/s000711 4511001620.

[96] Ninomiya, K., Matsuda, H., Shimoda, H., Nishida, N., Kasajima, N., Yoshino, T., Morikawa, T., and Yoshikawa, M. (2004). Carnosic acid, a new class of lipid absorption inhibitor from sage. *Bioorganic & Medicinal Chemistry Letters* 14 (8):1943-1946. doi: https://doi. org/10.1016/j.bmcl.2004.01.091.

[97] Romo Vaquero, M., Yáñez-Gascón, M.-J., García Villalba, R., Larrosa, M., Fromentin, E., Ibarra, A., Roller, M., Tomás-Barberán, F., Espín de Gea, J. C., and García-Conesa, M.-T. (2012). Inhibition of Gastric Lipase as a Mechanism for Body Weight and Plasma Lipids Reduction in Zucker Rats Fed a Rosemary Extract Rich in Carnosic Acid. *PLOS ONE* 7 (6):e39773. doi: 10.1371/journal.pone. 0039773.

[98] Fernando, W. I. T., Attanayake, A., Perera, H. K. I., Sivakanesan, R., Jayasinghe, L., Araya, H., and Fujimoto, Y. (2019). Isolation, identification and characterization of pancreatic lipase inhibitors from Trigonella foenum-graecum seeds. *South African Journal of Botany* 121:418-421. doi: 10.1016/j.sajb.2018.10.023.

[99] Hilles, A. R., and Mahmood, S. (2017). A review on phytochemistry and pharmacological effects of Trigonella foenum-graecum %J Advanced Herbal Medicine. *Advanced Herbal Medicine* 3 (3):61-67.

[100] Zhuo, R., Wang, L., Wang, L., Xiao, H., and Cai, S. (2010). Determination of trigonelline in Trigonella foenum-graecum L. by hydrophilic interaction chromatography. *Chinese Journal of Chromatography* 28 (4):379-382. doi: 10.3724/sp.j.1123.2010.00379.

[101] Villiger, A., Sala, F., Suter, A., and Butterweck, V. (2015). In vitro inhibitory potential of Cynara scolymus, Silybum marianum, Taraxacum officinale, and Peumus boldus on key enzymes relevant to metabolic syndrome. *Phytomedicine* 22 (1):138-144.

[102] Sohn, E. H., Kim, T., Jeong, Y. J., Han, H.-S., Lea, Y., Cho, Y. M., and Kang, S. C. (2015). Triglyceride control effect of Agrimonia eupatoria L. in oleic acid induced NAFLD-HepG2 model. *Korean Journal of Plant Resources* 28 (5):635-640.

[103] Kiselova, Y., Vankova, D., Tasinov, O., Feklicheva, I., Potoroko, I., and Ivanova, D. (2017). Changes in interleukin 6 and monocyte chemoattractant protein-1 expression levels in 3t3-l1 cells after exposure to *Agrimonia eupatoria* L. extract and subsequent oxidative stimulation with tert-buthyl hydroperoxide reveal possible adaptogenic potential. *Scripta Scientifica Pharmaceutica* 4 (2):7-13.

[104] Park, S.-J., Lee, I.-S., Lee, S.-P., and Yu, M.-H. (2013). Inhibition of adipocyte differentiation and adipogenesis by supercritical fluid extracts and marc from *Cinnamomum verum*. *Journal of Life Science* 23 (4):510-517.

[105] Hatano, T., Sameshima, Y., Kawabata, M., Yamada, S., Shinozuka, K., Nakabayashi, T., and Mizuno, H. (2014). St. John's Wort Promotes Adipocyte Differentiation and Modulates NF-κB Activation in 3T3-L1 Cells. *Biological and Pharmaceutical Bulletin* 37 (7):1132-1138. doi: 10.1248/bpb.b13-00989.

[106] Richard, A. J., Amini, Z. J., Ribnicky, D. M., and Stephens, J. M. (2012). St. John's Wort inhibits insulin signaling in murine and human adipocytes. *Biochimica et Biophysica Acta* 1822 (4):557-63. doi: 10.1016/j.bbadis.2011.12.005.

[107] Hong, Y., Kim, M.Y., and Yoon, M. (2011). The anti-angiogenic herbal extracts Ob-X from *Morus alba*, *Melissa officinalis*, and

Artemisia capillaris suppresses adipogenesis in 3T3-L1 adipocytes. *Pharmaceutical Biology* 49 (8):775-83. doi: 10.3109/13880209.2010.547208.

[108] Woo, S., Yoon, M., Kim, J., Hong, Y., Kim, M. Y., Shin, S. S., and Yoon, M. (2016). The anti-angiogenic herbal extract from Melissa officinalis inhibits adipogenesis in 3T3-L1 adipocytes and suppresses adipocyte hypertrophy in high fat diet-induced obese C57BL/6J mice. *Journal of Ethnopharmacology* 178:238-50. doi: 10.1016/j.jep. 2015.12.015.

[109] Yang, L., Li, X. F., Gao, L., Zhang, Y.-O., and Cai, G.-P. (2012). Suppressive Effects of Quercetin-3-O-(6″-Feruloyl)-β-D-Galactopyranoside on Adipogenesis in 3T3-L1 Preadipocytes Through Down-regulation of PPARγ and C/EBPα Expression. *Phytotherapy Research* 26 (3):438-444. doi: 10.1002/ptr.3604.

[110] Ahn, E.-K., and Oh, J. S. (2012). Inhibitory effect of galanolactone isolated from *Zingiber officinale* roscoe extract on adipogenesis in 3T3-L1 cells. *Journal of the Korean Society for Applied Biological Chemistry* 55 (1):63-68. doi: 10.1007/s13765-012-0011-6.

[111] Kiyofuji, A., Yui, K., Takahashi, K., and Osada, K. (2014). Effects of xanthohumol-rich hop extract on the differentiation of preadipocytes. *Journal of Oleo Science* 63 (6):593-7. doi: 10.5650/ jos.ess14009.

[112] Quintero-Fabián, S., Ortuño-Sahagún, D., Vázquez-Carrera, M., and López-Roa, R. I. (2013). Alliin, a garlic (*Allium sativum*) compound, prevents LPS-induced inflammation in 3T3-L1 adipocytes. *Mediators of Inflammation* 2013.

[113] Gray, A. M., and Flatt, P. R. (1998). Actions of the traditional anti-diabetic plant, Agrimony eupatoria (agrimony): effects on hyperglycaemia, cellular glucose metabolism and insulin secretion. *British Journal of Nutrition* 80 (1):109-114.

[114] Bratoeva, K., Bekyarova, G., Kiselova, Y., and Ivanova, D. (2010). Effect of Bulgarian herb extracts of polyphenols on metabolic disorders–induced by high fructose diet. *Trakia Journal of Sciences* 8 (2):56-60.

[115] Ivanova, D., Vankova, D., and Nashar, M. (2013). Agrimonia eupatoria tea consumption in relation to markers of inflammation, oxidative status and lipid metabolism in healthy subjects. *Archives of Physiology and Biochemistry* 119 (1):32-37.

[116] Chen, Y.-C., Kao, T.-H., Tseng, C.-Y., Chang, W.-T., and Hsu, C.-L. (2014). Methanolic extract of black garlic ameliorates diet-induced obesity via regulating adipogenesis, adipokine biosynthesis, and lipolysis. *Journal of Functional Foods* 9:98-108.

[117] Sukandar, E., Permana, H., Adnyana, I., Sigit, J., Ilyas, R., Hasimun, P., and Mardiyah, D. (2010). Clinical study of turmeric (*Curcuma longa* L.) and garlic (*Allium sativum* L.) extracts as antihyperglycemic and antihyperlipidemic agent in type-2 diabetes-dyslipidemia patients. *International Journal of Pharmacology* 6 (4):456-463.

[118] Kumar, R., Chhatwal, S., Arora, S., Sharma, S., Singh, J., Singh, N., Bhandari, V., and Khurana, A. (2013). Antihyperglycemic, antihyperlipidemic, anti-inflammatory and adenosine deaminase–lowering effects of garlic in patients with type 2 diabetes mellitus with obesity. *Diabetes, metabolic syndrome and obesity: targets and therapy* 6:49.

[119] Jung, Y.-M., Lee, S.-H., Lee, D.-S., You, M.-J., Chung, I. K., Cheon, W. H., Kwon, Y.-S., Lee, Y.-J., and Ku, S.-K. (2011). Fermented garlic protects diabetic, obese mice when fed a high-fat diet by antioxidant effects. *Nutrition research* 31 (5):387-396.

[120] Lee, H. S., Lim, W. C., Lee, S. J., Lee, S. H., Lee, J. H., and Cho, H. Y. (2016). Antiobesity Effect of Garlic Extract Fermented by *Lactobacillus plantarum* BL2 in Diet-Induced Obese Mice. *Journal of Medicinal Food* 19 (9):823-829. doi: 10.1089/jmf.2016.3674.

[121] Yang, C., Li, L., Yang, L., Lǚ, H., Wang, S., and Sun, G. (2018). Anti-obesity and Hypolipidemic effects of garlic oil and onion oil in rats fed a high-fat diet. *Nutrition & Metabolism* 15 (1):43.

[122] Lai, Y. S., Chen, W. C., Ho, C. T., Lu, K. H., Lin, S. H., Tseng, H. C., Lin, S. Y., and Sheen, L. Y. (2014). Garlic essential oil protects against obesity-triggered nonalcoholic fatty liver disease through

modulation of lipid metabolism and oxidative stress. *Journal of Agricultural and Food Chemistry* 62 (25):5897-5906.

[123] Kim, I., Kim, H. R., Kim, J. H., and Om, A. S. (2013). Beneficial effects of Allium sativum L. stem extract on lipid metabolism and antioxidant status in obese mice fed a high-fat diet. *Journal of the Science of Food and Agriculture* 93 (11):2749-2757.

[124] Fantini, N., Colombo, G., Giori, A., Riva, A., Morazzoni, P., Bombardelli, E., and Carai, M. A. (2011). Evidence of glycemia-lowering effect by a Cynara scolymus L. extract in normal and obese rats. *Phytotherapy Research* 25 (3):463-466.

[125] Magied, M. A., Hussien, S., Zaki, S. M., and Said, R. (2016). Artichoke (*Cynara scolymus* L.) leaves and heads extracts as hypoglycemic and hypocholesterolemic in rats. *Journal of Food and Nutrition Research* 4 (1):60-68.

[126] Kwon, E.-Y., Kim, S., and Choi, M.-S. (2018). Luteolin-Enriched Artichoke Leaf Extract Alleviates the Metabolic Syndrome in Mice with High-Fat Diet-Induced Obesity. *Nutrients* 10 (8):979.

[127] Loi, B., Fantini, N., Colombo, G., Gessa, G. L., Riva, A., Bombardelli, E., Morazzoni, P., and Carai, M. A. (2013). Reducing effect of a combination of *Phaseolus vulgaris* and *Cynara scolymus* extracts on food intake and glycemia in rats. *Phytotherapy Research* 27 (2):258-263.

[128] Rondanelli, M., Giacosa, A., Orsini, F., Opizzi, A., and Villani, S. (2011). Appetite control and glycaemia reduction in overweight subjects treated with a combination of two highly standardized extracts from *Phaseolus vulgaris* and *Cynara scolymus*. *Phytotherapy Research* 25 (9):1275-1282.

[129] Rondanelli, M., Opizzi, A., Faliva, M., Sala, P., Perna, S., Riva, A., Morazzoni, P., Bombardelli, E., and Giacosa, A. (2014). Metabolic management in overweight subjects with naive impaired fasting glycaemia by means of a highly standardized extract from Cynara scolymus: A double-blind, placebo-controlled, randomized clinical trial. *Phytotherapy Research* 28 (1):33-41.

[130] Misael, P. C. E., De Guadalupe, T., del CSGM, P., and Carlos, R. L. J. (2014). Effect of Cynara scolymus (artichoke) in homeopathic doses on body mass index in obese and overweight patients. *Biomedical and Pharmacology Journal* 7 (2):525-533.

[131] Kim, K., Kim, H., Kwon, J., Lee, S., Kong, H., Im, S.-A., Lee, Y.-H., Lee, Y.-R., Oh, S.-T., and Jo, T. H. (2009). Hypoglycemic and hypolipidemic effects of processed Aloe vera gel in a mouse model of non-insulin-dependent diabetes mellitus. *Phytomedicine* 16 (9):856-863.

[132] Devaraj, S., Yimam, M., Brownell, L. A., Jialal, I., Singh, S., and Jia, Q. (2013). Effects of Aloe vera supplementation in subjects with prediabetes/metabolic syndrome. *Metabolic Syndrome and Related Disorders* 11 (1):35-40.

[133] Shin, E., Shim, K. S., Kong, H., Lee, S., Shin, S., Kwon, J., Jo, T. H., Park, Y. I., Lee, C. K., and Kim, K. (2011). Dietary aloe improves insulin sensitivity via the suppression of obesity-induced inflammation in obese mice. *Immune Network* 11 (1):59-67.

[134] Cignarella, A., Nastasi, M., Cavalli, E., and Puglisi, L. (1996). Novel lipid-lowering properties of *Vaccinium myrtillus* L. leaves, a traditional antidiabetic treatment, in several models of rat dyslipidaemia: a comparison with ciprofibrate. *Thrombosis Research* 84 (5):311-322.

[135] Asgary, S., RafieianKopaei, M., Sahebkar, A., Shamsi, F., and Goli-malekabadi, N. (2016). Anti-hyperglycemic and anti-hyperlipidemic effects of *Vaccinium myrtillus* fruit in experimentally induced diabetes (antidiabetic effect of *Vaccinium myrtillus* fruit). *Journal of the Science of Food and Agriculture* 96 (3):764-768.

[136] Mykkänen, O. T., Huotari, A., Herzig, K.-H., Dunlop, T. W., Mykkänen, H., and Kirjavainen, P. V. (2014). Wild blueberries (*Vaccinium myrtillus*) alleviate inflammation and hypertension associated with developing obesity in mice fed with a high-fat diet. *PLoS One* 9 (12):e114790.

[137] Sidorova, Y., Shipelin, V., Mazo, V., Zorin, S., Petrov, N., and Kochetkova, A. (2017). Hypoglycemic and hypolipidemic effect of *Vaccinium myrtillus* L. leaf and *Phaseolus vulgaris* L. seed coat extracts in diabetic rats. *Nutrition* 41:107-112.

[138] Zagayko, A. L., Kolisnyk, T. Y., Chumak, O. I., Ruban, O. A., and Koshovyi, M. (2018). Evaluation of anti-obesity and lipid-lowering properties of *Vaccinium myrtillus* leaves powder extract in a hamster model. *Journal of Basic and Clinical Physiology and Pharmacology* 29 (6):697-703.

[139] Husain, G. M., Chatterjee, S. S., Singh, P. N., and Kumar, V. (2011). Hypolipidemic and antiobesity-like activity of standardised extract of *Hypericum perforatum* L. in rats. *ISRN Pharmacology* 2011. doi: 10.5402/2011/505247.

[140] Yatim, A., and Nor, N. M. (2011). Pink guava (*Psidium guajava*) puree supplement on high fat diet-induced obese rats: effects on antioxidant enzyme activities, kidney and liver functions. *Sains Malaysiana* 40 (2):147-154.

[141] Mamun, M. A. A., Faruk, M., Rahman, M. M., Nahar, K., Kabir, F., Alam, M. A., and Subhan, N. (2019). High Carbohydrate High Fat Diet Induced Hepatic Steatosis and Dyslipidemia Were Ameliorated by *Psidium guajava* Leaf Powder Supplementation in Rats. *Evidence-Based Complementary and Alternative Medicine* 2019:12. doi: 10.1155/2019/1897237.

[142] Díaz-de-Cerio, E., Rodríguez-Nogales, A., Algieri, F., Romero, M., Verardo, V., Segura-Carretero, A., Duarte, J., and Galvez, J. (2017). The hypoglycemic effects of guava leaf (*Psidium guajava* L.) extract are associated with improving endothelial dysfunction in mice with diet-induced obesity. *Food Research International* 96:64-71. doi: 10.1016/j.foodres.2017.03.019.

[143] Soliman, M. M., Attia, H. F., El-Shazly, S. A., and Saleh, O. M. (2012). Biomedical effects of cinnamon extract on obesity and diabetes relevance in Wistar rats. *American Journal of Biochemistry and Molecular Biology* 2 (3):133-45.

[144] Anderson, R. A., Qin, B., Canini, F., Poulet, L., and Roussel, A. M. (2013). Cinnamon counteracts the negative effects of a high fat/high fructose diet on behavior, brain insulin signaling and Alzheimer-associated changes. *PloS one* 8 (12):e83243.

[145] Joshi, S. C., Jain, P. K., and Sharma, P. (2017). Antiatherosclerotic and lipid lowering effect of *Cinnamomum verum* in cholesterol-fed rabbits *International Journal of Current Pharmacological Research* 9 (1):75-80.

[146] Camacho, S., Michlig, S., de Senarclens-Bezençon, C., Meylan, J., Meystre, J., Pezzoli, M., Markram, H., and Le Coutre, J. (2015). Anti-obesity and anti-hyperglycemic effects of cinnamaldehyde via altered ghrelin secretion and functional impact on food intake and gastric emptying. *Scientific Reports* 5:7919.

[147] Villalpando-Arteaga, E. V., Mendieta-Condado, E., Esquivel-Solís, H., Canales-Aguirre, A. A., Gálvez-Gastélum, F. J., Mateos-Díaz, J. C., Rodríguez-González, J. A., and Márquez-Aguirre, A. L. (2013). *Hibiscus sabdariffa* L. aqueous extract attenuates hepatic steatosis through down-regulation of PPAR-γ and SREBP-1c in diet-induced obese mice. *Food & Function* 4 (4):618-626. doi: 10.1039/c3fo30270a.

[148] Moyano, G., Sáyago-Ayerdi, S. G., Largo, C., Caz, V., Santamaria, M., and Tabernero, M. (2016). Potential use of dietary fibre from *Hibiscus sabdariffa* and *Agave tequilana* in obesity management. *Journal of Functional Foods* 21:1-9. doi: 10.1016/j.jff.2015.11.011.

[149] Kao, E. S., Yang, M. Y., Hung, C. H., Huang, C. N., and Wang, C. J. (2016). Polyphenolic extract from Hibiscus sabdariffa reduces body fat by inhibiting hepatic lipogenesis and preadipocyte adipogenesis. *Food & Function* 7 (1):171-82. doi: 10.1039/c5fo00714c.

[150] Lee, Y. S., Yang, W. K., Kim, H. Y., Min, B., Caturla, N., Jones, J., Park, Y. C., Lee, Y. C., and Kim, S. H. (2018). Metabolaid((R)) Combination of Lemon Verbena and Hibiscus Flower Extract Prevents High-Fat Diet-Induced Obesity through AMP-Activated Protein Kinase Activation. *Nutrients* 10 (9). doi: 10.3390/nu10091204.

[151] Ibrahim, K. G., Chivandi, E., Mojiminiyi, F. B. O., and Erlwanger, K. H. (2017). The response of male and female rats to a high-fructose diet during adolescence following early administration of *Hibiscus sabdariffa* aqueous calyx extracts. *Journal of Developmental Origins of Health and Disease* 8 (6):628-637. doi: 10.1017/s2040174417 00040x.

[152] Herranz-López, M., Olivares-Vicente, M., Boix-Castejón, M., Caturla, N., Roche, E., and Micol, V. (2019). Differential effects of a combination of *Hibiscus sabdariffa* and *Lippia citriodora* polyphenols in overweight/obese subjects: A randomized controlled trial. *Scientific Reports* 9 (1):2999. doi: 10.1038/s41598-019-39159-5.

[153] Boix-Castejon, M., Herranz-Lopez, M., Perez Gago, A., Olivares-Vicente, M., Caturla, N., Roche, E., and Micol, V. (2018). Hibiscus and lemon verbena polyphenols modulate appetite-related biomarkers in overweight subjects: a randomized controlled trial. *Food & Function* 9 (6):3173-3184. doi: 10.1039/c8fo00367j.

[154] Gurrola-Díaz, C. M., García-López, P. M., Sánchez-Enríquez, S., Troyo-Sanromán, R., Andrade-González, I., and Gómez-Leyva, J. F. (2010). Effects of *Hibiscus sabdariffa* extract powder and preventive treatment (diet) on the lipid profiles of patients with metabolic syndrome (MeSy). *Phytomedicine* 17 (7):500-505. doi: 10.1016/j.phymed.2009.10.014.

[155] Chang, H. C., Peng, C. H., Yeh, D. M., Kao, E. S., and Wang, C. J. (2014). *Hibiscus sabdariffa* extract inhibits obesity and fat accumulation, and improves liver steatosis in humans. *Food & Function* 5 (4):734-9. doi: 10.1039/c3fo60495k.

[156] Nazish, I., Ansari, S., Arora, P., and Ahmad, A. (2016). Antiobesity activity of *Zingiber officinale*. *Pharmacognosy Journal* 8 (5). doi: 10.5530/pj.2016.5.5.

[157] Misawa, K., Hashizume, K., Yamamoto, M., Minegishi, Y., Hase, T., and Shimotoyodome, A. (2015). Ginger extract prevents high-fat diet-induced obesity in mice via activation of the peroxisome

proliferator-activated receptor δ pathway. *The Journal of Nutritional Biochemistry* 26 (10):1058-1067. doi: 10.1016/j.jnutbio.2015.04.014.

[158] Ebrahimzadeh Attari, V., Ostadrahimi, A., Asghari Jafarabadi, M., Mehralizadeh, S., and Mahluji, S. (2016). Changes of serum adipocytokines and body weight following *Zingiber officinale* supplementation in obese women: a RCT. *European Journal of Nutrition* 55 (6):2129-36. doi: 10.1007/s00394-015-1027-6.

[159] Ebrahimzadeh Attari, V., Mahluji, S., Asghari Jafarabadi, M., and Ostadrahimi, A. (2015). Effects of Supplementation with Ginger (*Zingiber officinale* Roscoe) on Serum Glucose, Lipid Profile and Oxidative Stress in Obese Women: A Randomized, Placebo-Controlled Clinical Trial. *Pharmaceutical Sciences* 21 (4):184-191. doi: 10.15171/ps.2015.35.

[160] Ebrahimzadeh Attari, V., Asghari Jafarabadi, M., Zemestani, M., and Ostadrahimi, A. (2015). Effect of *Zingiber officinale* supplementation on obesity management with respect to the uncoupling protein 1-3826A> G and ß3-adrenergic receptor Trp64Arg polymorphism. *Phytotherapy Research* 29 (7):1032-1039. doi: 10.1002/ptr.5343.

[161] Miranda, C. L., Elias, V. D., Hay, J. J., Choi, J., Reed, R. L., and Stevens, J. F. (2016). Xanthohumol improves dysfunctional glucose and lipid metabolism in diet-induced obese C57BL/6J mice. *Archives of Biochemistry and Biophysics* 599:22-30. doi: 10.1016/j.abb.2016.03.008.

[162] Takahashi, K., and Osada, K. (2017). Effect of Dietary Purified Xanthohumol from Hop (Humulus lupulus L.) Pomace on Adipose Tissue Mass, Fasting Blood Glucose Level, and Lipid Metabolism in KK-Ay Mice. *Journal of Oleo Science* 66 (5):531-541. doi: 10.5650/jos.ess16234.

[163] Legette, L. L., Luna, A. Y., Reed, R. L., Miranda, C. L., Bobe, G., Proteau, R. R., and Stevens, J. F. (2013). Xanthohumol lowers body weight and fasting plasma glucose in obese male Zucker fa/fa rats. *Phytochemistry* 91:236-41. doi: 10.1016/j.phytochem.2012.04.018.

[164] Yui, K., Kiyofuji, A., and Osada, K. (2014). Effects of xanthohumol-rich extract from the hop on fatty acid metabolism in rats fed a high-fat diet. *Journal of Oleo Science* 63 (2):159-68. doi: 10.5650/jos.ess 13136.

[165] Mahli, A., Koch, A., Fresse, K., Schiergens, T., Thasler, W. E., Schonberger, C., Bergheim, I., Bosserhoff, A., and Hellerbrand, C. (2018). Iso-alpha acids from hops (*Humulus lupulus*) inhibit hepatic steatosis, inflammation, and fibrosis. *Laboratory Investigation* 98 (12):1614-1626. doi: 10.1038/s41374-018-0112-x.

[166] Everard, A., Geurts, L., Van Roye, M., Delzenne, N. M., and Cani, P. D. (2012). Tetrahydro iso-alpha acids from hops improve glucose homeostasis and reduce body weight gain and metabolic endotoxemia in high-fat diet-fed mice. *PLoS One* 7 (3):e33858. doi: 10.1371/journal.pone.0033858.

[167] Vroegrijk, I. O. C. M., van Diepen, J. A., van den Berg, S. A. A., Romijn, J. A., Havekes, L. M., van Dijk, K. W., Darland, G., Konda, V., Tripp, M. L., Bland, J. S., and Voshol, P. J. (2013). META060 protects against diet-induced obesity and insulin resistance in a high-fat–diet fed mouse. *Nutrition* 29 (1):276-283. doi: 10.1016/j.nut. 2012.05.004.

[168] Foca, G., Ulrici, A., Cocchi, M., Durante, C., Vigni, M. L., Marchetti, A., Sighinolfi, S., and Tassi, L. 2011. Chapter 76 - Seeds of Horse Chestnut (*Aesculus hippocastanum* L.) and Their Possible Utilization for Human Consumption. In *Nuts and Seeds in Health and Disease Prevention*, edited by Victor R. Preedy, Ronald Ross Watson and Vinood B. Patel, 653-661. San Diego: Academic Press.

[169] Harrold, J. A., Hughes, G. M., O'Shiel, K., Quinn, E., Boyland, E. J., Williams, N. J., and Halford, J. C. (2013). Acute effects of a herb extract formulation and inulin fibre on appetite, energy intake and food choice. *Appetite* 62:84-90. doi: 10.1016/j.appet.2012.11.018.

[170] Wootten, D., Simms, J., Koole, C., Woodman, O. L., Summers, R. J., Christopoulos, A., and Sexton, P. M. (2011). Modulation of the glucagon-like peptide-1 receptor signaling by naturally occurring and synthetic flavonoids. *Journal of Pharmacology and*

Experimental Therapy 336 (2):540-50. doi: 10.1124/jpet.110. 176362.

[171] Choo, C. Y., Sulong, N. Y., Man, F., and Wong, T. W. (2012). Vitexin and isovitexin from the Leaves of *Ficus deltoidea* with in-vivo alpha-glucosidase inhibition. *Journal of Ethnopharmacology* 142 (3):776-81. doi: 10.1016/j.jep.2012.05.062.

[172] Hong, J.S., Jin Woo Kim, Byung Il Yoon, Ki-Jong Rhee, Chang Six Rha, and Jung, a.B.D. (2017). Effect of *Ginkgo biloba* Extract (EGb 761) on Serum Cholesterol Levels in Wild-Type C57Bl/6 Mice. *Biomedical Science Letters* 23 (no. 2):80-88 doi: 10.15616/bsl.2017. 23.2.80.

[173] Weidner, C., Wowro, S. J., Freiwald, A., Kodelja, V., Abdel-Aziz, H., Kelber, O., and Sauer, S. (2014). Lemon balm extract causes potent antihyperglycemic and antihyperlipidemic effects in insulin-resistant obese mice. *Molecular Nutrition and Food Research* 58 (4):903-7. doi: 10.1002/mnfr.201300477.

[174] Jun, H. J., Lee, J. H., Jia, Y., Hoang, M. H., Byun, H., Kim, K. H., and Lee, S. J. (2012). *Melissa officinalis* essential oil reduces plasma triglycerides in human apolipoprotein E2 transgenic mice by inhibiting sterol regulatory element-binding protein-1c-dependent fatty acid synthesis. *Journal of Nutrition* 142 (3):432-40. doi: 10.3945/jn.111. 152538.

[175] Kim, J., Lee, H., Lim, J., Oh, J., Shin, S. S., and Yoon, M. (2017). The Angiogenesis Inhibitor ALS-L1023 from Lemon-Balm Leaves Attenuates High-Fat Diet-Induced Nonalcoholic Fatty Liver Disease through Regulating the Visceral Adipose-Tissue Function. *International Journal of Molecular Sciences* 18 (4). doi: 10.3390/ijms18040846.

[176] Kim, J., Lee, H., Lim, J., Lee, H., Yoon, S., Shin, S. S., and Yoon, M. (2017). The lemon balm extract ALS-L1023 inhibits obesity and nonalcoholic fatty liver disease in female ovariectomized mice. *Food Chemical and Toxicology* 106 (Pt A):292-305. doi: 10.1016/ j.fct.2017.05.059.

[177] Park, B. Y., Lee, H., Woo, S., Yoon, M., Kim, J., Hong, Y., Lee, H. S., Park, E. K., Hahm, J. C., Kim, J. W., Shin, S. S., Kim, M. Y., and Yoon, M. (2015). Reduction of Adipose Tissue Mass by the Angiogenesis Inhibitor ALS-L1023 from *Melissa officinalis*. *PLoS One* 10 (11):e0141612. doi: 10.1371/journal.pone.0141612.

[178] Kim, M. Y., Park, B. Y., Lee, H. S., Park, E. K., Hahm, J. C., Lee, J., Hong, Y., Choi, S., Park, D., Lee, H., and Yoon, M. (2010). The anti-angiogenic herbal composition Ob-X inhibits adipose tissue growth in obese mice. *International Journal of Obesity (Lond)* 34 (5):820-30. doi: 10.1038/ijo.2010.13.

[179] Yoon, M., and Kim, M. Y. (2011). The anti-angiogenic herbal composition Ob-X from *Morus alba, Melissa officinalis,* and *Artemisia capillaris* regulates obesity in genetically obese ob/ob mice. *Pharmacological Biology* 49 (6):614-9. doi: 10.3109/ 13880209.2010. 539617.

[180] Wang, T., Takikawa, Y., Satoh, T., Yoshioka, Y., Kosaka, K., Tatemichi, Y., and Suzuki, K. (2011). Carnosic acid prevents obesity and hepatic steatosis in ob/ob mice. *Hepatology Research* 41 (1):87-92. doi: 10.1111/j.1872-034X.2010.00747.x.

[181] Rau, O., Wurglics, M., Paulke, A., Zitzkowski, J., Meindl, N., Bock, A., Dingermann, T., Abdel-Tawab, M., and Schubert-Zsilavecz, M. (2006). Carnosic Acid and Carnosol, Phenolic Diterpene Compounds of the Labiate Herbs Rosemary and Sage, are Activators of the Human Peroxisome Proliferator-Activated Receptor Gamma. *Planta Medica* 72 (10):881-887. doi: 10.1055/s-2006-946680.

[182] Colker, C. M., Kaiman, D. S., Torina, G. C., Perlis, T., and Street, C. (1999). Effects of *Citrus aurantium* extract, caffeine, and St. John's Wort on body fat loss, lipid levels, and mood states in overweight healthy adults. *Current Therapeutic Research* 60 (3):145-153. doi: https://doi.org/10.1016/S0011-393X(00)88523-9.

[183] Haaz, S., Fontaine, K. R., Cutter, G., Limdi, N., Perumean-Chaney, S., and Allison, D. B. (2006). *Citrus aurantium* and synephrine alkaloids in the treatment of overweight and obesity: an update.

Obesity Reviews 7 (1):79-88. doi: 10.1111/j.1467-789X.2006.00195.x.

[184] Jones, D. 2001. Regulation of appetite, body weight and athletic function with materials derived from citrus varieties. United States: Google Patents.

[185] Hedrei, P., and Gougeon, R. (1997). Thermogenic effect of beta-sympathicomimetic compounds extracted from *Citrus aurantium. McGill Nutrition Food Science Center, Royal Victoria Hospital* Canadá.

[186] Pathak, B., Gougeon, R., and Center, M. (1999). Thermic effect of *Citrus aurantium* in obese subjects. *Current Therapeutic Research* 60:145-151.

[187] Chiba, H., Uehara, M., Wu, J., Wang, X., Masuyama, R., Suzuki, K., Kanazawa, K., and Ishimi, Y. (2003). Hesperidin, a Citrus Flavonoid, Inhibits Bone Loss and Decreases Serum and Hepatic Lipids in Ovariectomized Mice. *The Journal of Nutrition* 133 (6):1892-1897. doi: 10.1093/jn/133.6.1892.

[188] Mulvihill, E. E., Assini, J. M., Lee, J. K., Allister, E. M., Sutherland, B. G., Koppes, J. B., Sawyez, C. G., Edwards, J. Y., Telford, D. E., Charbonneau, A., St-Pierre, P., Marette, A., and Huff, M. W. (2011). Nobiletin Attenuates VLDL Overproduction, Dyslipidemia, and Atherosclerosis in Mice With Diet-Induced Insulin Resistance. *Diabetes* 60 (5):1446-1457. doi: 10.2337/db10-0589.

[189] Lim, J., Lee, H., Ahn, J., Kim, J., Jang, J., Park, Y., Jeong, B., Yang, H., Shin, S. S., and Yoon, M. (2018). The polyherbal drug GGEx18 from *Laminaria japonica, Rheum palmatum,* and *Ephedra sinica* inhibits hepatic steatosis and fibroinflammtion in high-fat diet-induced obese mice. *Journal of Ethnopharmacology* 225:31-41. doi: https://doi.org/10.1016/j.jep.2018.06.034.

In: Phytochemicals from Mexican … ISBN: 978-1-53617-397-0
Editors: J. Basilio Heredia et al.

Chapter 4

THE METABOLIC SYNDROME

M. G. Sánchez-Otero[1], A. Alexander-Aguilera[1], A. Ramírez-Higuera[2], M. Juárez-Arias[2] and R. M. Oliart-Ros[2,*]

[1]Facultad de Bioanálisis, Universidad Veracruzana, Veracruz, México
[2]UNIDA-Tecnológico Nacional de México/IT Veracruz, Veracruz, México

ABSTRACT

Metabolic syndrome (MetS) is one of the major public health challenges worldwide; this syndrome is a constellation of metabolic abnormalities such as hypertension, central obesity, insulin resistance, and atherogenic dyslipidemia. MetS is related with an increased risk for developing cardiovascular diseases. Some medicinal plants or their natural compounds have been used in traditional medicine to treat or prevent some of the manifestations of MetS. In this chapter, we summarize existing available information regarding studies about the use of Mexican medicinal plants in MetS parameters treatment.

---

* Corresponding Author's Email: roliart@itver.edu.mx.

Keywords: metabolic syndrome, medicinal plants

INTRODUCTION

The metabolic syndrome (MetS) is defined as the set of simultaneous alterations related to cardiovascular risks, such as obesity, insulin resistance, hypertension, hyperglycemia, reduced level of HDL-C and hypertriglyceridemia; the pathogenesis of MS involves both genetic and acquired factors that play a role in the final pathway of inflammation that leads to cardiovascular diseases which, along with Type 2 diabetes mellitus, have become the main causes of morbidity and mortality in the world. MetS is associated with a twofold increased risk to develop cardiovascular disease and a fivefold risk to develop diabetes mellitus, so early diagnosis is important in order to employ lifestyle and risk factors modifications effectively; management of MetS involves a dual approach that combines pharmacological therapy and lifestyle changes, the first is aimed at treating the individual components of MetS with drugs such as metformin and antihypertensives; the latter is based on the use of nutraceuticals, defined as dietary supplements that provide health benefits. In this sense, botanical extracts that ameliorate pathogenic mechanisms involved in MetS parameters represent an alternative and a complementary approach for its prevention or treatment [1, 2].

Various natural compounds derived from plant extracts, spices, herbs, and essential oils, such as curcumin, allicin from garlic [3, 4], resveratrol from grapes [5, 6], and polyunsaturated fatty acids from fish oils, among others [7, 8] have demonstrated benefits in the management of the manifestations of MetS; however, the positive and/or negative effects of these nutraceuticals are still under investigation in order to generate viable, non-toxic and affordable pharmaceutical preparations.

METABOLIC SYNDROME

Definition

Metabolic syndrome (MetS), also known as Syndrome X, insulin resistance syndrome, hypertriglyceridemic waist, and the deadly quartet, is recognized as an important cardiovascular risk factor. The diabetes consultation group of the World Health Organization realized the first

Table 1. Definitions of metabolic syndrome

World Organization of Health [10]	European Group for the Study of Insulin Resistance [11]	ATP III actualization 2005 AHA [12]	International Federation of Diabetes (IDF), 2005 [13]
Two or more of the following criteria:	Insulin resistance or hyperinsulinemia	Three or more of the following criteria:	Central Obesity: Waist circumference: > 80 cm women > 94 cm men
BMI> 29.9 kg / m2 and / or waist / hip (men> 0.9 and women> 0.85	Besides, two or more of the following criteria:	Obesity: Perimeter waist > 88 cm in women and > 102 cm in men	Also two of the following criteria:
Arterial hypertension (140/90 mm / Hg) or treatment	Central obesity: ≥ 80 cm women ≥ 94 cm men	Arterial hypertension (130/85 mm / Hg)	Arterial hypertension (130/85 mm / Hg) or treatment
Hypertriglyceridemia (> 150 mg / dL) and / or HDL cholesterol <35 mg / dL men or <40 women	Arterial hypertension (140/90 mm / Hg) or treatment	Hypertriglyceridemia (> 150 mg / dL)	Hypertriglyceridemia (> 150 mg / dL) or treatment
Microalbuminuria> 20 μg / min	Dyslipidemia: Triglycerides > 2.0 mmol / l or cHDL <1.0	HDL cholesterol <40mg /dL men <50mg / dL women	HDL-cholesterol <40mg / dL men <50 mg / dL women or treatment
Also, DM2 or glucose intolerance or resistance to insulin	Fasting blood sugar ≥ 6,1 mmol/L	Diabetes or glucose anomalous baseline (≥100 mg / dL)	Diabetes or glucose anomalous baseline (≥100 mg / dL)

internationally recognized definition of MetS in 1988 [9]. This organization pointed MetS as the presence of insulin resistance which is related to impaired fasting glucose, impaired glucose tolerance, or type 2 diabetes mellitus, together with two of the following risk factors: obesity obtained for waist-hip ratio or body mass index, hyperlipidemia (hypertriglyceridemia, low levels of the high-density lipoprotein (HDL) cholesterol), hypertension or microalbuminuria. Various definitions have been proposed, according to the criteria of the organization (Table 1).

Epidemiology

MetS is a problem of public health in the world. It affects 20%–30% of the population in developed countries. Its worldwide constant increase represents a risk factor for type 2 diabetes (T2D), cardiovascular diseases (CVD) [14, 15], and premature mortality [16]. There is no global data on metabolic syndrome prevalence since it is hard to differentiate from T2D, obesity, or CVD, but since this condition is about three times more common than diabetes, global prevalence could be estimated to be about one quarter of the world's population, that would represent around a billion people affected by MetS [16].

Pathogenesis

Along with the prevalence of MetS, our understanding of the mechanisms involved in the pathogenesis of this disease has increased, which has permitted the development of effective prevention strategies and adequate medical interventions in order to decrease the prevalence and limit its comorbidities, particularly in young population; the understanding of adipose tissue role in the overall physiology has made clear that it is not only an "energy storage depot" but an active tissue, which secrets hormones, like leptin and adiponectin, that have important and sometimes opposite effects on appetite, satiety, and energy metabolism regulation [16].

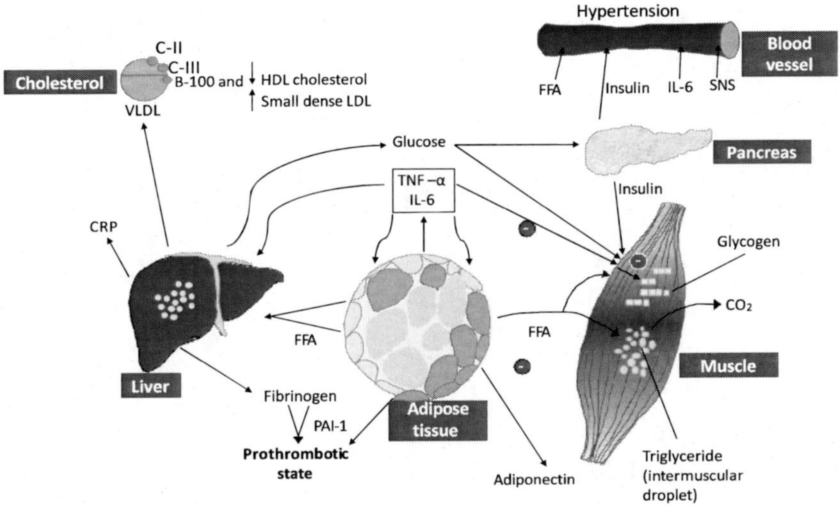

Figure 1. Pathophysiology of MetS [31 adapted].

There are several mechanisms to explain the pathophysiology of MetS; the most important is related to the presence of insulin resistance associated with an increased fatty acid flux, inflammation, and oxidative stress. These mechanisms are complex and have been partially elucidated (Figure 1). Whether the individual components of MetS represent separate pathologies or manifestations of an only central mechanism is still debated. Visceral adiposity is one of the most important conditions involved in MetS, stressing the influence of high caloric intake and lack of physical activity as major causative etiological factors. Of all the proposed mechanisms, insulin resistance, neurohormonal activation, and chronic inflammation emerge as responsible for the initiation and development of MetS [9, 17].

Insulin Resistance

Insulin plays a crucial role in the pathogenesis of MetS since, according to Rochliani et al. [2] 'it increases glucose uptake in muscle and liver and inhibits lipolysis and hepatic gluconeogenesis. Insulin resistance in adipose tissue impairs insulin-mediated inhibition of lipolysis, leading to an increase in circulating' free fatty acids (FFAs) [18, 19]. High levels of FFAs inhibit protein kinase activation in the muscle, decreasing glucose uptake, and increase activation in the liver, promoting gluconeogenesis and lipogenesis. In addition, a hyperinsulinemic state is developed due to a compensatory secretion of insulin that, along with the lipotoxicity of FFAs, damage pancreatic beta cells functionality in the long term resulting in a decrease in insulin secretion and finally to the development of type 2 DM [11, 19]. Insulin resistance also contributes to the development of arterial hypertension due to the loss of the vasodilator effect of insulin and the vasoconstriction caused by FFA [12, 20]. Additionally, increased sympathetic activation and reabsorption of sodium in the kidneys provoke an increase in serum viscosity, the induction of a prothrombotic state, and the release of proinflammatory cytokines from adipose tissue, which altogether rise the risk for CVD [13, 21].

Inflammation

Low grade or chronic inflammation is mainly triggered by nutrients and metabolic disorders, characterized by abnormal cytokine production, activation of a network of inflammatory signaling pathways, and increased production of acute-phase reactants and other mediators [3, 21]. MetS and abdominal obesity are recognized to be a proinflammatory and prothrombotic physiological state. Adipose tissue is an endocrine and paracrine organ with many biological functions. Adipocytes undergo hypertrophy and hyperplasia in response to caloric excess, which leads cells to an hypoxic state and to cell necrosis with macrophage infiltration, and the production of adipocytokines such as the pro-inflammatory mediators interleukin-6 (IL-6) and tumor necrosis factor-alpha (TNF-α), and the prothrombotic mediator plasminogen activator inhibitor-1 (PAI-1) [14-16, 22]. Lipids also regulate metabolic, inflammatory and innate immune processes, through the activation of transcription factors, such as the peroxisome-proliferator activated receptor (PPAR) and liver X receptor (LXR) families; the ligands to members of PPAR family suppress production of proinflammatory cytokines, mainly by suppressing the inhibitor of nuclear factor-κB (NF-κB), lipid chaperone proteins, such as adipocyte/macrophage fatty-acid-binding proteins (FABPs), and influence inflammatory pathways, including the activation of c-Jun amino-terminal kinase (JNK) and insulin action target (IKK) in cells [21].

Oxidative Stress

MetS is characterized by the increase of many chemical and immunological compounds associated with cellular stress, but particularly plasma-free fatty acids (FFAs) that activate reactive oxygen species (ROS) production [17]. Even though the mechanism is still poorly understood, elevated levels of FFA have shown to enhance $^{\bullet}OH$, $O_2^{-\bullet}$, and H_2O_2 in endothelial and vascular smooth muscle cells (VSMC) [18, 23]. Moreover, fat accumulation constitutes a trigger for ROS augmentation in adipose tissue, likely by stimulating the homologs molecules of nicotinamide adenine dinucleotide phosphate oxidase (NOX) and lowering antioxidative enzymes [19, 24]. The exposure of adipose tissues to OxS under MetS conditions is a determining factor for a decrease in anti-inflammatory adiponectin and an increase in inflammatory cytokines [20, 21, 25, 26], that lead to alterations in insulin signaling and glucose transporter 4 (GLUT4) translocation [3, 27].

Additionally, under the MetS pathophysiological conditions, suppression of the transcription factor p53 downregulates basal glutathione peroxidase enzyme (GPX1) and superoxide dismutase (SOD2) expression, which induces ROS production in cells, leading to oxidative damage, whereas re-establishing physiological levels of p53 enhances the antioxidant enzymes and lowers ROS formation [24-26, 28, 29].

Metabolic Syndrome Management

Strategies for the prevention and treatment of MetS include the increase in physical activity and exercise, weight reduction and maintenance of body weight, pharmacological interventions, and dietary approaches that include calories-restriction, low-carb or low-protein diets, a high intake of omega-3 polyunsaturated fatty acids, amongst others; these strategies might be enriched with the inclusion of bioactive compounds, such as polyphenols and antioxidant vitamins from vegetal sources, that have shown benefits as antioxidants, vasodilatory, anti-atherogenic, antithrombotic, and anti-inflammatory; moreover, safe extracts of edible plants represent a rich source of structurally diverse molecules that can effectively interfere with multifactorial diseases as MetS [2, 16, 30].

MEXICAN MEDICINAL PLANTS AGAINST METABOLIC SYNDROME

Worldwide, there is an aging population and a growing generation of children and adolescents that suffer from MetS due to unhealthy diets and lifestyle factors; these patients present insulin resistance, chronic inflammation and oxidative stress that are conditions to CVD and diabetes development [32]. Currently, there are synthetic drugs used to ameliorate the cardiovascular risk and other MetS manifestations; however, some of them present side effects or are out of reach for poor population; in this sense, traditional medicine is an important part of the health care system for developing countries, in some Asian and African countries the wide availability of herbal medicines to the general public favors their use, so up to 80% of the population depends on them for primary health care and for prevention and management of the onset of many chronic diseases; the traditional healer might prescribe them, or they are self-medicated. On the other side, in many developed countries, most people have used some form of alternative or complementary medicine in the form of teas, commercial plant extracts, essential oils, or interventions such as acupuncture [33-36]. The use of medicinal plants is a time-honored tradition used since the ancient era for the treatment of various ailments in human beings. Ayurveda therapy is based in herbal medicine and has been used since 1500 BC. Traditional medicine of different world regions has allowed the identification of several nutrients, food supplements, herbs, and spices which exhibit effects against several diseases including MetS, many of them are produced as secondary metabolites in the plant; therefore, they hold great public and medical interest worldwide as sources of novel compounds for low-cost drugs or low-cost herbal preparations [33, 35].

The scientific community around the globe is carrying out studies focused on the molecular action of bioactive compounds from herbal species with proved positive effects on MetS manifestations. Among the studied species, the Mexican medicinal plants are the main target in search of molecules or herbal preparations with antioxidant, vasodilatory, anti-atherogenic, antithrombotic, or anti-inflammatory effects that might be used against MetS. In this chapter, we review the most recent pharmacological studies *in vitro* and *in vivo* about the effect of Mexican plants preparations or its bioactive compounds on MetS biomarkers and manifestations.

Agrimonia eupatoria L.

Agrimonia eupatoria L. (AE) is indigenous of middle and northern Europe, temperate Asia and North America, is commonly known as agrimony, stickle wrote, cocklebur, or stitchwort [37]. This species has been used as antibacterial, antiviral, antitumor, analgesic, antioxidant, immunomodulatory, antidiabetic, gastrointestinal and hepatoprotective, as a wound healer, amongst other pharmacological applications [38]. AE is well known for its beneficial effects in various diseases, and all the parts of the plant are used in folk medicine; that information has encouraged the studies on aqueous and aqueous-alcoholic extracts directed to characterize its antioxidant, antidiabetic and anti-inflammatory activity [37, 39].

Phytochemical Constituents

Agrimonia eupatoria phytochemical analysis has revealed the presence of carbohydrates, glycosides, tannins, terpenoids, phenolic compounds, flavonoids, agrimony furanocoumarins, agrimony-lactone, oils, amino acids and many other bioactive chemical groups [40].

In Vitro Studies

The antioxidant activity of agrimony is associated with its high polyphenolic content, and the AE extracts significantly modify the expression of pro-inflammatory factors and enzymes from glutathione metabolism in cell culture and animal models [41]. AE extract contains epicatechin, thought to be responsible for modulating aorta vasoconstriction, by elevating nitric oxide (NO) concentration, a well-known endothelial vasodilator [39]. Agrimonolide and desmethylagrimonolide are the main bioactive polyphenols in agrimony; both of them possess considerable protective activity from oxidative DNA damage and have been used to ameliorate oxidative stress in HepG2 cells mice liver [42].

In Vivo Studies

The antioxidant and anti-inflammatory effects of one-month consumption of AE tea was evaluated in healthy volunteers. Significant elevation of plasma total antioxidant capacity was observed, parallel to a significant lowering of interleukin 6 (IL-6), and an increase in the high-density lipoprotein cholesterol levels (HDL) at the end of the intervention [43].

The hepatoprotective effects of AE against chronic ethanol-induced liver injury were evaluated in a murine model; rats were fed a Lieber–DeCarli liquid diet for eight weeks. After chronic consumption of ethanol, there was a markedly increase in several hepatic markers such as aminotransferases and cytochrome P450 activities, elevated Toll-receptor 4 (TLR4), lipid peroxidation, mRNA expression, nuclear translocation of NF-κB, and pro-inflammatory cytokines production; all of these increases were attenuated in the group treated with AE water extract. The authors suggest that the observed protective action is due to the suppression of oxidative stress and TLR-mediated inflammatory signaling [44].

Allium sativum L.

The common name of *Allium sativum* L. (AS) is garlic, this species is native from Central Asia and then was spread to China, the Near East, and the Mediterranean region and eventually was distributed and cultivated in Central and Southern Europe, Northern Africa and North América [45]. AS has been considered as a good source of nutrients and is known to be part of a group of functional foods which have antimicrobial, antifungal antioxidant, anti-inflammatory, anticholesterol, hepatoprotective, anticancer, anti-aging and wound healing properties; also, contributes to reduce the risk factors associated to CVD, as high blood pressure and glucose concentrations, which suggest that AS is a potential source of compounds and preparations to use in the management of Metabolic Syndrome [46, 47]. In diabetic patients, AS reduce fasting plasma glucose and TG levels; this effect is probably related to a kind of insulin-sensitizing effects exerted by garlic extract [48].

Phytochemical Constituents

The principal active compounds present in AS cleaves are sulfur compounds and high concentrations of trace mineral content. Most of the sulfur compounds found in whole garlic cloves are of two equally distributed classes: the s-alkylcysteine sulfoxides (alliin) and the γ-glutamyl-s-alkylcysteines [49].

In Vivo Studies

Garlic cleaves extracts have mainly a significant blood pressure-lowering effect due to an inhibitory activity on angiotensin-converting enzyme (ACE) and as calcium channel blocker, which reduces the sensitivity to catecholamines; extract also increases the levels of bradykinin and nitric oxide and consequently improves arterial compliance [50]. This extract also exerts variable effects on human lipid metabolism, reducing the serum level of apolipoprotein B and increasing HDL-cholesterol [51].

Poonam et al. [52] reported that 'AS cleaves alone, as well as in combination with metformin, improved body weight, whilst the combination therapy was more effective in reducing blood glucose levels', highlighting that garlic extract potentiates the hypoglycemic effect of metformin [52]. However, it is important to adjust the metformin dosage when co-administered with garlic since the later increased the peak plasma concentration (Cmax) and the area under the curve of metformin in a murine model [53].

A report by Ashraf et al. [54] in a clinical trial, evaluated 60 diabetic patients with fasting blood sugar levels above 126 mg/dl, they were randomly divided into two groups to receive AS cleaves tablets (300 mg thrice daily) and metformin (500 mg twice daily), or placebo and metformin over 24 weeks. The authors reported a significantly greater reduction in blood glucose level (3–12%) was found in the group with co-treatment of garlic and metformin when compared to that of the placebo and metformin group (0.59%), indicating an enhancement effect [54].

Aloe vera L. Burm

Aloe vera L. Burm (AVLB), also known as sabila [55], is worldwide distributed in diverse habitats and altitudes, from sea level up to 2700 m above sea level, it is found in closed canopy forests or desert shrublands [56]. AVLB presents multiple biological properties, including antibacterial and antimicrobial, antitumoral, anti-inflammatory, anti-arthritic, anti-rheumatoid, anti-cancer, and anti-diabetic activities; it is also used to treat constipation, flushing out toxins and wastes from the body, promoting digestion on gastrointestinal disorders treatment, and reducing illnesses likelihood, as well as for immune system deficiencies [57].

Phytochemical Constituents

Among the nutrients and phytochemicals identified in AVLB are vitamins, minerals, enzymes, simple and complex polysaccharides, fatty acids, indoles, alkanes, pyrimidines, aldehydes, dicarboxylic acids, ketones, phenolic compounds, phytosterols, and alkaloids with potential biological activities [58, 59].

In Vitro Studies

In order to study the role of AVLB leaves' gel on the production of inflammatory cytokines, a study with lipopolysaccharide-activated (LPS) human THP-1 cells and monocyte-derived macrophages treated with AVLB leaves gel was conducted by Budai and co-workers in 2013 [60]. AVLB significantly reduced the production of IL-8, TNF-a, IL-6 and IL-1 cytokine; this phenomenon was substantially more elevated in the primary cells, highlight the fact that LPS-induced activation of signaling pathways like NF-αB were inhibited, demonstrating a significant reduction of cytokine secretion, and providing a new therapeutic approach to regulate inflammation-mediated responses. AVLB has also demonstrated antioxidant and chelating activities of AVLB *in vitro* [61].

In Vivo Studies

The effects of lophenol and cycloartanol, minor phytosterols of AVLB on peroxisome proliferator-activated receptors (PPAR) was evaluated using a luciferase assay. DNA microarray and real-time quantitative RT-PCR (qPCR) analyses were also performed in a diet-induced obesity mouse model. Authors reported that the AVLB treatment provoked an activation of PPAR in a dose-dependent manner The expression levels of many PPAR target genes, such as Fatp1, Acox1, Cpt1, and Hmgcs2 were significantly increased in the *Aloe* phytosterol treatment compared with those in the control high-fat diet (HFD) group; in contrast, the expression level of ApoCIII was significantly decreased in the *Aloe* phytosterol group [62].

The compared effect of AVLB extract (10 ml/kg) and glimepiride (10 mg/kg), was studied using female albino rats. The AVLB extract significantly decreased serum glucose levels and increased serum insulin levels as compared to the glimepiride group; this hypoglycemic effect could be useful in the MetS and Type 2 DM treatment [63].

Calendula officinalis L.

This plant, native to Europe, is widely cultivated in North America, Balkans, Eastern Europe, Germany, and India. It is also known as Marigold or Pot Marigold [64]. *Calendula officinalis* L. (COL) extract is consumed for its anti-inflammatory, antioxidant, antitumor, antigenotoxic, chemoprotective, hepatoprotective anti-HIV, cytotoxic, spasmolytic and spasmogenic properties [65].

Phytochemical Constituents

Phytochemical studies have underlined the presence of various classes of compounds as a wide variety of terpenoids, including sitosterols, stigmasterols, oleanolic acid saponins; flavonoids such as quercetin, isoquercetin, narcissin, calendoflaside, calendoflavoside, calendoflavobioside, rutin; coumarines such as scopoletin, umbelliferone and esculetin; quinones such as plastoquinone, phylloquinone and ubiquinone; carotenoids, varying in composition in differnts part of the plan but mainly neoxanthin, 9Z-neoxanthin, violaxanthin, luteoxanthin in all of them; and free amino acids as arginine or glutamic acid. The essential oil of COL flower contains several monoterpenes and sesquiterpenes (sabinene, β-pienene, limonene, 1,8-cineol, p-cymene) [65].

In Vitro Studies

The COL cardioprotective activity against ischemic heart disease was evaluated in a rat's hearts study; rat's hearts were perfused with a COL extract solution at 50 mM in KHB buffer for 15 min prior to subjecting the heart to ischemia, while the control group was perfused with only buffer. COL solution achieved cardioprotection by stimulating left ventricular pressure and aortic flow as well as by reducing myocardial infarct size and cardiomyocytes apoptosis. Ray et al. [66] mentioned that 'cardioprotection appears to be achieved by changing ischemia reperfusion-mediated death signal into a survival signal by modulating antioxidant and anti-inflammatory pathways as evidenced by the activation of Akt and Bcl2 and depression of TNFα'. The results further strengthen the concept of using natural products in chronic and degenerative diseases like ischemic heart disease [66].

In Vivo Studies

There are inflammation experimental models to measure anti-inflammatory activities; for example, the model that uses inflammation inducers: carrageenan and dextran to induce paw edema. This model was used to assess the potential anti-inflammatory activity of ethanolic extract of COL; carrageenan induces edema through the release of histamine, kinins, and prostaglandins, and dextran mediates serotonin and histamine in the inflamed tissue. Oral administration of 250 and 500 mg/kg body weight of COL extract produced a significant inhibition of more than the 40% in paw edema induced by carrageenan and dextran in BALB/c mice; the anti-inflammatory activity of the compounds present in COL extract might be inhibiting inflammation mediators like cytokines and prostaglandins [67].

Cinnamomum verum J. Presi

Cinnamomum verum J. Prezi (CVJP) is indigenous to Sri Lanka, which is the largest producer and exporter of the quill and oil of the best quality that is also called Sri Lankan or Ceylon cinnamon [68]. There are several species of the Cinnamomun genus that are industrially exploited, due mainly to its fragrance, which can be incorporated into different products in food, cosmetic and pharmaceutical industry; also, the biological activities associated with the consumption of CVJP make its different preparation (powder, extract, essential oil, tea) a good candidate to be used as antioxidant, anti-inflammatory, antidiabetic, antimicrobial, anticancer, lipid-lowering, and against neurological disorders, such as Parkinson's and Alzheimer's diseases [69].

Phytochemical Constituents

The most important constituents of CVJP are cinnamaldehyde and trans-cinnamaldehyde. In the essential oil, these compounds are the main contributors to fragrance and thought to be responsible of the diverse biological activities observed; it also contains cinnamaldehyde, cinnamic acid, ferulic acid, cinnamate, procyanidins and catechins, and many other important [69].

In Vitro Studies

A study performed in rat epididymal fat cells and adipocytes demonstrated that water-soluble polymeric compounds isolated from cinnamon have insulin-enhancing biological activity *in vitro* [70].

In Vivo Studies

The consumption of an aqueous cinnamon extract at a dose of 30 and 300 mg/kg bw in rats during three weeks potentiated the insulin effect through the up-regulation of the glucose uptake in adipocytes and induced glucose utilization, the increased insulin-receptor substrate (IRS)-1 tyrosine phosphorylation levels, the skeletal-muscle insulin receptors stimulation, and the association of the IRS-1 with phosphatidylinositol (PI) 3-kinase [71].

In another study, the cardiac and hepatic antioxidant enzymes, the levels of lipid conjugated dienes, and glutathione concentrations of rats increased after 90 days of administration of the bark powder of CVJP (10%) [72].

The antihypertensive action of a cinnamon extract was successfully demonstrated in both spontaneously hypertensive rats [73] and patients with MetS (22 men and women treated daily with a 500-mg extract for 12 weeks) [74]. Triglyceride-lowering effects with cinnamon were observed in Wistar rats fed a high-fructose diet, and in some clinical trials [75].

Citrus aurantium L.

Citrus aurantium L., known as bitter orange, is a tree original from Asia and eastern Africa, which is predominantly found in the Spanish Mediterranean basin. Its health properties have been described since the time of the ancient Greeks and Romans [76] and its nutraceutical activities are supported by several scientific and clinical studies [77].

The aqueous or ethanolic extracts that are obtained from the immature fruit (peel and juice) are used in traditional medicine for the treatment of digestive problems such as nausea, indigestion or constipation. They are also widely used for weight management products, as a thermogenic agent, for sports performance products, in appetite control, to improve energy, mental focus, and cognition. Fruit, peel, leaves, flowers, seeds, and the essential oil of *C. aurantium* are utilized in food, perfumes, cosmetics and confectionery industries. The essential oil, known as neroli oil, is widely used in aromatherapy for its suggested effects as central nervous system stimulator, and as sedative, analgesic, anti-inflammatory, antispasmodic, carminative, digestive, and diuretic, and blood pressure regulator [78, 79].

Phytochemical Constituents

C. aurantium peel and juice contain several bioactive compounds, including alkaloids, flavonoids (especially hesperidin and naringin), and polyphenols. The main bioactive compound of *C. aurantium* extracts is the phenyl-ethylamine protoalkaloid p-synephrine, and small amounts of other components such as octopamine, hordenine, m-methylturamine, and tyramine. These are mainly adrenergic agents with beta-agonistic actions that stimulate lipolysis and increase resting metabolic rate [78, 80, 81].

To the best of our knowledge, there are no studies in animal models or humans, related to the use of *C. aurantium* extracts for metabolic syndrome treatment or prevention. However, several reports have addressed its efficacy for obesity management, and its beneficial effects on lipids metabolism and adipose tissue function, which are intimately associated with metabolic syndrome origin and development. *C. aurantium* extracts, alone or in combination with caffeine, St. John's wort, green tea, ginkgo, ginseng, guarana, yerba mate, caloric restriction and/or exercise programs, have been extensively studied and used in weight management and fat loss programs, due to the beta-agonistic actions of p-synephrine on adrenergic receptors of the central nervous system, that stimulate lipolysis and increase resting metabolic rate and energy expenditure. It is interesting to note that p-synephrine alone or in combination with other herbal products, has no adverse side effects on cardiovascular or central nervous system stimulation at commonly used doses, and in consequence, has substituted the banned compound ephedrine in weight and fat loss formulations [78, 81-86].

The beneficial effects of *C. aurantium* flavonoids found in fruit juice extracts on adipocyte functional capacity have been studied also, demonstrating their capacity to suppress lipid accumulation in cultured pre-adipocytes [87] and to improve insulin-induced glucose uptake [77] with a positive impact on insulin resistance amelioration.

In addition, the isolated flavonoids hesperidin and naringin have emerged as promising therapeutic agents for the prevention and treatment of metabolic syndrome, obesity, and cardiovascular disease. In several studies in humans and animal models, it has been demonstrated that these flavonoids possess strong antioxidant, anti-inflammatory, hypoglycemic, cardioprotective and antihypertensive activities, and that are able to promote insulin sensitivity, to attenuate dyslipidemia, and to protect the liver from steatosis [88-94].

Cynara scolymus L.

This plant is commonly known as artichoke, is native to the Mediterranean region, there are wild forms, but some varieties are cultivated for food [95]. Generally, they are used in folk medicine against common colds, asthma, kidney problems, skin inflammations, liver diseases, hypertension, astringent, blood cleanser, cardiotonic, detoxifier, digestive stimulant, diuretic and hypoglycaemic [96]. Pharmacologically *Cynara scolymus* L. (CSL) leaf extracts have proved to have hepatoprotective, anticarcinogenic, antioxidant, anti-inflammatory, antibacterial, and anti-HIV activities, as well as the ability to inhibit cholesterol biosynthesis and LDL oxidation [97].

Phytochemical Constituents

Cynara scolymus L. contains fibers, inulin, minerals (potassium, sodium, and phosphorus), and vitamin C. The major bioactive components of head and leaves are polyphenolic compounds, including quinic, mono and dicaffeoylquinic acids (e.g., chlorogenic acid and cynarin) and flavonoids [98].

In Vitro Studies

Two organic acids (caffeic and quinic) from CSL have proven to exert hepatoprotective activity against CCl_4 toxicity in isolated rat hepatocytes. The CCl_4 model is widely used to mimic several aspects related to liver pathologies, such as increased lipid peroxidation and oxidative stress cytotoxicity; the activity was compared with some polyphenolic compounds, such as cynarin, isochlorogenic acid, chlorogenic acid, and luteolin-7-Oglucoside [99].

CSL aqueous extracts reduce lipid peroxidation (measured as production of malondialdehyde) and cytotoxicity (measured as lactate dehydrogenase leakage) in cultures of rat primary hepatocytes exposed to tert-butyl hydroperoxide (t-BHP). Furthermore, artichoke extracts prevented the corresponding loss of intracellular glutathione caused by t-BHP, which in turn induces lipid peroxidation [100].

Endothelial nitric oxide synthase (eNOS) is responsible for nitric oxide (NO) synthesis, NO is a signaling molecule that participates in important physiological processes such as the modulation of vascular tone, insulin secretion, and peristalsis. It has been reported that cynarin increases eNOS activity above 100μM concentration in human endothelial cells [101]. While eNOS has vascular protective effects, nitric oxide synthase (iNOS) has a proinflammatory effect; is has been proved that cynarin reduced iNOS activity with 3 μM concentration in human endothelial cells [102].

Compounds that are able to reduce the activity of digestive enzymes have been used as drugs in obesity-related disease treatments; lipase and α-glucosidase inhibitors are currently the most used [103]. CSL leaf extract has weak α-glucosidase and pancreatic lipase inhibitory activities that could be applied in weight control strategies [104].

In Vivo Studies

In regard of digestive enzymes and digestive system functionality, there are evidences that suggests that CSL compounds not only inhibits digestive enzymes, but increases the bile secretion, inhibits inflammation and ROS production, improves liver function, enhances lipolysis and lipid metabolism, and reduces blood glucose in preclinical and clinical studies [105], which are all relevant keys to prevent and treat MetS. The ability of CLS extracts to inhibit digestive enzymes may explain the beneficial effects on blood glucose levels or satiety signals observed in patients with obesity [106].

Elettaria cardamomum L. Maton

Elettaria cardamomum L. Maton, commonly known as cardamom or small cardamom, is a perennial, herbaceous plant, from the Zingiberaceae family; this species is native of tropical evergreen forests of southern India. Small cardamom oil is used in beauty products, and it was commonly used as an antiputrefactive agent during embalmment process; it is known as the "Queen of spices," is one of the most expensive spices in the world; it has been used in traditional medical applications as antioxidant, gastroprotective and anti-inflammatory agent [107, 108].

Phytochemical Constituents

The composition of *Elettaria cardamomum* L. Maton (ECLM) essential oil might vary due to seasonal and geographical factors and the soil conditions; the major components are monoterpenes such as 1,8-cineole also called eucalyptol, α-pinene, sabinene, linalool, α-terpineol and nerol; sesquiterpenes as elemene and nerolidol; and esters as a-terpinyl acetate and ocimenyl acetate (Marongiu et al. 2004; Ashokkumar et al. 2018). Some of these compounds have demonstrated pharmacological activities, such as the gastroprotective activity of α-pinene [109].

In Vivo Studies

Although there is no abundant literature regarding cardamom as an effective agent in preventing or treating metabolic syndrome, improvements in some parameters have been published. With respect to a cardioprotective effect, the administration of ECLM fruits as crude powder on blood pressure, lipid profile, fibrinolysis, fibrinogen, and total antioxidant was evaluated in a long-term study by Verma and coworkers [110]; the study was conducted in primary hypertension diagnosed patients, that were treated with two capsules of cardamom powder twice a day. ECLM powder was able to reduce systolic and diastolic blood pressure values significantly, but it was not able to normalize them; changes in total cholesterol, triglycerides, VLDL-C and LDL-C levels were statistically not significant, and there was a significant enhancement of fibrinolysis and in the overall antioxidant state in treated subjects; the combined descent of blood pressure values and the improved fibrinolysis make ECLM powder a candidate as cardioprotective in metabolic disorders.

In a rat model of diet-induced metabolic syndrome, the treatment with ECLM powder (whit α-terpinyl acetate as main constituent) for 16 weeks decreased plasma insulin and liver enzymes (ALT and AST) levels; however, ECLM provoked an increase in adiposity, a decrease in liver function and worsened cardiovascular structure and function [111].

ECLM has also been studied in combination with other medicinal plants in order to explore combined bioactivities; in a recent studio, seeds of ECLM along with different parts of three other medicinal species were evaluated for their synergistic cardioprotective potential and antioxidant activities. The preparation with ECLM was able to ameliorate cardiotoxicity in a surgical-induced myocardial infarction in dogs by keeping cardiac markers as CK-MB, SGOT, and LDH near to normal levels. Authors proposed that phytochemical constituents present in this herbal preparation as responsible for its cardioprotective potential, resulting as an alternative product for the management of cardiovascular diseases [36].

Hibiscus sabdariffa L.

Hibiscus sabdariffa L. (HSL), commonly known as roselle, florida cranberry, jamaica sorrel, has been used in food and medicinal applications in Egypt, Sudan, Trinidad and Tobago, Mexico, China, Thailand, Malaysia, Indonesia, India, and South America. HSL can be used as food and fiber. The calyces or flowers are prepared into different beverages as tea, wine, and cocktail, sauces or as jam, marmalade, candy, pickles, or liquors. It has a distinctive aroma, a scarlet red color with a sour taste that is very appreciated. Many biological and pharmacological activities have been associated with the consumption of HSL extracts, such as antipyretic, anti-inflammatory, antioxidant, hepatoprotective, nephroprotective, cancer-preventive, anti-anemic, anti-obesity and anti-hypertensive, amongst others [112-114].

Phytochemical Constituents

HSL constituents, specially those found in calyces from the flowers, include organic acids such as hibiscus, malic, tartaric, citric, hydroxycitric, oxalic and ascorbic acid; polyphenols and flavonoids, mostly anthocyanins such as delphinidin-3-glucoside, sambubioside, and cyanidine-3-sambubioside, flavonoids such as gossypetine, hibiscetin and their respective glycosylated forms; protocatechuic acid, catechin, gallocatechins, eugenol, and ergoesterol. HSL is also rich in minerals, especially Ca and Fe [112, 114, 115-118]. HSL seeds have been the focus of attention since they might be a good source of unsaturated fatty acids, as oleic and linoleic acids, saturated acids, like palmitic acid, as well as of lipid-soluble antioxidants as alpha-tocopherol [113].

Particularly, polyphenols found in HSL have been of interest in the study of prevention strategies for SM manifestations since they have multi-targeted effects on obesity, regulation of energy metabolism, oxidative stress and inflammatory pathways, acting through transcription factors, hormones, peptides, digestive enzymes, as well as by epigenetic modifications [106].

In Vitro Studies

In a study performed in 2003 by Kim and co-workers [119], the effects of an HSL extract on 3T3-L1 cells adipogenic differentiation at molecular and morphological levels were evaluated; the pre-adipocyte cell culture was treated with hibiscus extract (100 mg/mL). After treatment, a significant reduction in the accumulation of cytoplasmic fat was observed, and even in differentiated adipocytes, hibiscus extract could attenuate the accumulation of intracellular triglycerides. At the molecular level, a lower expression of the adipogenic transcription factors CCAAT element-binding protein (C/EBP) and peroxisome proliferator-activated receptor PPARγ was found.

In another *in vitro* study, 3T3-L1 mouse adipocytes were treated with an HSL aqueous extract, a *Hibiscus sabdariffa* purified extract (HSPE), and three fractions enriched with different compounds such as delphinidin-3-sambubioside, cyanidin-3-sambubioside, or glycosylated flavonols, such as quercetin-3-sambubioside. HSPE and the enriched fractions significantly inhibited adipogenesis, decreasing triglyceride content in hypertrophied and insulin-resistant adipocytes. They also proved that the aqueous extract is more active in inhibiting TG accumulation and modifying adipokine secretion when the fiber and polysaccharides are removed [120].

In Vivo Studies

Increased body weight and obesity are closely associated with MetS. Studies on the effect of HSE consumption in murine models have been conducted during the last years in this regard; for example, the effect of standardized HSL aqueous extract containing 28% of anthocyanins was evaluated in METSG obese mice. The authors demonstrated that extract consumption inhibited body weight gain and reduced glycemia in obese mice [121]. The effect of the consumption of an HSL ethanolic extract was evaluated in rats with a diet-induced atherogenic condition and demonstrated a lower weight gain in the treated groups and differences in the excreted fatty acids which suggested that some components could inhibit pancreatic amylase, preventing polysaccharide unfolding and absorption [122]. Pérez-Torres and co-workers reported similar results in 2014 [33], in a sucrose-induced metabolic syndrome rats model treated with an HSL calyx's infusion, and reported a significant decrease in weight gain, in systolic blood pressure, in HOMA index, in intra-abdominal fat, and in serum triglycerides and insulin levels.

In another study, Huang and coworkers in 2015 [123] evaluated the effect of an HSL extract in a high-fat diet-induced obese hamsters; the extract contained the polyphenols protocatechuic acid, catechin, epigallocatechin, and caffeic acid, and the anthocyanins delphinidin and cyanidin, and significantly decreased the levels of serum and hepatic lipids (total cholesterol and triglycerides) and the LDL-C /HDL-C ratio. The authors suggested that this could be attributed to the inhibition of lipid absorption in the gut and concluded that the consumption of polyphenolic- and flavonoid-rich HSL extract is a safe therapy for obesity and metabolic syndrome.

Humulus lupulus

The genus *Humulus,* native of China, belongs to the family of Cannabinaceae, and it has three known species. *Humulus lupulus Linneus*, is the best known and is commonly named as hops; currently, is cultivated in central Europe and several temperate regions of North and South America, South Africa, and Australia. Besides its use in beer brewing, this plant has been used in traditional medicine all over the world to treat tension headache, toothache, earache, and neuralgia; it is also used for its diuretic antispasmodic, sedative and analgesic activities [124].

Phytochemical Constituents

More than 100 volatile compounds have been identified in hops essential oil including acids and terpenes such as formic acid, acetic acid, isopentyl acetate, 1,3-nonadiene, glycerol, R-terpinene, phenol, isooctanol, beta-ocimene, alpha terpineol, beta- caryophyllene, farnesene and humulene [125].

In hop mature cones different phytochemicals are found, such as terpenes and chalcones, flavonol glycosides such as kaempferol, quercetin, and rutin; and some catechins as catechin and epicatechin gallate [124]. Particularly, bitter acids are a group of lipophilic compounds that define beer quiality (humulones or alpha-acids, cohumulone, adhumulone, and the corresponding beta-acids as lupulone, colupulone and adlupulone) [126]. In contrast, in the female inflorescences there is a group of compounds called prenylflavonids [127]. Both isohumulone and xanthohumol have been tested in pharmacological studies focused on cancer-preventive activity and metabolic disorders.

In Vitro Studies

FXR is the physiological receptor for bile acids, which plays an important role in cholesterol, lipid, carbohydrates and lipoprotein metabolism and is considered a molecular target for disorders associated to metabolic syndrome amelioration. In 2005 [128], Nozawa reported that xanthohumol significantly activated FXR in a transient transfection assay in HepG2 cells.

In Vivo Studies

Another molecular target of interest is peroxisome proliferator-activated receptors (PPARs) α and γ. The isomerized hop extract (IHE) mostly composed of isohumulones was administered in C57BL/6NCrj and PPARα-/-/129S4/SvJae mice that have an altered response to compounds that induce peroxisome proliferation and hepato-carcinogenesis. The authors reported that IHE intake significantly diminished final body weight, regardless of whether or not the diet was supplemented with cholesterol or if the individuals were PPAR a-deficient or not. It also reduced serum triglycerides and FFA and regulated genes involved in microsomal N-oxidation and peroxisomal and mitochondrial oxidation in C57BL/6N mice [129].

Obara and coworkers in 2009 [130] demonstrated in a human study that an oral administration of at least 32 mg of isohumulones per day for 12 weeks, decreased fasting blood glucose level, LDL-cholesterol, and triglycerides levels, but had no effect on HDL-cholesterol. The treatment also led to a significant decrease in body weight, BMI, and abdominal fat area.

In another study in humans with elevated LDL cholesterol and metabolic syndrome was conducted to evaluate the effect of the oral administration of rho iso-alpha acids from *Humulus lupulus* L. combined with proanthocyanidins from *Acacia nilotica*. After 12 weeks, significant decreases from baseline in total cholesterol, LDL-C, non-HDL-C, and cholesterol/HDL-C were observed. There was no change in serum HDL-C, fasting glucose or HbA1c concentrations; the authors proposed this combination as a potential nutritional alternative to the cardiovascular disease risk management of patients with both metabolic syndrome and elevated LDL-C [131].

***Hypercurium perforatum* L.**

Hypericum is a genus of the family Hypericaceae, which includes almost 500 species of different herbs, shrubs, and trees. Hypercurium perforatum L. (HPL) is widely known as St. John's wort, "hipérico" or "corazoncillo"; is a herb indigenous to Europe, Africa and Western Africa and it has been valued for its medicinal applications for centuries and known by Greek physicians like Hippocrates and Dioscorides; this plant has been used to treat liver disorders, depression, rheumatism, gastric diseases, and duodenal ulcer, as regenerative and anti-inflammatory agent in digestive tract, and antioxidant [132-134]. The study of HPL flavonoids suggests that it could help in reducing lipid oxidation due to their high antioxidant capacity, and they might be used in the treatment for menopause-related chronic disease including the cardiovascular system, central nervous system and vegetative nervous system [135]. Evidence from animal, preclinical, and clinical studies suggest that HPL preparations may modulate Cytochrome P450, which is relevant since many drugs may act as inhibitors or inducers of Cytochrome P450, causing interference with other treatments [132, 134].

Phytochemical Constituents

The main phytochemical compounds found in HPL are flavonoids: flavones, flavonols, isoflavones, flavonones, and chalcones; amongst them, flavonols are considered the most active, and they include rutin, hyperoside, quercitrin and quercetin [135]. Pharmacologically active ingredients of *H. perforatum* include naphthodianthrones such as hypericin and its derivatives and phloroglucinols derivatives like hyperforin. This compound is an inhibitor of the reuptake of different neurotransmitters like serotonin; therefore, hyperforin and hypericin are the standardized molecules in commercial preparations [132]. *H. perforatum* essential oil contains hydrocarbons, monoterpenoids, and sesquiterpenoids [134].

In Vivo Studies

Most of the pharmacological studies of HPL are related to its antidepressant activity and other CNS related conditions. Hernández-Saavedra and coworkers [136] evaluated the effect of aqueous leaf extracts from HPL and other Mexican herbs in diet-induced insulin resistance using obese Sprague–Dawley rats. After a 12-week treatment, they found a significant reduction in adipocyte's volume and fat content, an increase in adiponectin levels and a decrease in serum leptin. The extract was also able to regulate the expression of IL-6 and TNF-α compared with the obese group.

Melissa officinalis L.

Melissa officinalis L. (MOL), commonly known as lemon balm, melissa, toronjil or hierba limonera according to the geographical region, belongs to the Family *Lamiaceae* (nom. alt. *Labiatae*) Is an aromatic herb whose fragrant leaves have been widely used in cooking for over 2000 years, and has wide applications in the treatment of several conditions in the traditional medicine, especially in the European Traditional Medicine, the Iranian Traditional Medicine and in some regions of America, where it is used for digestive system illness as stomach ache and colic, to treat insomnia, anxiety, migraines, to modulate mood, cognition and memory; and in the treatment of hypertension and cancer; its essential oil has been used as antimicrobial and antiviral agent [137-139].

Phytochemical Constituents

The main phytochemicals of MOL are monoterpenes, sesquiterpenes and triterpenes, and diverse phenolic compounds such as flavonoids, phenolic acids, and tannins; its essential oil posses geranial, citronellal, geraniol, ursolic and oleanolic acid, luteolin, naringin and hesperidin [138].

In Vivo Studies

Using male Swiss albino hyperlipidemic rats as murine model, the effect of daily administration of 2 g/kg of a MOL extract for 28 days in morphological and biochemical parameters was assessed. Authors reported a reduction of total cholesterol, total lipids, alanine transaminase, aspartate transaminase and alkaline phosphatase levels in serum, and of the levels of tissue lipid peroxidation, and increased glutathione levels in liver tissue; at morphological level, all the degenerative modifications observed in hyperlipidemic rats such as vacuolization, mononuclear cell infiltration or damage in the endothelium of the central veins in the hepatocytes was minimal or absent in the animals treated with the MOL extract; authors suggest that MOL extract showed hypolipidemic and protective effect on the liver of hyperlipidemic rats [140]. A very similar study performed by Zarei and co-workers in 2014 demonstrated that the intraperitoneal administration of MOL extract to hypercholesterolemic rats was able to reduce total cholesterol as well as some hepatic damage signaling enzymes in serum as alanine transaminase, aspartate transaminase, and alkaline phosphatase. In this study, MOL extract application exerted a hypolipidemic effect similar to Atorvastatin and showed a protective effect on the liver of hyperlipidemic rats [141].

In a study conducted in 2014 [142], Weidner and coworkers evaluated the effect of the consumption of an ethanolic MOL leaf extract in C57BL/6 insulin-resistant and dyslipidemic mice. A dose of 200 mg/kg/day was supplied for six weeks, and their results showed an activation of the peroxisome proliferator-activated receptors (PPARs), which have key roles in the regulation of whole-body glucose and lipid metabolism. They also observed a significant reduction of hyperglycemia and insulin resistance, plasma triacylglycerol, non-esterified fatty acids, and LDL/VLDL cholesterol levels.

In a more recent studio, ovariectomized wild-type C57BL/6J mice in high-fat diet were treated with a MOL fraction denominated ALS-L1023 for 15 weeks. The ALS-L1023 fraction was standardized with rosmarinic acid and caffeic acid as references; mice in high-fat diet had increases in mass and inflammation in visceral adipose tissue, angiogenesis, body weight, while mice treated with the standardized fraction showed a significant reduction of all these parameters, along with an improvement of serum lipids and lowering of hepatic lipid accumulation, inflammatory cells, and collagen levels. Importantly, ALS-L1023 showed a tendency to normalize hepatic expression of genes involved in lipid metabolism and fibrosis. The authors suggest that this fraction might be a possible candidate to treat NAFLD in post-menopausal women with obesity [140].

***Mentha X piperita* L.**

Mentha piperita L. (MPL), commonly known as peppermint, is one of the aromatic herbs most frequently used for therapeutic, culinary, and flavoring purposes [144]. Peppermint leaves (fresh and dried) and the essential oil extracted from them are used in many food, cosmetic, and pharmaceutical products [145].

Phytochemical Constituents

The chemical components of peppermint leaves and oil vary with plant maturity, variety, geographical region, and processing conditions [146-148]. The fatty acid composition of the non-polar lipid fraction of peppermint leaves is dominated by palmitic, linoleic and linolenic acids. The main volatile components identified in the essential oil of peppermint are menthol, menthone, isomenthone, 1,8-cineole (eucalyptol), menthyl acetate, menthofuran, limonene, β-myrcene, β-caryophyllene, pulegone, and carvone; it also contains the polyphenols eugenol, thymol, terpenoids, eriotricin, luteolin, hesperidin, rosmarinic, cinnamic and caffeic acids and other compounds that demonstrate antioxidant activity. The total polyphenolic content of peppermint leaves is approximately 19–23% (total flavonoids 12%), which includes eriocitrin and rosmarinic acid, luteolin 7-O-rutinoside, hesperidin, 5,6-dihydroxy- 7,8,3′,4′-tetramethoxyflavone, pebrellin, gardenin B and apigenin [146, 147, 150-152].

In Vivo Studies

Several health benefits have been reported for MPL consumption in the form of teas, tisanes, or as essential oil for the treatment of respiratory and gastrointestinal problems, as well as for its antioxidant and antiperoxidant effects [145].

Due to the high content of flavonoids with antioxidant properties, various studies have analyzed the use *Mentha piperita* extracts (teas, tisanes, juices) as an alternative for the treatment of some parameters of the metabolic syndrome, particularly with respect to glycemia and dyslipidemia associated with insulin dysfunction. Such is the case of the report of the fact that the oral administration of a leaves' aqueous extract of *Mentha piperita* (100 and 250 mg/Kg) to fructose-induced insulin resistant male Sprague Dawley rats for 21 days significantly reduced serum total cholesterol, triglycerides and LDL-c, associated with a significant increase in HDL-c levels and a decrease in the atherogenic index, indicating its potent antihyperlipidemic and antiatherogenic activity. The improvement of liver functions and the subsequent increase in the uptake and utilization of blood glucose might be the mechanism of action of this extract as glucose lowering and hypolipidemic agent [153].

In another investigation, Barbalho and co-workers [154], compared the effect *Mentha pipertita* tea and juice on plasma lipids of Wistar rats fed a diet supplemented with saturated fats. Rats received a dose of 0.29 g/kg twice a day by intra-gastric route (gavage) for 30 days. Both tea and juice provoked a significant reduction in LDL-c values and weight gain percentage when administered concomitantly with normal or saturated fat-diet, and in cholesterol levels only in rats fed normal diet. In addition, the HDL-c levels were increased, and triacylglycerol values were diminished significantly, even in animals fed with saturated fats.

In a similar study, there were benefits in glucose and lipid profiles (significant reduction in plasma levels of cholesterol, triglycerides, and LDL-c and significant increased HDL-c values) after administering peppermint juice at a dose of 0.29 g/kg once a day for 30 days in normoglycemic Wistar rats [155].

In humans, Barbalho and co-workers [155] examined the effect of *Mentha piperita* juice on glycemic, lipid, and anthropometric profiles and blood pressure in university students. The daily dose was 200 mL of juice twice a day for 30 days. After administration period, significant reductions in glycemia, total cholesterol, triacylglycerides, LDL-c, blood pressure, body weight, and body mass index, and an increase in HDL-c levels, were observed in the studied population, without alterations in hepatic enzymes (GOT and GPT) nor urea levels, which could indicate attack on the liver or kidney.

Origanum vulgare

Origanum vulgare is an aromatic herb that commonly occurs throughout Asia, Europe, and northern Africa [156]. It is the most recognized herb as "oregano" around the world. In folk medicine is used to treat respiratory disorders, dyspepsia, painful menstruation, general infections, inflammation-related illnesses, diabetes, rheumatoid arthritis, scrofulosis, and urinary tract disorders [156-158]. It is also used as a culinary herb in gastronomy, and its essential oil can be used to preserve food [156, 159] Anti-bacterial, antioxidant and antithrombotic activities have also been reported for oregano [160].

Phytochemical Constituents

The bioactive compounds of *Oregano vulgare* can be found either in the essential oil or in the phenolic fraction that is obtained from stems, leaves, and flowers. The main types of phenolic compounds are flavonoids and phenolic acids, being the most abundant rosmarinic, caffeic, chicoric and p-coumaric acids, luteolin-7-O-glucoside, apigenin-7-O-glucoside, eriodictyol, apigenin, and naringenin [161]. The essential oil is mainly composed of carvacrol, p-cymene, thymol, γ-terpinene, linalool and terpinen-4-ol [162].

Various biological activities related to metabolic syndrome have been reported for *O. vulgare* flavonoids and phenolic acids, as well as for its essential oil.

In Vitro Studies

For *O. vulgare* flavonoids and phenolic acids, *in vitro* antioxidant and anti-inflammatory properties have been demonstrated [163, 164]. With respect to the anti-inflammatory capacity, it has been reported that the aqueous infusion of *O. vul*gare promoted the significant reduction of the pro-inflammatory and cancer-promoting cytokine IL-8 concentration on the medium in HT-29 and PC3 cancerous cell lines [165]. The anti-IL-8 activity was attributed to caffeic acid present in the infusion.

O. vulgare essential oil has also demonstrated *in vitro* antioxidant activity [162], and an anti-inflammatory capacity determined by the reduced synthesis of the pro-inflammatory adipokines TNF-alpha, IL-1, and IL-6, and the increased synthesis of the anti-inflammatory cytokine IL-10 in a human adipocytes model of inflammation [166], and the inhibition of the levels of inflammatory biomarkers (MCP-1, VCAM-1 and ICAM-1) on activated-primary human neonatal fibroblasts [167].

In Vivo Studies

It has also been demonstrated that the administration of an aqueous extract of *O. vulgare* leaves to rats exerts a hypoglycemic effect and promotes the increase of CAP and adiponectin levels in adipose tissue, resulting in the improvement of insulin sensibility in adipocytes and skeletal myocytes. These effects have been related to Biochanin A (BCA) contained in the aqueous extract, which is known as a PPARγ agonist [168]. The peroxisome proliferator-activated receptor PPARγ is a key regulator of adipocyte differentiation, plays an important role in lipid metabolism, glucose homeostasis, and insulin sensitivity, modulates metabolism and inflammation in immune cells, and controls cell proliferation. In consequence, it is considered as a therapeutic drug target for metabolic syndrome treatment and prevention [169].

The essential oil of some species of oregano has been extensively studied as hypoglycemic agents [162, 170, 171]. Sarikurkcu and co-workers [172] showed that the essential oil of *O. vulgare* subsp. *vulgare* and subsp. *hirtum* possess alpha-amylase and alpha-glucosidase inhibitory activity; since these enzymes have an important role in dietary carbohydrates metabolism, its inhibition is used as a therapeutic target in metabolic syndrome and diabetes treatment to help retard post-prandial blood glucose increase. Interestingly, when *O. vulgare* essential oil was combined with rosmarinic acid, the inhibition of the a-amylase was higher, suggesting that the combination of phenolic compounds with essential oil potentiates the inhibitory activity.

Psidium guajava L.

Psidium guajava L. (PGL), commonly known as guava, is a tropical species native of Central América, its fruit stands out for its high nutritional value, and is consumed as fresh fruit and in different processed foods such as jams, nectars, and juices [173, 174]. The guava leaves are commonly used in folk medicine in the treatment of different diseases, particularly in type 2 diabetes mellitus [175-177].

Phytochemical Constituents

PGL fruit contains vitamins and minerals such as vitamin C, vitamin A, calcium, phosphorus and potassium, it is a good fiber source, and its soluble fiber can be used to enrich functional foods [174]. The composition and metabolites in guava fruits and leaves varies with factors such as geographical region, culture conditions, maturity of fruits and leaves [173, 178, 179]. PGL leaves contain high concentrations of phenolic compounds such as procyanidin B isomer stenophyllanin A, catechin, morin, hyperin, isoquercitin, reynoutrin, guajaverin, avicularin, quercetin, myricetin, kaempferol and luteolin [176, 180], including flavonoid glycosides such as strictinin, isostrictinin and pedunculagin [175].

Carotenoids are also present in guava fruits; the main carotenoids are phytofluene β-carotene and lycopene [178]. The main volatile compounds in guava fruits are 3Z-hexenyl acetate, 3-phenylpropyl acetate, (E)-cinnamyl, acetate, and hexanal and in leaves, 1,8-cineole and transcaryophyllene are amongst the most frequently encountered [173].

In Vitro Studies

Since inflammation is an important component of SM, the flavonoid fraction of PGL leaves extract (GLE) was used to investigate its effect on lipopolysaccharide-induced inflammatory response in head-kidney (HK) macrophages of *Labeo rohita,* a fish from the carp family; they found that the compounds present in this fraction caused an anti-inflammatory effect since they inhibited the expression of TNF-α, IL-1b, iNOS and COX-2 via down-regulation of the NF-kB pathway. Those results suggest that the active compounds may be used for the treatment of chronic inflammatory diseases such as SM treatment [176].

In Vivo Studies

Different murine models studies have been conducted in order to asses the effect of the consumption of GLE in some SM manifestations; for example, in spontaneous hyperglycemic, obese, hypertensive, hyperleptinemic, (SHRSP/ZF) MS rats, oral treatment with a *Psidium guajava* ethanolic extract obtained from leaves produced an improvement in glucose metabolism, reduced body weight gain, and increased insulin sensitivity in skeletal muscles; therefore authors conclude that this extract may be useful in the design of new therapies for the prevention and treatment of insulin resistance-related MS [181].

A similar study with SHRSP/ZF rats, evaluated the effect of the administration of a 2 g/kg/day GLE by forced administration for 6 weeks, after this time they found a diminished fat-drop formation and triglyceride (TG) in the liver; they evaluated gene expression of proteins involved in liver lipid metabolism, and concluded that the administration of GLE may ameliorate hepatic insulin resistance-related METS by enhancing the adiponectin-beta-oxidation system [182].

In a more recent study, diet-induced obese Male C57BL/6J mice were treated with an oral dose of 5 mg/kg GLE, for 7 weeks. The administration of GLE significantly ameliorated serum lipid markers such as low-density lipoprotein (LDL)-cholesterol and high-density lipoprotein (HDL)-cholesterol, and therefore, they increased LDL:HDL ratio, suggesting that this improvement in the lipid metabolism can have a positive impact on vascular dysfunction associated with obesity, thus preventing the development of atherosclerosis and the subsequent cardiovascular events. They also quantified 57 phenolic compounds in GLE, flavan-3-ols, flavonols, and ellagic acid derivatives amongst them, which could be responsible for the effects [180].

The anti-inflammatory activity of guava leaves essential oil has been evaluated in a male swiss mice model; the eosinophil migration in the pleural lavage was inhibited by the EO and by their terpene-enriched fractions; to a lesser extent, neutrophil and mononuclear cell migration was inhibited too; the effect was correlated to the presence of β-pinene and β-caryophyllene [183].

There are some studies based on surveys that report that the consumption of guava leaves' tea (GLT) is widely accepted in different regions of the world, such as East Asia, Africa, and North America, due to medicinal properties such as anti-inflammatory, antimalarial, antitussive, hepatoprotective, antitumor, antidiabetic and cardiovascular activity [177, 184-186].

In another study, the effect of a 20 days consumption of arepas in 75 students was analyzed. Arepas, a traditional food in South America, were prepared with corn dough or with precooked cornmeal added with 10% pectin extracted from *Psidium guajava* L fruit. Results showed a significant decrease in serum levels of total cholesterol, LDL cholesterol and glucose in students after arepas consumption [174].

Rosmarinus officinalis

Rosmarinus officinalis L., commonly known as rosemary, is a medicinal plant native to the Mediterranean region. It is a perennial and aromatic plant cultivated all around the world. Rosemary has been used in folk medicine, cosmetics, and in food industry. It is widely consumed as a food ingredient, particularly as a culinary spice and in the conservation of food products, and as an ornamental and medicinal plant. It is traditionally used in various illnesses, such as asthma, atherosclerosis, cataract, renal colic, hepatotoxicity, peptic ulcer, inflammatory diseases, ischemic heart disease, hypercholesterolemia, oxidative stress, physical and mental fatigue, hypertension, diabetes, cancer, hyperglycemia and depression [187, 188].

Phytochemical Constituents

Several bioactive compounds with pharmacological activities have been isolated from essential oils and aqueous/organic extracts of *R. officinalis* L. The compounds most reported include flavonoids, phenolic acids and phenolic diterpenes and triterpenes, such as caffeic acid, carnosic acid, chlorogenic acid, monomeric acid, oleanolic acid, rosmarinic acid, ursolic acid, betulinic acid, alpha-pinene, camphor, carnosol, eucalyptol, rosmadial, rosmanol, rosmaquinones A and B, secohinokio, and derivatives of eugenol and luteolin [188, 189]. Among the isolated phenolic compounds, carnosic acid and rosmarinic acid have been shown to possess the predominant effects.

Several bioactivities have been reported for rosemary, namely antiinflammatory, antioxidant, chemopreventive, anti-cancer, anti-proliferative, antimicrobial, antidiarrheic, antiseptic, antirheumatic, insecticidal, lipid peroxidation reduction, antiangiogenic, neuroprotective, antiviral, anti-spasmodic in renal colic and dysmenorrhea, anti-rheumatic, carminative, cholagogue, diuretic, expectorant, anti-convulsant, hepatoprotective, nephroprotective [187, 188, 190, 191].

With respect to metabolic syndrome physiological disorders, several therapeutic and prophylactic effects of *R. officinalis* L. have been reported *in vivo* and *in vitro*. In that respect, the main reported effects are antioxidant, anti-inflammatory, hypoglycemic, hypolipidemic, hypotensive, and hepatoprotective. For some of them, the responsible bioactive compounds has been described: carnosic acid (anti-inflammatory, inhibitory of digestive enzymes (lipase, α-amylase, and α-glucosidase), lipogenesis suppression); chlorogenic acid (antioxidant); oleanolic acid (antioxidant); ursolic acid (insulin sensitivity inductor, weight gain and atherosclerosis reduction); camphor (hypoglycemic); carnosol (anti-inflammatory, antidiabetic); eucalyptol (anti-inflammatory); eugenol (anti-inflammatory and antioxidative) [187, 188].

In Vitro Studies

Antioxidant activity of rosemary extracts has been extensively reported. For example, Hegazy and co-workers [192] described the antioxidant activity of aqueous extracts using the 1,1-diphenyl-2-picrylhydrazyl assay and the radical scavenging activity, and Ibarra et al. [193] reported the antioxidant activity *in vitro* by oxygen radical absorbance capacity and ferric-reducing/antioxidant power assays of a carnosic acid-rich rosemary extract.

The anti-inflammatory activity of *R. officinalis* extracts has also been demonstrated. Arranz and co-workers [194] reported that rosemary extract rich in carnosic acid and carnosol inhibited the secretion of pro-inflammatory cytokines TNF-alpha (tumor necrosis factor-a), IL-1b (interleukin-1b) and IL-6 in THP-1 human macrophages activated with lipopolysaccharides and human oxidized low-density lipoproteins.

The hypoglycemic action has also been addressed. In cultured HepG2 cells, methanolic extracts increased glucose consumption by means of the increase of liver glycolysis and fatty acid oxidation, by the activation of AMPK and PPAR pathways [195] and the suppression of the cAMP responsiveness of gluconeogenic genes, PEPCK-C and G6Pase promoters [196]. The enzymes alpha-amylase and alpha-glucosidase, which participate in carbohydrates breakdown previous to their absorption, have also been inhibited by aqueous and methanolic extracts [197, 198]. Ibarra and co-workers [199] reported the inhibitory effects of a carnosic acid-rich extract on pancreatic lipase, and the activation of PPAR-g activity, a nuclear receptor protein that controls glucose transport and metabolism and the transcription of the proteins and enzymes responsible for glucose and fatty acid cellular uptake. Also, Harach and co-workers have reported inhibition of the pancreatic lipase [200]. In addition, an increase in glucose uptake in 3T3-L1 adipocytes treated with *R. officinalis* extracts [201], and in L6 myotubes through increasing AMPK and acetyl-CoA carboxylase (ACC) phosphorylation, has been reported [202].

With respect to obesity, working with 3T3-L1 adipocytes, Gaya et al. [203] reported that an acetone extract inhibited preadipocyte differentiation, exerted antiadipogenic effects, and blocked mitotic clonal expansion and expression of PPAR-γ and CCAAT/enhancer-binding protein α (C/EBPα), which are critical for acquiring the adipocyte phenotype. The differentiation of the adipocytes and increase in their size are considered as therapeutic targets for the treatment of obesity [187].

The combination of an aqueous extract with cranberry inhibited the Angiotensin I-converting enzyme ACE-I activity. The inhibition of ACE-I exerts antihypertensive and cardioprotective effects [204].

In Vivo Studies

The analysis of *in vivo* effects of *R. officinalis* extracts on metabolic syndrome conditions has been undertaken in various murine models. Hegazy and co-workers [192] evaluated the protective effect of rosemary aerial parts aqueous extracts on the gentamicin (GM)-induced hepatotoxicity and dyslipidemia rat model. After 10 days of coadministration of GM and the extract, an inhibition of body weight gain and a reduction in liver weight was observed. Plasma liver enzymes (AST and ALT), bilirubin level, and total protein concentration were normalized, suggesting hepatoprotective actions. A hypolipidemic effect was observed by the significant reductions in total cholesterol, phospholipids, triacylglycerols, and atherogenic index and pancreatic lipase activity; raised hydrogen peroxide levels and DNA fragmentation were ameliorated. The authors suggest that the antioxidant activity of rosemary polyphenols may be responsible for the observed results.

In obesity and metabolic syndrome mice models induced by a high-fat diet, the preventive effects of a *R. officinalis* extract rich in carnosic acid (CA) was evaluated. After 16 weeks on the experimental protocol, the extract supplementation significantly reduced body weight gain, epididymal fat weight, percent of fat, plasma ALT, AST, free fatty acids, glucose, insulin and cholesterol levels, liver weight and liver triglycerides. The enriched extract also decreased the levels of plasma and liver malondialdehyde, advanced glycation end products (AGEs), the liver expression of the receptor for AGE (RAGE), and the lipid accumulation in hepatocytes, as well as enhanced fecal lipid excretion, which inhibits lipid absorption [199, 205]. In the same model, Harach et al. [200] reported that the administration of an ethanolic leaf extract for 50 days induced a reduction in weight and fat mass gain, associated with an increase of fecal lipid excretion, suggesting inhibition of pancreatic lipase activity and therefore limiting lipid absorption. They also observed a decrease in hepatic triglyceride levels and a reduction in liver steatosis in treated mice. In high-cholesterol fed mice, the administration of an aqueous extract for 36 days resulted in the reduction of plasma triacylglycerols, total cholesterol and LDL levels, and an increase in HDL levels [206].

The ability of *R. officinalis* extracts to moderate weight gain and to improve the plasma lipids profile was also observed in insulin-resistant/obese Zucker rats [207]. The administration of an ethanolic extract rich in carnosic acid for 64 days moderately reduced body weight gain in both lean (fa/+) and obese (fa/fa) animals. Serum triglycerides, cholesterol, and insulin levels were also markedly decreased in lean animals. Since lipase activity was significantly inhibited in the stomach of the supplemented animals, the authors suggest that these effects may be caused, at least in part, by a significant inhibition of gastric lipase and subsequent reduction in fat absorption. In the same experimental rats, Romo Vaquero et al. [208] reported changes in the microbiota composition, and a reduction in b-glucosidase activity in the caecum, which may contribute to reduce dietary energy extraction from non-digested polysaccharides and to moderate body weight gain, as deduced by the increased fecal fiber excretion in lean and obese animals. In addition, *R. officinalis* extract showed anti-inflammatory effects in the Zucker rats: decreased the pro-inflammatory tumor necrosis factor alpha, IL-1β, which contribute to insulin resistance and diabetes, and leptin, upregulated adiponectin that has anti-inflammatory effects improves insulin resistance, and reduces serum TG levels and weight, induced phase I and phase II gene expression and the peroxisome proliferator-activated receptor gamma coactivator 1-alpha (PGC1α) that improves fatty acid oxidation and insulin-stimulated glucose transport decreased adipose phosphorylated AMP-activated protein kinase. These results demonstrate that the extract exerts anti-inflammatory effects and affects hepatic metabolism. In mice, oral administration of a distilled extract resulted in the reduction of plasma glucose levels, due to the inhibition of intestinal α-glucosidase (sucrase) activity [209].

The hypolipidemic and antioxidant effects of an aqueous leaf extract (major compound rosmarinic acid) and a non-esterified phenolic fraction (major compound carnosic acid) was evaluated in rats with diet-induced hypercholesterolemia. After four weeks, lower serum total cholesterol, non-HDL-c, and tissue thiobarbituric acid reactive substance (TBARS) levels, and increased activity of tissue antioxidant enzymes (superoxide dismutase, catalase, and glutathione peroxidase) were observed, lowering the susceptibility to the peroxidative damage elicited by a cholesterol-rich diet. *R. officinalis* extracts also showed remarkable antioxidant activity *in vitro* in the DPPH assay and the β-carotene/linoleic acid system. The authors suggest that phenolic compounds, mainly rosmarinic and carnosic acids, might be responsible for the observed effects [210].

The effects of *R. officinalis* on blood lipid level and lipid peroxidation was also studied in a hyperlipidemic model of rats fed a high-fat diet, using an aqueous extract (major compound rosmarinic acid), essential oil, and crude plant powder (a mixture of hydrophilic and lipophilic compounds). The three samples enhanced the activities of superoxide dismutase (SOD) and total antioxidant capacity (T-AOC), and alleviated the peroxidation injury in hyperlipidemic rats. The decrease in serum cholesterol and triglycerides levels, the increase HDL-c, and the improvement in the anti-oxidant capacity were exerted in different magnitude by the three rosemary samples [211].

Vanithadevi and co-workers studied the effect of *R. officinalis* on insulin sensitivity [212], in mice fed a high fructose diet, as a murine model of insulin resistance. Extract administration for 60 days prevented fructose-induced body weight gain, fasting glucose concentration, and AUNglucose and AUNinsulin increase. Treatment also reduced the levels of glucose, insulin, fructosamine, glycated hemoglobin, and advanced glycated end products (AGE), HOMA-homeostatic model assessment, ISI0, 120-insulin sensitivity index, and QUICKI-quantitative insulin check index, to near normal values. The extract significantly lowered the oxidative stress markers in mitochondria, levels of non-enzymatic antioxidants in plasma, and the activities of SOD, CAT, GPX, and GST in the liver were increased, demonstrating that *R. officinalis* extract can mitigate metabolic and redox disturbances in this dietary model. The authors suggest that the antioxidant action of the extract could be one of the contributing mechanisms for the improvement of insulin sensitivity.

Thymus vulgaris L.

Thymus vulgaris L., commonly known as thyme, is a spicy herb native to Southern Europe It has a worldwide distribution, particularly in the Mediterranean and neighboring countries, Northern Africa and parts of Asia [213].

It is used as a preservative, odorant and flavoring agent in foods, and as a culinary herb and in herbal medicine [214]. Thyme possesses potent antibacterial, antifungal, sedative, antibiotic, antiseptic, antioxidative, antitussive, expectorant, antispasmodic, astringent, antifungal, antiviral, antihelminthic, carminative and diaphoretic effects [215-217]. The plant contains terpenoids, flavonoids, glycosides, and phenolic acids, and the essential oil contains thymol and carvacrol as their main bioactive components [218, 219]. The polyphenolic compounds that are responsible for its antioxidant effects are rosmarinic acid, quercetin, eriocitrin, luteolin, and apigenin [220].

Phytochemical Constituents

In traditional medicine, thyme has been used as an infusion for the treatment of asthma, bronchitis, diabetes, and cold and chest infections, and as a syrup for digestive problems, such as intestinal infections and infestations. It is also used in liver function improvement, appetite stimulant effect, treatment of cartilaginous tube, bronchial, and urinary infections, and treatment of laryngitis and inflammation, rheumatic aches, and neuralgia [221].

The beneficial effects of thyme in the amelioration of MetS-related markers have been reported.

In Vitro Studies

The antioxidant activity of an aqueous extract of the aerial parts of *T. vulgaris* was reported by [222], as determined by the 1,1-diphenyl-2-picrylhydrazyl assay and the radical scavenging activity.

Regarding the effects on lipids and carbohydrates metabolism, and El-Houri et al. [223] reported that methanol and dichloromethane extracts from the aerial parts of T. vulgaris induced PPARγ transactivation in a mouse embryonic fibroblasts assay, without stimulation of adipocyte differentiation. The extracts also increased insulin-stimulated glucose uptake in primary porcine myotubes and in 3T3-L1adipocytes. The authors suggest that carvacrol, a major component in thyme that has been previously reported to activate PPARγ [224], could be responsible for the observed bioactivity.

Similar results were previously reported by Christensen et al. [225] using the same screening platform, but the authors also reported that thyme extract did not affect cell viability as assayed in macrophages and endothelial cells.

In Vivo Studies

With respect to lipids metabolism and obesity, El-Houri et al. [223] reported that a methanol/dichloromethane extract from the aerial parts of T. vulgaris reduced fat accumulation in the worm *C. elegans*, making it a promising source of natural products with an effect on metabolic syndrome and type-2 diabetes management.

Hegazy et al. [222] evaluated the protective effect of an aqueous extract of thyme aerial parts on the gentamicin (GM)-induced hepatotoxicity and dyslipidemia rat model. After 10 days of coadministration of GM and the extract, inhibition of the body weight gain and a reduction in liver weight was observed. Plasma liver enzymes (AST and ALT), bilirubin level and total protein concentration were normalized, suggesting hepatoprotective actions. A hypolipidemic effect was observed by the significant reductions in total cholesterol, phospholipids, triacylglycerols, LDL-c, VLDL-c and atherogenic index; raised hydrogen peroxide levels, and DNA fragmentation were ameliorated. The authors suggest that the antioxidant activity of thyme polyphenols may be responsible of the observed results.

Urtica dioica **L.**

Urtica dioica L., also known as stinging nettle, is a medicinal herbaceous perennial flowering plant, native to Eurasia. It grows in wild form in most of the world and has been widely used in traditional medicine since ancient times to treat a variety of diseases including diabetes, hypertension, prostate cancer, osteoarthritis, gastrointestinal, respiratory and hepatic disorders, inflammation, arthritis rheumatoid, allergic rhinitis, benign prostatic hyperplasia, urologic diseases, anemia, bladder stones. It is also used in food as a leaf vegetable, primarily in soups, vegetable pies, and salads without reported side effects [226-230].

Phytochemical Constituents

The plant is rich in biologically active compounds, such as the phenylpropanoids and flavonol glycosides: 5-O- caffeoylquinic acid (chlorogenic acid), 3-O-rhamnosylglucoside (rutin), p-coumaroyl malic acid, caffeic acid, 2-O-caffeoyl malic acid, and 3-O-glucoside (isoquercitrin), kaempferol 3-O-rutinoside, isorhamnetin 3-O-rutinoside, and isorhamnetin hexoside. The leaves contain the highest concentration of active ingredients, and the root is rich in sterols and sterolglycosides, phenylpropanes, and coumarins, being the most abundant compounds quinic acid and p-coumaric acid [231-233]. The reported bioactivities are antifungal and anti-microbial, stimulation of the central nervous system, metabolism enhancement, antioxidant, anti-inflammatory, anti-apoptotic, immunomodulatory, antiplatelet, hypoglycemic, hypocholesterolemic, diuretic, natriuretic and hypotensive [228, 230, 231, 234].

In vitro and *in vivo* studies have demonstrated several beneficial effects of *U. dioica* on metabolic syndrome associated abnormalities, particularly with respect to its antioxidant, anti-inflammatory, hypoglycemic, antihyperlipidemic, antihypertensive, antiplatelet, hypocholesterolemic and insulin-sensitizing properties.

In Vitro Studies

It has been demonstrated that *U. dioica* leaf extracts can inhibit inflammatory processes. Barnes et al. [235] reported that an aqueous extract was able to inhibit the production of the pro-inflammatory IL-2 and interferon-y (IFNy), and to stimulate the anti-inflammatory IL-4 in a culture of mononuclear cells. Riehemann and Schulze-Osthoff [236] suggested that some of the anti-inflammatory effects of *U. dioica* could be attributed to its inhibitory effect on TNF-kappa B activity. Obertreis et al. [237] reported that a leaf alcoholic extract reduced significantly the pro-inflammatory cytokines Tumor Necrosis Factor (TNF-a) and interleukin-1 (IL-l), after 24 h exposure on healthy humans' blood. Francišković et al. [223] demonstrated that an aqueous/methanol extract of the aerial parts and roots can inhibit COX-1 and 12-LOX pathways ex vivo on IEC18 human platelets. The enzymes COX and LOX catalyze the production of inflammatory mediators associated mostly with inflammatory conditions and certain types of cardiovascular diseases.

The antioxidant properties of *U. dioica* have also been addressed. [238] demonstrated that leaf extracts obtained by an ultrasound-assisted extraction in aqueous/methanol media possess a greater antioxidant capacity than BHT and a high metal chelating ability.

With respect to adipocytes metabolism and insulin sensitivity, Obanda et al. [239] reported that an ethanolic extract of the aerial part of *U. dioica* enhanced of the expression of the insulin-sensitizing hormone adiponectin and the ceramidase activity in cultured 3T3-L1 adipocytes treated with excess FFAs as a model of metabolic dysfunction; in consequence the ceramide accumulation was abated, and insulin sensitivity was increased via enhanced insulin-stimulated Akt serine phosphorylation.

Hypoglycemic activity *in vitro* was reported by Domola et al. [240], who showed that leaf aqueous extracts increased glucose uptake in rat L6-GLUT4myc-myoblasts, without affecting insulin secretion in the clonal insulin-secreting mouse cell line MIN6. The authors propose the existence of several distinct isoforms of cyclic peptides that are in part responsible for the hypoglycemic effect, which may form a pore through which glucose can pass, facilitating cellular uptake.

In Vivo Studies

The antioxidant capacity of *U. dioica* has also been evaluated *in vivo*. The administration of leaf extracts to mice and rats has resulted in the increase of SOD and CAT activity in erythrocytes, in the increase of plasma antioxidant capacity, and in the reduction of plasma lipid peroxidation, increasing the endogenous antioxidative defense and decreasing the oxidative stress in experimental rats [238] and in numerous organs of mice [241], probably due to the presence of rutin in the extracts.

With respect to hypertension management, Tahri et al. [242] reported a hypotensive effect of an aqueous leaf extract in rats; in a similar way, Qayyum et al. [243] using a crude methanolic rhizome extracts, reported an antihypertensive effect after intravenous administration in normotensive and salt-induced hypertensive rats, probably due to the NO-mediated vasorelaxation and the calcium channel blockade exerted by U. dioica extracts observed on rat and rabbit thoracic aortae. The phytochemical analysis indicated the presence of alkaloids, tannins, flavonoids, saponins, reducing sugars, cardiac glycosides, steroids and terpenoids in the methanolic crude extract. Testai et al. [244] evaluated the effect of a root extract obtained by a water infusion and a methanol extraction in Wistar rats and Dunkin Hartley guinea-pigs, and concluded *that U. dioica* roots can produce hypotensive responses, through a vasorelaxant effect mediated by the release of endothelial nitric oxide and the opening of potassium channels, and through a negative inotropic action, as observed in rat's aorta with and without endothelium.

Vajic et al. [238] evaluated the effect of different doses of an aqueous/methanol leaf extract in spontaneously hypertensive rats (SHR), an experimental model of essential human hypertension. After 4 weeks of extract administration, a significant reduction in systolic and diastolic blood pressure and an improvement of cardiac performances (cardiac index, cardiac output, stroke volume) was observed, probably as a result of calcium channel blockade and to the increase in endogenous antioxidative defense and the decrease in oxidative stress.

Several studies have also shown the capacity of *Urtica dioica* to decrease blood glucose in human and animal studies through both pancreatic and extrapancreatic pathways. Farzami et al. [245] observed a reduction in glucose and an increase in insulin levels in normal and diabetic rats after intraperitoneal injection of U. dioica leaf extracts, as well as a marked increase in insulin secretion on pancreatic islets of Langerhans exposed to several fractions of *U. dioica* leaf extracts. The study of Bnouham et al. [246] also showed that an aqueous extract of aerial plant material had a significant antihyperglycemic effect when administered to rats 30 min before an oral glucose tolerance test, which may be caused in part by the reduction of intestinal glucose absorption, as observed in situ in a segment of rats jejunum. In addition, the ability of leaf aqueous extracts to inhibit alpha-amylase activity was demonstrated *in vitro* by Rahimzadeh et al. [247] as another mechanism to reduce circulating levels of glucose.

The beneficial effects of *U. dioica* have also been published on metabolic syndrome murine models. Ahangarpour et al. [248] described the effect of a leaf hydroalcoholic extract on fructose-induced insulin resistance rats. The injection of the extract for two weeks significantly decreased serum glucose, insulin, LDL, leptin, LDL/HDL ratio, and FIRI. It also significantly increased serum TG, VLDL, and AST, but did not change serum ALP. Daher et al. [249] analyzed the effect of an aqueous and petroleum ether extracts of aerial parts administered to rats fed a high-fat diet for one month. Significant decreases in total cholesterol, LDL cholesterol, LDL/HDL cholesterol ratio and plasma total apo B were observed. The activity of the liver enzymes GOT, GPT and LDH diminished, showing that no liver damage occurred during the study period. Nassiri-Asl et al. [250] reported the effect of an ethanolic leaf extract in a rat model of hypercholesterolemia induced by a high-cholesterol diet. After 28 days of extract administration, total cholesterol and LDL-c were reduced to normal, as well as plasma AST and ALT activities, and a reversion of liver steatosis was observed. The authors suggest that *U. dioica* sterols could mediate the hypocholesterolemic effect of the extract and that the hepatoprotective effect may be related to the antioxidant and anti-inflammatory activities of the extract. Ethanolic and aqueous extracts were also assayed in mice fed with a cholesterol-rich diet; a decrease in serum LDL-c and an increase in serum HDL concentration were observed, with no changes in serum total cholesterol, triglycerides, glucose, nor AST and ALT activities [251].

Vaccinium myrtillus L.

Vaccinium myrtillus L. (VML), also known as bilberry or blueberry, grows wildly in northern Europe and North America. VML is a popular ingredient in different food products as juices and other fruit preparations, and it has been reported that VML fruits consumption protects against oxidative stress and free radicals [252-254].

Phytochemical Constituents

The nutritional quality of VLM fruit depends on environmental factors such as origin, climate, fertilization, and soil characteristics; the profile of the different bioactive compounds, such as anthocyanins, flavonols, and tocopherols might be variable, but in general, they have a high content of phenolic compounds including anthocyanidins and anthocyanins, and other potent antioxidants such as flavonols and phenolic acids, stilbenes (resveratrol) and ascorbic acid [255]. The oil extracted from VML seeds contains PUFAs such as linoleic acid (18:2 n-6) and α-linolenic acid (18:3 n-3) and the MUFA oleic acid (18:1 n-9) and vitamin E [254]. Consumption of VML has demonstrated beneficial effects for prevention and treatment of MS due to the high content of anthocyanins and the ability of this group of compounds to improve insulin resistance, impaired glucose tolerance, and to prevent free radical production, amongst other activities [256].

In Vitro Studies

In 3T3-L1 adipocytes culture, it was demonstrated that a blueberry extract inhibited adipocyte differentiation in a dose-dependent manner by diminishing lipid accumulation with concomitant down-regulation in the mRNA expression levels of PPARγ and C/ebp α, key-role molecules in adipocyte differentiation [257]. Tsuda et al. [258] demonstrated that the treatment of rat adipocytes with the anthocyanins cyanidin 3-glucoside and cyanidin resulted in the up-regulation of genes involved in lipid metabolism and signal transduction. In particular, an increase in hormone-sensitive lipase gene expression and enhancement of the lipolytic activity by the treatment of adipocytes with anthocyanins were observed. These genes have important functions in adipocytes related to obesity and diabetes.

In Vivo Studies

Blueberry juice anthocyanins attenuated obesity development, elevated serum leptin, and diabetes in mice fed with an obesogenic diet; also, its consumption attenuated insulin resistance in high-fat diet-fed mice [259, 260]. Blueberry extracts enriched in anthocyanins attenuated cardiac injury induced by cyclophosphamide via reduction of arterial blood pressure and increased cardiac enzyme activities and heart rate in rats [261]. Consumption of fresh blueberries for 75 days in guinea pigs fed a high cholesterol diet decreased the accumulation of cholesterol and oxidative stress in aorta and liver [253].

In addition to their antiglycemic and anti-obesogenic effects, blueberries have also been investigated for their hypotensive potential. Salt-sensitive, spontaneously hypertensive stroke-prone rats (SHRSP) were fed a diet enriched with 3% blueberry powder during eight weeks; blood pressure was significantly reduced in the blueberry-treated group compared with the control group [262]. Blueberry extracts have been observed to increase vasodilation consistently in rat models through an endothelium-mediated pathway involving nitric oxide (NO) metabolism and the production/activity of COX-derived products [263].

Zingiber officinale Roscoe

The rhizome of *Zingiber officinale Roscoe* (ZOR), commonly known as ginger is one of the most popular spices in the world; its essential oil is used worldwide in the food and cosmetics industries [264]. There are various reviews about the use and biological activities of ginger and ginger extracts, namely antioxidant, anti-inflammatory, antimicrobial, anticancer, anti-emetic, anti-metabolic syndrome, and neuroprotective [265-268].

Phytochemical Constituents

ZOR essential oil is rich in sesquiterpenes, carbonyl compounds, alcohols, and monoterpene hydrocarbons [264]. The fresh fruit possess gingerols; a group of lipophilic compounds that impart the characteristic flavor and have many of the pharmacological activities of fresh ginger [267].

In Vitro Studies

Since inflammation is an important component of SM, the murine macrophage cell line Raw 264.7 stimulated with lipopolysaccharides (LPS) was used to evaluate the effect of various concentrations of pure 12-dehydrogingerdione (DHGD) from ZOR rhizome in pro-inflammatory cytokine production. Authors report a strong inhibition of the expression of LPS-stimulated IL-6, NO, prostaglandin E2, iNOS, and cyclooxygenase-2 in this cell line, making DHGD a potential compound as an anti-inflammatory drug [269].

Peroxisome proliferator reactivated receptor gamma co-activator 1-alpha (PGC-1α), is a receptor closely related to the pathogenesis of insulin resistance. In an interesting *in vitro* studio, rat myoblast culture was treated with a ZOR fraction containing (S)-(6)-gingerol 94% to investigate its effect on muscle metabolism, particularly in the PGC-1α. The authors reported an increase in the expression of PCG-1α and propose that skeletal muscle cells improve their metabolic capacity when consuming ginger extract by effectively lowering insulin resistance [270].

In Vivo Studies

Murine models have been useful in studying the effects of the administration of ginger extract in some metabolic syndrome parameters. Administration of an ethanolic ginger extract to adult male Sprague–Dawley rats fed a high-fat and high-simple-carbohydrate diet, similar to a cafeteria-style western diet, was found to be effective in preventing insulin resistance by enhancing the metabolic capacity of skeletal muscle [270].

Male Golden Syrian hamsters in high-cholesterol diets were used to evaluate the effect of gingerol- and shogaol-enriched ginger extract (GSE). GSE significantly decreased total fatty acids in the liver and increased the excretion of fecal fatty acids in a dose-dependent manner. Authors also report that GSE helped to modulate lipoprotein profile in plasma by increasing cholesterol excretion, specifically by down-regulating the protein that participates in cholesterol absorption, favoring its excretion in the intestine [271].

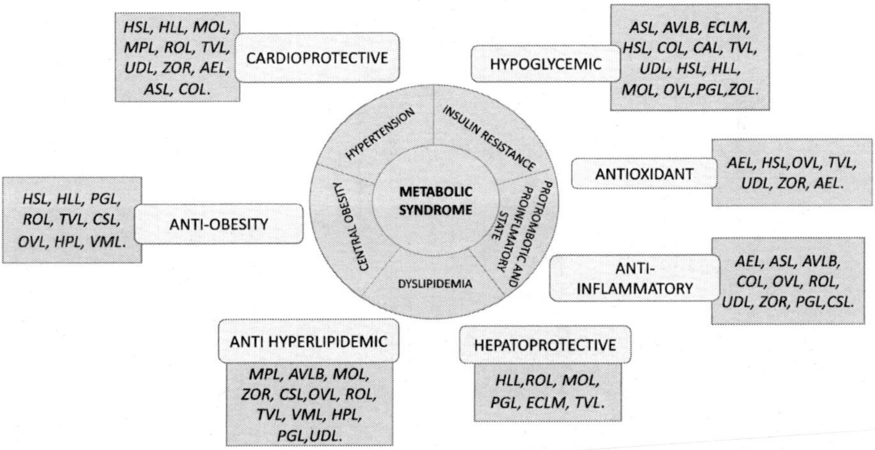

Figure 2. Mexican medicinal plants with positive effects on MetS manifestations: *AEL- Agrimonia eupatoria* L. *ASL- Allium sativum* L. *AVLB- Aloe vera* L. *Burm COL- Calendula officinalis* L. *CVJP- Cinnamomum verum J Prezi CAL-Citrus aurantium* L. *CSL- Cynara scolymus* L. *ECLM- Elettaria cardamomum* L. *Maton HSL- Hibiscus sabdariffa* L. *HLL-Humulus lupulus* L. *HPL- Hypercorium perforatum* L. *MOL- Melissa officinalis* L. *MPL- Mentha piperita* L. *OVL- Origanum vulgare* L. *PGL- Psidium guajava* L. *ROL- Rosmarinus officinalis* L. *TVL- Thymus vulgaris* L. *UDL- Urtica dioica* L. *VML- Vaccinium myrtillus* L. *ZOR- Zingiber officinale R.*

A mice model of obesity fed with a high-fat diet, with chronic inflammation and prothrombotic state was used as an early vascular model in order to evaluate the effect of the consumption of Tongqiaohuoxue in SM associated parameters; Tongqiaohuoxue decoction (TD) is commonly used in East Asian medicine, and is prepared by combining eight different types of herbal plants including *Zingiber officinale* Roscoe. Mice treated in TD showed an ameliorated prothrombotic state, since PAI-1and fibrins, which are markers of thrombosis, decreased significantly. A significant increase in adiponectin levels in blood and adipose tissue were also found, with no effects on liver or total body weigh [272].

Table 2. Mexican medicinal plants with positive effects on MetS

Biological effect	Mechanism	Mexican medicinal plant
Anti-inflammatory	↓Pro-inflammatory cytokines	*Agrimonia eupatoria L* (AE), *Allium sativum L, Aloe vera L. Burm, Calendula officinalis L, Origanum vulgare, Rosmarinus officinalis* L, *Urtica dioica L, Zingiber officinale Roscoe*
	↓ Via NF-kB pathway down-regulation	*Psidium guajava L*
	↓ iNOS proinflammatory activity	*Cynara scolymus L*
Anti hyperlimidemic	↑HDL-c↓ LDL-c	*Mentha piperita L*
	↑Expression of PPAR taget gene	*Aloe vera L. Burm, Melissa officinalis L, Zingiber officinale Roscoe*
	↓ Lipid peroxidation	*Cynara scolymus L*
	PPAR γ agonist	*Origanum vulgare, Rosmarinus officinalis* L, *Thymus vulgaris L, Vaccinium myrtillus L*
	↑ Adiponectin ↓ Serum leptin	*Hypercurium perforatum L, Psidium guajava L, Urtica dioica L, Zingiber officinale Roscoe*
	↑ Fatty acid oxidation	*Rosmarinus officinalis* L
Antioxidant	↑Proteccion against DNA damage	*Agrimonia eupatoria L* (AE)
	↓ Oxidative stress	*Agrimonia aupatoria L* (AE), *Hibiscus sabdariffa L, Origanum vulgare; Thymus vulgaris L, Urtica dioica L, Zingiber officinale Roscoe*
	↑ Plasma antoxidant capacity	*Agrimonia eupatoria L* (AE)
	↓COX-1 and 12-LOX pathways	*Urtica dioica L*
Anti-obesity	↓ Adipogenic transcription factors	*Hibiscus sabdariffa L*
	↓ Weight gain and glycemia in obesity	*Hibiscus sabdariffa L, Humulus lupulus Linneus, Psidium guajava L, Rosmarinus officinalis* L, *Thymus vulgaris L*
	Digestive enzymes inhibitory activity	*Cynara scolymus L, Origanum vulgare*
	↑Liver glycolysis and fatty acid oxidation	*Rosmarinus officinalis* L
	↓ Pre-adipocyte differentiation and antiadipogenic effect	*Hypercurium perforatum L, Rosmarinus officinalis* L, *Vaccinium myrtillus L*
	↑ Insulin activity	*Calendula officinalis L, Citrus aurantium L, Thymus vulgaris L; Urtica dioica L*
	↓ HOMA	*Hibiscus sabdariffa L*

Biological effect	Mechanism	Mexican medicinal plant
	↓Fasting glucose levels	*Humulus lupulus Linneus, Urtica dioica L*
	↑Insulin sensitivity in adipocytes and skeletal muscle	*Melissa officinalis L, Origanum vulgare, Psidium guajava L, Urtica dioica L, Zingiber officinale Roscoe*
	↓Alpha-amylase activity	*Urtica dioica L*
Hipoglycemic	↓ Blood glucose levels	*Allium sativum L, Aloe vera L. Burm*
	↑ Serum insulin levels	*Aloe vera L. Burm, Elettaria cardamomum L Maton, Hibiscus sabdariffa L*
Hepatoprotective	↓ Hepatic lipid synthesis	*Humulus lupulus Linneus; Rosmarinus officinalis* L
	Via ↑Glutathione expression	*Melissa officinalis L*
	↓ Adipogénesis	*Psidium guajava L*
	↓ Serum liver enzymes	*Elettaria cardamomum L Maton*
	↓ Liver weight and liver triglycerides	*Rosmarinus officinalis* L, *Thymus vulgaris L*
Cardioprotective	↓ Serum TC, TG, LDL-c and atherogenic index	*Hibiscus sabdariffa L; Humulus lupulus Linneus, Melissa officinalis L. Mentha piperita L, Rosmarinus officinalis* L, *Thymus vulgaris L, Urtica dioica L, Zingiber officinale Roscoe*
	↑ HDL-c	*Agrimonia eupatoria L* (AE), *Allium sativum L, Mentha piperita L*
	↓ iNOS proinflammatory activity	*Calendula officinalis L*
	↑ Arterial compliance	*Allium sativum L*
	↓ Systolic and diatolic pressure	*Cinnamomum verum J Prezi; Elettaria cardamomum L Maton, Urtica dioica L, Vaccinium myrtillus L*
	↑ Fibrinolysis	*Elettaria cardamomum L Maton*
	↓ Atherogenic index	*Hibiscus sabdariffa L, Mentha piperita L, Psidium guajava L*
	ACE-I inhibition	*Rosmarinus officinalis* L

Increased oxidative stress is a major factor of obesity-related metabolic syndrome; therefore thirty-two obese men with a body mass index of 30 (18-30 years old) were volunteers to participate in a study in which they consumed 250 mg of a ZOR rhizome powder four times a day for 10 weeks along with a resistance training; the results showed that either resistance training or ginger supplementation can be an effective therapeutic strategy to reduce oxidative stress in obese individuals [273].

In this chapter, it has been established that many of the Mexican medicinal plants possess at least one proved activity against MetS manifestations; many preparations and compounds of these medicinal species may provide a wide spectrum of biological activities such as anti-inflammatory, anti-hypertensive, ant-hyperlipidemic, against oxidative stress, enhancement of insulin sensitivity, hepatoprotective and cardioprotective activities among many others, potentially useful in MetS prevention or treatment (Figure 2).

All the above-mentioned actions are accomplished through diverse molecular mechanisms that include the inhibition of the production of pro-inflammatory cytokines, the modification in the secretion of hormones like leptin and adiponectin, the inhibition of digestive enzymes' activity, and the modification of the expression of many genes involved in lipid synthesis and glucose homeostasis. The different effects of Mexican medicinal plants preparations and their bioactive compounds with beneficial effects on MetS are summarized in Table 2.

CONCLUSION

MetS is a disease of a multifactorial and complex origin; it is, by far, an epidemy of global proportions, constituting one of the major public-health challenges in many countries. Continuous efforts have been done in order to clearly define and identify MetS clinical markers so the early identification of people at risk and the detection in early stages could be more opportune; in this sense, MetS prevention strategies are as necessary as treatment ones, and the development of personalized dietary strategies to ameliorate the consequences in morbidity and mortality due to this disease is still pending. In consequence, the use of Mexican medicinal plants and their isolated compounds has enormous potential, since their use is a common practice in many regions of the world, and they are culturally and economically closer to the general population than drugs.

REFERENCES

[1] Engin, Atilla. 2017. "The Definition and Prevalence of Obesity and Metabolic Syndrome. Obesity and Lipotoxicity." *Advances in Experimental Medicine and Biology 960,* Accessed July 19, 2019. doi 10.1007/978-3-319-48382-5_1.

[2] Rochlani, Yogita. Naga, Venkata, Pothineni. Swathi, Kovelamudi. and Jawahar, L. 2017. "Metabolic syndrome: pathophysiology, management, and modulation by natural compounds." MehtaTherapeutic Advances in Cardiovascular Disease, Vol. 11(8) 215–225. Accessed July 19, 2019. doi: 10.1177/1753944717711379.

[3] Aggarwal, Bharat B. 2010. "Targeting inflammation-induced obesity and metabolic diseases by curcumin and other nutraceuticals." *Annu Rev Nutr; 30: 173–199.* Accessed July 19, 2019 doi: 10.1146/annurev.nutr.012809.104755.

[4] Padiya, Raju. Tarak, N, Khatua. Pankaj, K, Bagul. Madhusudana, Kuncha. and Sanjay, K, Banerjee. 2011. "Garlic improves insulin sensitivity and associated metabolic syndromes in fructose fed rats." *Nutr Metab (Lond); 8: 53, 7075–8–53.* Accessed July 19, 2019. doi: 10.1186/1743-7075-8-53.

[5] Bremer, Andrew. 2014. "Resveratrol use in metabolic syndrome." *Metabolic Syndrome and Related Disorders; 12(10): 493–495.* Accessed July 19, 2019. doi: 10.1089/met.2014.1505.

[6] Hou, Chih Yao. Tain, You-Lin. Yu Hong-Ren, Huang, Li-Tung. 2019. "The Effects of Resveratrol in the Treatment of Metabolic Syndrome." *International Journal of Molecular Science, 20(3). 535.* Accessed July 19, 2019. doi: 10.3390/ijms20030535.

[7] Herrera, Meza, Ma, Del Socorro. Ma, Remedios, Mendoza, López. Oscar, García-Barradas. Ma. Guadalupe, Sanchez-Otero. Erick, R, Silva-Hernández. Jesús, Ofelia, Angulo. & Rosa, María. Oliart-Ros. 2013. "Dietary anhydrous milk fat naturally enriched with conjugated linoleic acid and vaccenic acid modify cardiovascular risk biomarkers in spontaneously hypertensive rats," *International Journal of Food Sciences and Nutrition, Vol.64 575-586.* Accessed July 19, 2019. doi:10.3109/09637486.2013.763908.

[8] Guo, Xiao-fei. Li, Xin. Shi, Meiqi. and Li, Duo. 2017. "n-3 Polyunsaturated Fatty Acids and Metabolic Syndrome Risk." *A Meta-Analysis. Nutrients 2017, 9, 703*. Accessed July 19, 2019 doi:10.3390/nu907070.

[9] Grundy, Scott M. Barbara, Hansen. Sidney, C. Smith. Jr. James, I. Cleeman. Richard, A. Kahn. Lung. 2004 "Blood Institute, American Diabetes Association. Clinical management of metabolic syndrome: report of the American Heart Association" *American Diabetes Association conference on scientific issues related to management.* 24(2): e19–e24. Accesed June 15,2019. doi:10.1161/01.ATV. 0000112379.88385.67.

[10] WHO Consultation. 1999. "OMS World Health Organization. Definition, Diagnosis and Classification of Diabetes Mellitus and Its Complications:. Geneva." *World Health Organization,*. Accessed June15, 2019.

[11] (Adults Treatment Panel III). Expert Panel on detection, evaluation and treatment of high blood cholesterol in adults. 2001. *Executive Summary of the Third Report of the National Cholesterol Education Program (NCEP).JAMA; 285: 2486-2497.* Accessed June15, 2019.

[12] Grundy, Scott. M. James, Cleeman. Stephen, Daniels. Karen, Donato. Robert, Eckel. Barry, Franklin. David, Gordon. Ronald, Krauss. Peter, Savage. Sidney, SmithJr. John, Spertus. and Fernando, Costa. 2005. "Scientific statement. Diagnosis and management of the metabolic syndrome." *Circulation;* 112: 2735-2752. Accessed June 15, 2019. doi:10.1161/CIRCULATIONAHA. 105.169404 The IDF consensus worldwide definition of the

metabolic syndrome http://www.idf.org/ webdata/docs/IDF_ Metasyndrome_definition. Accessed June,15, 2019.

[13] Ingelsson, Erik. Lisa, M. Sullivan. Joanne, M. Murabito. Caroline, S. Fox. Emelia, J. Benjamin. Joseph, F. Polak. James, B. Meigs. Michelle, J. Keyes. Christopher, J. O'Donnell. Thomas, J. Wang. Ralph, B. D'Agostino. Sr. Philip, A. Wolf. and Ramachandran, S. Vasan. 2007. "Prevalence and prognostic impact of subclinical cardiovascular disease in individuals with the metabolic syndrome and diabetes." *Diabetes* 56: 1718–1726 Accessed June 15, 2019. .doi: 10.2337/db07-0078.

[14] McNeill, Ann Marie. Ronit, Katz. Cynthia, J. Girman. Wayne, D. Rosamond. Lynne, E. Wagenknecht. Joshua, I. Barzilay. Russell, P. Tracy. Peter, J. Savage. Sharon, A. Jackson. 2006. "Metabolic syndrome and cardiovascular disease in older people: the cardiovascular health study." *J Am Geriatr Soc* 54: 1317–1324, Accessed June 15 2019 doi: 10.1111/j.1532-5415.2006.00862.x.

[15] Patel, Tushar P. Komal, Rawal. Ashim, K. Bagchi. Gauri, Akolkar. Nathalia, Bernardes. Danielle, da Silva, Dias. Sarita, Gupta. Pawan, K. Singal. 2016. "Insulin resistance: An additional risk factor in the pathogenesis of cardiovascular disease in type 2 diabetes." *Heart Fail Rev 21: 11–23.* Accessed June 15,2019. doi: 10.1007/s10741-015-9515-6.

[16] Saklayen, Mohammad G.. 2018. "The Global Epidemic of the Metabolic Syndrome." *Current Hypertension Reports 20: 12.* Accessed July 19, 2019. doi:10.1007/s11906-018-0812-z.

[17] Boden, Gebers. and Shulman, Gerald. 2002 "Free fatty acids in obesity and type 2 diabetes: defining their role in the development of insulin resistance and beta-cell dysfunction." *Eur J Clin Invest 32.(Suppl. 3): 14–23.* Accessed July 19, 2019. doi: 10.1046/j.1365-2362.32.s3.3.x.

[18] Tooke, John E. and Michael, Hannemann. 2000. "Adverse endothelial function and the insulin resistance syndrome." *J Intern Med* 247(4): 425–431. Accessed June 15, 2019.

[19] Tripathy, Devjit. Priya, Mohanty. Sandeep, Dhindsa. Tufail, Syed. Husam, Ghanim. Ahmad, Aljada. and Paresh, Dandona. 2003. "Elevation of free fatty acids induces inflammation and impairs vascular reactivity in healthy subjects." *Diabetes* 52(12): 2882–2887. Accessed June 15, 2019. doi: 10.2337/diabetes.52.12.2882.

[20] Juhan-Vague, Irene. Marie-Christine, Alessi. Pierre, Emmanuel, Morange. Riyadh Saif-Ali, Khaled A Ahmed Sekaran Muniandy. 2003. "Plasminogen activator inhibitor-1, inflammation, obesity, insulin resistance and vascular risk." *J Thromb Haemost*. 1(7): 1575–1579. Accessed June 15, 2019. Doi: 10.1186/1475-2840-10-23.

[21] Hotamisligil, Gökhan S. 2016. "Inflammation and metabolic disorders." *NATURE Vol 444|14*. Accessed July 19, 2019. doi: 10.1038/nature05485.

[22] Matsuzawa, Yuji. Tohru, Funahashi. Tadashi, Nakamura. 2011. "The concept of metabolic syndrome: contribution of visceral fat accumulation and its molecular mechanism." *J Atheroscler Thromb*; 18(8): 629–639. Accessed June 15, 2019. doi: 10.5551/jat.7922.

[23] Inoguchi, Toyoshi. Ping, Li. Fumio, Umeda. Hai, Yan, Yu. Maiko, Kakimoto. Minako, Imamura. Tsuyoshi, Aoki. Takashi, Etoh. Toshihiko, Hashimoto. Masaichi, Naruse. Hiroaki, Sano. Hideo, Utsumi. Hajime, Nawata. 2000. "High glucose level and free fatty acid stimulate reactive oxygen species production through protein kinase C—dependent activation of NAD(P)H oxidase in cultured vascular cells." Accessed June 16, 2019. doi: 10.2337/diabetes. 49.11.1939.

[24] Furukawa, Shigetada. Takuya, Fujita. Michio, Shimabukuro. Masanori, Iwaki. Yukio, Yamada. Yoshimitsu, Nakajima. Osamu, Nakayama. Makoto, Makishima. Morihiro, Matsuda. Iichiro, Shimomura. 2004. "Increased oxidative stress in obesity and its impact on metabolic syndrome." *J Clin Invest 114: 1752–1761,*. Accessed June 16, 2019. doi: 10.1172/JCI21625.

[25] Soares, Anisio-Francesco. Delpfine, Cozzone. Nathalie, Bernoud-Hubac. Nabile, Bouzaidi-Tiali. Michel, Lagarde. Alain, Geloen.

2004."Effects of oxidative stress on adiponectin secretion and lactate production in 3T3-L1 adipocytes." *Free Radic Biol Med 38: 882–889*, Accessed June 16, 2019. Doi: 10.1016/j.freeradbiomed. 12.010.

[26] Otani, Hajime. 2011. "Oxidative stress as pathogenesis of cardiovascular risk associated with metabolic syndrome." *Antioxid Redox Signal 15: 1911–1926.* Accessed June 16, 2019. doi: 10.1089/ars.2010.3739.

[27] Bloch-Damti, Asnat. Bashan, Nava. 2005. "Proposed mechanisms for the induction of insulin resistance by oxidative stress." *Antioxid Redox Signal 7: 1553–1567.* Accessed June 16, 2019. doi: 10.1089/ars.2005.7.1553.

[28] Sablina, Anna. Andrei, Budanov. Galina, Ilyinskaya. Larissa, Agapova. Julia, Kravchenko. Peter, Chumakov. 2005. "The antioxidant function of the p53 tumor suppressor." *Nat Med 11: 1306–1313*, Accessed June 16, 2019. doi: 10.1038/nm1320.

[29] Kruiswijk, Flore. Christiaan, Labuschagne. Karen, Vousden. 2015. "p53n survival, death and metabolic health: A lifeguard with a licence to kill." *Nat Rev Mol Cell Biol 16: 393–405*. Accessed June 16, 2019. doi: 10.1038/nrm4007.

[30] De la Iglesia, Rocio. Viviana, Loria-Kohen. Maria Angeles, Zulet. Jose Alfredo, Martinez. Guillermo, Reglero. Ana, Ramirez de Molina. 2016. "Dietary Strategies Implicated in the Prevention and Treatment of Metabolic Syndrome." *Int. J. Mol. Sci. 17, 1877.* Accessed July 19, 2019. doi:10.3390/ijms17111877.

[31] Jameson, J. L., A. S. Fauci, D. L. Kasper, S. L. Hauser, D. L. Longo J. Loscalzo. *Harrison's Principles of Internal Medicine,* 20th Edition McGraw. Hill Education. Accessed July 19, 2019.

[32] Jungbauer Alois, Svjetlana Medjakovic. 2012. "Anti-inflammatory properties of culinary herbs and spices that ameliorate the effects of metabolic syndrome." *Maturitas 71 (2012) 227– 23*. Accessed July 19, 2019. doi: 10.1016/j.maturitas.2011.12.009.

[33] Pérez-Torres, Israel. Alejandra, Zúñiga Muñoz. Ulises, Beltrán-Rodríguez, Díaz. Raúl, Martínez-Memije. Verónica, Guarner Lans.

2014. "Modification of the liver fatty acids by *Hibiscus sabdariffa Linnaeus (Malvaceae)* infusion, its possible effect on vascular reactivity in a metabolic syndrome model." *Clin Exp Hypertens, 2014; 36(3): 123–131.* Accessed june 15, 2019. doi: 10.3109/10641963.2013.789046.

[34] Priyanka, M. Jadhav. Nilima, A. Kshirsagar. 2015. "Innovative Approach for Classification of Traditional System of Medicine." *Nat Prod Chem Res 3: 191.* Accessed July 19, 2019. doi:10.4172/2329-6836.1000191.

[35] Kandikattu, Hemanth, Kumar. P. Rachitha. Jayashree. Krupashree, Sukhith. Abdul, Majid. Narayanappa, Amruta. Farhath, Khanum. 2017. "Anti-inflammatory and anti-oxidant effects of *Cardamom (Elettaria repens (Sonn.) Baill)* and its phytochemical analysis by 4D GCXGC TOF-MS." *Biomedicine & Pharmacotherapy 91 (2017) 191–201.* Accessed July 19, 2019 doi:10.1016/j.biopha.2017.04.049.

[36] Afsheen, Nadia. Khalil-ur-Rehman. Nazish, Jahan. Misbah, Ijaz. Asad, Manzoor. Khalid, Mahmood, Khan. Saman, Hina. 2018. "Cardioprotective and Metabolomic Profiling of Selected Medicinal Plants against Oxidative Stress." *Oxidative Medicine and Cellular Longevity Volume 2018, Article ID 9819360, 17 pages.* Accessed July 19, 2019 doi:10.1155/2018/9819360.

[37] Ali, Al-Snafi, Esmail. 2015. "The pharmacological and therapeutic importance of *Agrimonia eupatoria* a review." *Asian Journal of Pharmaceutical Science and Technology 5(2):112-117.* Accessed June 19, 2019. doi: 10.1155/2014/282698.

[38] Mikołaj, Kostryco. Mirosława, Chwil. 2017. "Biologically active compounds in *Agrimonia eupatoria L.* and their therapeutic effects." *World Scientific News 89: 90-97.* Accessed Jume 19, 2019. doi: 10.1016/j.jfda.2016.02.007.

[39] Kuczmanová, Anika. Andrea, Balažová. Eva, Racˇanská. Miroslava, Kameníková. Silvia, Fialová. Jaroslav, Majerník. Milan, Nagy. Peter, Gál. Pavel, Mucaji. 2016. "*Agrimonia eupatoria L.* and Cynara cardunculus L. Water Infusions: Comparison of Anti-

Diabetic Activities." *Molecules* 21: 2-12. Accessed June 19, 2019. doi: 10.3390/molecules21050564.

[40] Mirjana, Muruzović. Katarina, Mladenović. Ljiljana, Čomić. Aleksandar, Glišović. 2016. "Extracts of *Agrimonia eupatoria L.* as sources of biologically active compounds and evaluation of their antioxidant, antimicrobial, and antibiofilm activities." *Journal of Food and Drug Analysis 24(3): 539-547.* Accessed Jue 19, 2009. doi:10.5937/KgJSci1941097G.

[41] Ivanova, Diana. Oskan, Tasinov. Deyana, Vankova. Yoana, Kiselova, Kaneva. 2011. "Antioxidative potential of *Agrimonia eupatoria L.*" *Science & Technologies 1(1):20-24.* Accessed June 19, 2019.

[42] Chen, Lei. Hui, Teng. Kalin, Yanbo, Zhang. Krystyna, Skalicka, Wozniak. Milen, I, Georgiev. and Jianbo, Xiao. 2017. "Agrimonolide and Desmethylagrimonolide Induced HO-1 Expression in HepG2 Cells through Nrf2-Transduction and p38 Inactivation." *Front. Pharmacol 7:513.* doi: 10.3389/fphar. 2016.00513. Accessed june 20, 2019. doi: 10.3389/fphar. 2016.00513.

[43] Ivanova, Diana. Deyana, Vankova. Milka, Nasha. 2013. "*Agrimonia eupatoria* tea consumption in relation to markers of inflammation, oxidative status and lipid metabolism in healthy subjects." *Archives of Physiology and Biochemistry 119(1): 32–37.* Accessed June 21, 2019. doi:10.3109/13813455.2012.729844.

[44] Seong, Jin Yon. Kho, Eun, Ji. Kim, Chang, Soo. Zee, Ok, Pyo. Kwak Jong, Hwan. Jeong, Won, Jin. Kim, Jee, Hyun. and Lee, Sun, Mee. 2012. "*Agrimonia eupatoria* protects against chronic ethanol-induced liver injury in rats." *Food and Chemical Toxicology 50: 2335–2341.* Accessed June 21, 2019. doi: 10.1016/j.fct.2012.04.005.

[45] Londhe, Vikas. Avinash, Gavasane. Sonali, Nipate. Deepti, Dinesh Bandawane. Praveen, Chaudhari. 2011. "Role of garlic (*Allium sativum*) in various diseases: an overview." *Journal of Pharmaceutical Research and Opinion 1(4): 129 – 134.* Accessed June 21, 2019.

[46] Lanzotti, Virginia. Felice,Scala. Giuliano,Bonanomi. 2014. "Compounds from *Allium* species with cytotoxic and antimicrobial activity." *Phytochemistry reviews 13(4):769-791.* Accessed June 21, 2019. doi:10.1007/s11101-014-9366-0.

[47] Hosseini, Azar. Hossein, Hosseinzadeh. 2015. "A review on the effects of *Allium sativum* (Garlic) in metabolic síndrome." *J. Endocrinol Invest 38(11):1147-57.* Accessed June 21, 2019. doi:10.1007/s40618-015-0313-8.

[48] Sobenin, Igor A., Lyudmila V. Nedosugova, Lyudmila V. Filatova, MikhailI Balabolkin, Tatiana V. Gorchakova. 2008. "Metabolic effects of time-released garlic powder tablets in type 2 diabetes mellitus: the results of double-blinded placebo-controlled study." *Acta Diabetol 45 (1):1–6.* Accessed June 21, 2019. doi:10.1007/s00592-007-0011-x.

[49] Murat, Ali, Turan. Taban, Suleyman. Taban, Nilgun. Ersan, Lutfiye, Yilmaz. 2017. "Characterization of garlic (*Allium sativum l.*) According the geographical origin by analysis of minerals." *Fresenius Environmental Bulletin 26(6): 4292:4298.* Accessed June 21, 2019.

[50] Ried, Karin. Peter, Fakler. 2014. "Potential of garlic (*Allium sativum*) in lowering high blood pres- sure: mechanisms of action and clinical relevance." *Integr. Blood Press Control 7: 71–82* . Accessed June 21, 2019. doi: 10.2147/IBPC.S51434.

[51] Jung, Eun-Soo. Soo-Hyun Park. Eun-Kyung Choi. Beun-Ho Ryu. Byung-Hyun Park. Dal-Sik Kim.Young-Gon Kim. Kim Soo-Wan Chae. 2014. "Reduction of blood lipid parameters by a 12-wk supplementation of aged black garlic: a randomized controlled trial." *Nutrition 30 (9):1034–1039.* Accessed June 21, 2019. doi: 10.1016/j.nut.2014.02.014.

[52] Poonam, Tripathi. Prem Prakash Gupta, Vijay Kumar Lal. 2013. "Effect of co-administration of *Allium sativum* extract and Metformin on Blood glucose of Streptozotocin induced diabetic rats." *J Intercult Ethnopharmacol 2:81–4.* Accessed June 21, 2019. doi:10.5455/jice.20130530123748.

[53] Shikha Chourey, Shikha. Tamanna, Narsinghani. and Love, Kumar Soni. 2011. "Effect of *Allium Sativum* on the pharmacokinetic of Metformin in rat plasma: a herb–drug interaction study." *Der Pharm Chem 3:287–91.* Accessed June 21, 2019.

[54] Ashraf, Rizwan. Rafeep, Alam Khan. Imran, Ashraf. 2011. "Garlic (*Allium sativum*) supplementation with standard antidiabetic agent provides better diabetic control in type 2 diabetes patients." *Pak J Pharm Sci 24(4):565–70.* Accessed June 21, 2019. PMID: 21959822.

[55] Animesh, Datta, Kumar. Jaime A., Teixeira da Silva. Aditi Saha. Paul, Rita. Sonali, Sengupta, Priyanka, Kumari Dubey. Sandip, Halder. 2012. "An Updated Overview on *Aloe vera (L.) Burm.* f." *Medicinal and Aromatic Plant Science and Biotechnology* 6 (1):1-11. Accessed June 22, 2019.

[56] Salehi, Bahare. Sevil, Albayrak. Hubert, Antolak. Dorota, Kręgiel. Ewelina, Pawlikowska. Mehdi, Sharifi. Yadav, Uprety. Patrick, Velere Tsouh, Fokou. Zubaida, Yousef. Zainul, Amiruddin, Zakaria. Elena Maria, Varoni. Farukh, Sharopov. Natália, Martins. Marcello, Iriti. and Javad, Sharifi, Rad. 2018. "*Aloe* Genus Plants: From Farm to Food Applications and Phytopharmacotherapy." *Int. J. Mol. Sci 19:2843.* Accessed June 22, 2019. doi: 10.3390/ijms19092843.

[57] Radha, Maharjan. Nampoothiri, Laxmipriya. 2015. "Evaluation of biological properties and clinical effectiveness of *Aloe vera*: A systematic review." *J. Tradit. Complement. Med.5:21–26.* Accessed June 23, 2019. doi: 10.1016/j.jtcme.2014.10.006.

[58] Boudreau, Mary. Frederick Beland. 2006. "An evaluation of the biological and toxicological properties of *Aloe Barbadensis (miller)*, Aloe vera." *J. Environ. Sci. Health C 24:103–154.* Accessed June 22, 2019. doi: 10.1080/10590500600614303.

[59] Nejatzadeh, Barandozi, Fatemeh. 2013. "Antibacterial activities and antioxidant capacity of *Aloe vera*." *Org. Med. Chem. Lett. 3(1):5.* Accessed June 22, 2019. doi: 10.1186/2191-2858-3-5.

[60] Budai, Marietta M., Aliz Varga, Sándor Milesz, József T″ozsér, Szilvia Benk″o. 2013. "*Aloe vera* downregulates LPS-induced

inflammatory cytokine production andexpression of NLRP3 inflammasome in human macrophages" *Molecular Immunology 56 (2013) 471–479.* Accessed July 5, 2019.doi: 10.1016/j.molimm. 2013.05.005.

[61] Hęś, Marzanna. Krzysztof, Dziedzic. Joanna, Mai, Le Thanh-Blicharz. Dominik, Kmiecik. 2016. "Antioxidant activity of true aloe (*Aloe vera*) extract in model systems." *Nauka Przyr. Technol 10(4).* Accessed June 23, 2019. doi:10.17306/J.NPT.2016.4.53.

[62] Kouji Nomaguchi. Miyuki Tanaka, Eriko Misawa, Muneo Yamada, Tomohiro Toida, Keiji Iwatsuki, Tsuyoshi Goto, Teruo Kawada. 2011. "*Aloe vera* phytosterols act as ligands for PPAR and improve the expression levels of PPAR target genes in the livers of mice with diet-induced obesity." *Obesity Research y Clinical Practice 5:190-201.* Accessed June 23, 2019. doi: 10.1016/j.orcp.2011.01.002.

[63] Amira, Hussein, Abo-Youssef, Mourad. Basim, Anwar, Shesheta Messiha. 2013. "Beneficial effects of *Aloe vera* in treatment of diabetes: Comparative in vivo and in vitro studies." *Bulletin of Faculty of Pharmacy Cairo University 51:7-11.* Accessed June 23, 2019. doi: 10.1016/j.bfopcu.2012.03.002.

[64] Vrish, Ashwlayan. Amrish, Amrish Kumar, Mansi Verma, Vipin Kumar Garg, SK Gupta. 2018. "Therapeutic Potential of *Calendula officinalis*." *Pharm Pharmacol Int J 6(2):149–155.* Accessed June 24, 2019. doi: 10.15406/ppij.2018.06.00171.

[65] Mullaicharam, Bhupathyraaj. Nirmala, Amaresh. 2014. "Phytochemistry and Pleiotropic Pharmacological Properties of *Calendula officinalis* - A Review." *Journal of pharmacognosy and phytochemistry 2(4).* Accessed June 24, 2019. doi: 10.18502/ pbr.v4i2.214.

[66] Ray, Diptarka. Subhendu, Mukherjee. Mario, Falchi. Aldo, Bertelli. Pier, Carlo, Braga. Dipak, Das. 2010. "Amelioration of myocardial ischemic reperfusion injury with *Calendula officinalis*." *Curr Pharm Biotechnol 11(8):849–854.* Accessed June 25, 2019. doi:10.2174/ 138920110793261971.

[67] Preethi, Korengath, Chandran. Girij Kuttan. Ramadasan, Kuttan, R. 2009. "Anti-inflammatory activity of flower extract of *Calendula officinalis Linn* and its possible mechanism of action." *Indian J Exp Biol 47(2):113–120.* Accessed June 25, 2019. PMID: 19374166.

[68] Pei, Chen., Jianghao, Sun., and Paul, Ford. 2014. "Differentiation of the Four Major Species of *Cinnamons (C. burmannii, C. verum, C. cassia, and C. loureiroi)* Using a Flow Injection Mass Spectrometric (FIMS) Fingerprinting Method." *J. Agric. Food Chem 62: 2516−2521.* Accessed June 25, 2019. doi:10.1021/jf405580c.

[69] Jayaprakasha, Guddarangavvanahally. Jagan, Mohan, Rao. 2011. "Chemistry, biogenesis, and biological activities of *Cinnamomum zeylanicum.*" *Crit Rev Food Sci Nutr 51:547–562.* Accessed June 25, 2019. doi: 10.1080/10408391003699550.

[70] Anderson, Richard. Broadhurst, Leigh. Marilyn Polansky. Walter, Schmidt. Alam, Khan, Vincent, Flanagan. 2004. "Isolation and characterization of polyphenol type-A polymers from *cinnamon* with insulin-like biological activity." *J Agric Food Chem 52:65-70.* Accessed June 25, 2019. doi: 10.1021/jf034916b.

[71] Qin, Bolin. Masaru, Nagasaki. Ming, Ren. Gustavo, Bajotto. Yoshiharu Oshida. Yuzo, Sato. 2003. "*Cinnamon* extract (traditional herb) potentiates in vivo insulin-regulated glucose utilization via enhancing insulin signaling in rats." *Diabetes Res Clin Pract 62:139-148.* Accessed June 25, 2019. doi:10.1016/S0168-8227(03)00173-6.

[72] Hamid, Mollazadeh. Hosseinzadeh, Hossein. 2016. "*Cinnamon* effects on metabolic syndrome: a review based on its mechanisms." *Iran J Basic Med Sci 19:1258-1270.* Accessed June 25, 2019. doi:10.22038/ijbms.2016.7906.

[73] Preuss, Harry. Bobby Echard. Marilyn, Polansky. Richard, Anderson. 2006. "Whole *cinnamon* and aqueous extracts ameliorate sucrose- induced blood pressure elevations in spontaneously hypertensive rats." *J Am Coll Nutr 25(2):144–150.* Accessed June 26, 2019. doi:10.1080/07315724.2006.10719525.

[74] Ziegenfuss, Tim. Jennifer, Hofheins. Ronald, Mendel. Jamie Landis. Richard, Anderson. 2006. "Effects of a water-soluble *cinnamon* extract on body composition and features of the metabolic syndrome in pre-diabetic men and women." *J Int Soc Sports Nutr* 3:45–53. Accessed June 26, 2019. doi:10.1186/1550-2783-3-2-45.

[75] Qin, Bolin. Marilyn, Polansky. Richard, Anthony, Anderson. 2010. "*Cinnamon* extract regulates plasma levels of adipose-derived factors and expression of multiple genes related to carbohydrate metabolism and lipogenesis in adipose tissue of fructose-fed rats." *Horm Metab Res 42(3):187–193.* Accessed June 26, 2019. doi: 10.1055/s-0029-1242746.

[76] Karimi Ehsan, Oskoueian, Ehsan. Hendra, Rudi. Oskoueian Armin. Jaafar, Hawa Z. 2012. "Phenolic compounds characterization and biological activities of *Citrus aurantium bloom.*" *Molecules. 2012; 17(2): 1203–1218.* Accessed July 2, 2019. doi:10.3390/molecules 17021203.

[77] Raciti, Gregory, Alexander. Francesca, Fiory. Michele, Campitelli. Antonella, Desiderio. Rosa, Spinelli. Michele, Longo. Cecilia, Nigro. Giacomo, Pepe. Eduardo, Sommella. Pietro, Campiglia. Pietro, Formisano. Francesco, Beguinot. Claudia, Miele. 2018. "*Citrus aurantium L.* dry extracts promote C/ebpβ expression and improve adipocyte differentiation in 3T3-L1 cells." *PLoS ONE 13(3): e0193704.* Accessed July 2, 2019. doi:10.1371/journal. pone.0193704.

[78] Stohs Sidney J. 2017. "Studies Regarding *Citrus aurantium* (Bitter Orange) Extract and p-Synephrine." *Phytother Res. 2017 Oct; 31(10): 1463–1474.* Accessed July 2, 2019. doi: 10.1002/ptr.5879.

[79] Mannucci, Carmen. Fabrizio, Calapai. Luigi, Cardia. Giuseppina, Inferrera. Giovanni, D'Arena. Martina, Di Pietro. Michele, Navarra. Sebastiano, Gangemi. Elvira, Ventura, Spagnolo. Gioacchino, Calapai. 2018. "Clinical Pharmacology of *Citrus aurantium* and *Citrus sinensis* for the Treatment of Anxiety." *Evid Based Complement Alternat Med. 2018 Dec 2;2018:3624094.* Accessed July 2, 2019. doi: 10.1155/2018/3624094.

[80] Karoui, Jabri, Iness. Brahim, Marzouk. 2013. "Characterization of bioactive compounds in Tunisian bitter orange (*Citrus aurantium L.*) peel and juice and determination of their antioxidant activities." *Biomed Res Int. 2013;2013:345415.* Accessed July 3, 2019. doi:10.1155/2013/345415.

[81] Preuss, Harry. Donna, DiFerdinando. Mihir, Bagchi. Debasis, Bagchi. 2002. "*Citrus aurantium* as a thermogenic, weight-reduction replacement for ephedra: an overview." *J Med. 2002;33(1-4):247-64.* Accessed July 3, 2019. PMID:12939122.

[82] Colker, Carlon. Douglas, Kalman. Georgean, Torina. Theresa, Perlis. Chris, Street. 1999. "Effects of *Citrus aurantium* extract, caffeine, and St. John's wort on body fat loss, lipid levels, and mood states in overweight healthy adults." *Curr Ther Res 60(3): 146-153, 1999.* Accessed July 3, 2019. doi:10.1016/S0011-393X(00)88523-9.

[83] Haaz, Steffany. Kevin R. Fontaine. David, Allison. 2006. "*Citrus aurantium* and synephrine alkaloids in the treatment of overweight and obesity: an update." *Obes Rev 2006, 7:79-88.* Accessed July 3, 2019. doi:10.1111/j.1467-789X.2006.00195.x.

[84] Mercader, Josep. Estelle, Wanecq. Jian, Chen. Christian, Carpéné. 2011. "Isopropylnorsynephrine is a stronger lipolytic agent in human adipocytes than synephrine and other amines present in *Citrus aurantium.*" *J Physiol Biochem 2011, 67(3):443-452.* Accessed July 3, 2019. doi:10.1007/s13105-011-0078-2.

[85] Stohs, Sidney, J. Harry, G, Preuss. Mohd, Shara. 2011. "The safety of *Citrus aurantium* (bitter orange) and its primary protoalkaloid p-synephrine." *Phytother Res. 2011 Oct;25(10):1421-8.* Accessed July 3, 2019. doi:10.1002/ptr.3490.

[86] Stohs, Sidney. Harry, G, Preuss. Mohd, Shara. 2012. "A review of the human clinical studies involving *Citrus aurantium* (bitter orange) extract and its primary protoalkaloid p-synephrine." *Int J Med Sci. 2012;9(7):527-38.* Accessed July 3, 2019. doi:10.7150/ijms.4446.

[87] Kim Gon-Sup. Hyoung, Joon, Park. Jong-Hwa, Woo. Mi-Kyeong, Kim. Phil-Ok, Koh. Wongi, Min. Yeoung-Gyu, Ko. Chung-Hei,

Kim. Chung-Kil, Won. Jae-Hyeon, Cho. 2012. "*Citrus aurantium* flavonoids inhibit adipogenesis through the Akt signaling pathway in 3T3-L1 cells." *BMC Complement Altern Med. 2012 Apr 3;12:31.* Accessed July 3, 2019. doi:10.1186/1472-6882-12-31.

[88] Rizza, Stefano. Ranganath Muniyappa, Micaela Iantorno, Jeong-a Kim, Hui Chen, Philomena Pullikotil, Nicoletta Senese, Manfredi Tesauro, Davide Lauro, Carmine Cardillo, Michael J. Quon. 2011. "*Citrus* polyphenol hesperidin stimulates production of nitric oxide in endothelial cells while improving endothelial function and reducing inflammatory markers in patients with metabolic syndrome." *J Clin Endocrinol Metab. 2011 May;96(5):E782-92.* Accessed July 4, 2019. doi:10.1210/jc.2010-2879.

[89] Assini, Julia. Erin, Mulvihill. Murray, Huff. 2013. "*Citrus* flavonoids and lipid metabolism." *Curr Opin Lipidol. 2013 Feb;24(1):34-40.* Accessed July 4, 2019. doi:10.1097/MOL. 0b013e32835c07fd.

[90] Alam M-Ashraful. Nusrat, Subhan. Mahbubur, Rahman. Shaikh, Uddin. Hasan, Reza. Satyajit, Sarker. 2014. "Effect of *citrus* flavonoids, naringin and naringenin, on metabolic syndrome and their mechanisms of action." *Adv Nutr. 2014 Jul 14;5(4):404-17.* Accessed July 3, 2019. doi:10.3945/an.113.005603.

[91] Amiot Maie, Josephe. Catherine, Riva. Agnès, Vinet. 2016. "Effects of dietary polyphenols on metabolic syndrome features in humans: a systematic review." *Obes Rev. 2016 Jul;17(7):573-86.* Accessed July 3, 2019. doi:10.1111/obr.12409.

[92] Mulvihill, Erin. Amy, Burke. and Murray, W. Huff. 2016. "*Citrus* Flavonoids as Regulators of Lipoprotein." *Metabolism and Atherosclerosis. Annu. Rev. Nutr. 36: 275-299.* Accessed July 3, 2019. doi:10.1146/annurev-nutr-071715-050718.

[93] Burke, Amy. Brian, Sutherland. Dawn, Telford. Marisa, Morrow. 2018. "Intervention with *citrus* flavonoids reverses obesity and improves metabolic syndrome and atherosclerosis in obese Ldlr-/- mice." *J Lipid Res. 2018 Sep;59(9):1714-1728.* Accessed July 3, 2019. doi:10.1194/jlr.M087387.

[94] Mosqueda-Solís Aandrea. Juana, Sánchez. Bárbara, Reynés. Mariona, Palou. María, Portillo. Andreu, Palou. Catalina, Picó. 2018. "Hesperidin and capsaicin, but not the combination, prevent hepatic steatosis and other metabolic syndrome-related alterations in western diet-fed rats." *Sci Rep. 2018 Oct 10;8(1):15100.* Accessed July 4, 2019. doi: 10.1038/s41598-018-32875-4.

[95] Sepideh, Miraj. Kiani, Sadegh. 2016. "Study of therapeutic effects of *Cynara scolymus L.*: A review." *Der Pharmacia Lettre 8 (9):168-173.* Accessed June 21, 2019.

[96] Lattanzio, Vincenzo. Paul, Kroon. Vito, Linsalata. Angela, Cardinali. 2009. "Globe artichoke: a functional food and source of nutraceutical ingredients." *J Funct Foods 1(2):131–144.* Accessed June 21, 2019.

[97] Bundy, Rafe. Ann, Walker. Richard, Middleton. Carol, Wallis. Hugh Simpson. 2008. "Artichoke leaf extract (*Cynara scolymus*) reduces plasma cholesterol in otherwise healthy hypercholesterolemic adults: a randomized, double blind placebo controlled trial." *Phytomedicine 15(9): 668-75.* Accessed June 21, 2019. doi: 10.1016/j.phymed.2008.03.001.

[98] Ben, Maryem. Hanen, Affes. 2017. "Chemicals Compositions, Antioxidant and Anti-Inflammatory Activity of *Cynara scolymus Leaves* Extracts, and Analysis of Major Bioactive Polyphenols by HPLC." *2017:4951937.* Accessed June 21, 2019. doi: 10.1155/2017/4951937.

[99] Emine, Colak. Ustuner, Mehmet. Tekin, Neslihan. Colak, Ertugrul. Burukoglu, Dilek. Degirmenci, Irfan. Gunes, Hasan. 2016. "The hepatocurative effects of Cynara scolymus L. leaf extract on carbón tetrachloride-induced oxidative stress and hepatic injury in rats." *Springer Plus 5:216.* Accessed June 21, 2019. doi: 10.1186/s40064-016-1894-1.

[100] Jiménez, Escrig, Antonio. Lars Ove Dragsted Bahram Daneshvar Raquel Pulido Fulgencio Saura-Calixto. 2003. "In vitro antioxidant activities of edible artichoke (*Cynara scolymus L.*) and effect on

biomarkers of antioxidants in rats." *J Agric Food Chem 51(18): 5540-5545*. Accessed June 21, 2019. doi: 10.1021/ jf030047e.

[101] Li, Huige. Ning Xia, Isolde Brausch, Ying Yao and Ulrich Förstermann. 2004. "Flavonoids from artichoke (*Cynara scolymus L.*) up-regulate endothelial- type. Nitric-oxide synthase gene expression in human endothelial cells." *J Pharm Exp Ther 310(3):926-932*. Accessed June 21, 2019. doi: 10.1124/jpet. 104.066639.

[102] Xia, Ning. Andrea Pautz, Ursula Wollscheid, Gisela Reifenberg, Ulrich Förstermann and Huige Li. 2014 "Artichoke, cynarin and cyanidin Downregulate the Expression of Inducible." *Molecules 19:3654-3668.* Accessed June 21, 2019. doi: 10.3390/molecules 19033654.

[103] Tucci, Sonia. Emma. Boyland. Jason, Halford. 2010. "The role of lipid and carbohydrate digestive enzyme inhibitors in the management of obesity: a review of current and emerging therapeutic agents." *Diabetes, Metab. Syndrome Obesity: Targets Ther, 3:125-143.* Accessed June 22, 2019. doi:10.2147/dmsott. s7005.

[104] Villiger, Angela. Filippo, Sala. Andy, Suter. Veronika, Butterweck. 2015. "In vitro inhibitory potential of *Cynara scolymus, Silybum marianum, Taraxacum officinale, and Peumus boldus* on key enzymes relevant to metabolic syndrome." *Phytomedicine 22:138-144.* Accessed June 22, 2019. doi: 10.1016/j.phymed.2014.11.015.

[105] Mohaddese, Mahboubi. 2018. "*Cynara scolymus* (artichoke) and its efficacy in management of obesity." *Bulletin of Faculty of Pharmacy, Cairo University 56: 115–120.* Accessed June 22, 2019. doi:10.1016/j.bfopcu.2018.10.003.

[106] Ben, Salem, Maryem. Rihab Ben Abdallah Kolsi, Raouia Dhouibi, Kamilia Ksouda, Slim Charfi, Mahdi Yaich, Serria Hammami, Zouheir Sahnoun, Khaled Mounir Zeghal, Kamel Jamoussi, Hanen Affes. 2017. "Protective effects of *Cynara scolymus* leaves extract on metabolic disorders and oxidative stress in alloxan-diabetic rats."

BMC Complementary Altern. Med. 17:328. Accessed June 22, 2019. doi: 10.1186/s12906-017-1835-8.

[107] Marongiu, Bruno. Alessandra, Piras. Silvia, Porcedda. 2004. "Comparative Analysis of the Oil and Supercritical CO_2 Extract of *Elettaria cardamomum (L.) Maton.*" *J. Agric. Food Chem. 2004, 52, 6278−6282.* Accessed July 16, 2019. doi: 10.1021/jf034819i.

[108] Ashokkumar, Kaliyaperuma. Muthusamy, Murugan. Dhanya, Surya Raj. 2019. "Phytochemical variations among four distinct varieties of Indian cardamom *Elettaria cardamomum (L.) Maton.*" *Natural Product Research.* Accessed July 16, 2019. doi: 10.1080/14786419. 2018.1561687.

[109] Polo Cristian. Thiago, de Mello, Moraes. Clélia, Akiko, Hiruma-Lima. 2012. "Gastric ulcers in middle-aged rats: The healing effect of essential oil from *Citrus aurantium L.* (Rutaceae)." *Evid Based Complement Alternat Med. 2012:509451.* Accessed July 16, 2019. doi: 10.1155/2012/ 509451.

[110] Verma, Sanjeev. Vikyath Jain, Satyender Katewa. 2009 "Blood pressure lowering, fibrinolysis enhancing and antioxidant activities of Cardamom (*Elettaria cardamomum*)" *Indian Journal of Biochemistry & Biophysics Vol 46, December 2009, pp 503-506.* Accessed July 16, 2019. PMID: 20361714.

[111] Bhaswant, Maharshi. Hemant Poudyal. Michael L. Matha., Leigh C.Ward. Peter Mouatt. Lindsay Brown. 2019. "Green and Black Cardamom in a Diet-Induced Rat Model of Metabolic Syndrome." *Nutrients 2015, 7, 7691-7707*; Accessed July 16, 2019. doi:10.3390/ nu7095360.

[112] Da-Costa-Rocha Ines, Bernd Bonnlaender, Hartwig Sievers, Ivo Pischel, Michael Heinrich. 2014. "*Hibiscus sabdariffa L.* A phytochemical and pharmacological review." *Food Chemistry 165, 424–443.* Accessed June 28, 2019. doi: 10.1016/j.foodchem. 2014.05.002.

[113] Patel, Seema. 2014. "*Hibiscus sabdariffa*: An ideal yet under-exploited candidate for nutraceutical applications." *Biomedicine &*

Preventive Nutrition 4, 23–27. Accessed June 28, 2019. doi:10.1016/j.bionut.2013.10.004.

[114] Riaz, Ghazala. Rajni, Chopra. 2018. "A review on phytochemistry and therapeutic uses of *Hibiscus sabdariffa L.*" *Biomedicine & Pharmacotherapy 102, 575–586.* Accessed June 28, 2019. doi:10.1016/j.biopha.2018.03.023.

[115] Alarcon-Alonso Javier. Alejandro, Zamilpa. Francisco, Alarcón, Aguilar. Maribel, Herrera-Ruiz. Jaime, Tortoriello. Enrique Jimenez-Ferrer E. 2012. "Pharmacological characterization of the diuretic effect of *Hibiscussabdariffa Linn (Malvaceae*) extract." Accessed June 28, 2019. doi:10.1016/j.jep.2011.12.005.

[116] Camelo-Mendez Gustavo, Adolfo. Juan Arturo, Ragazzo-Sanchez. Antonio, Ruperto, Jimenez-Aparicio. Pablo, Vanegas-Espinoza. 2013. "Comparative study ofanthocyanin and volatile compounds content of four varieties of Mexicanroselle (*Hibiscus sabdariffa L.*) by multivariable analysis." *Plant Food Hum Nutr2013. PMID 57882371.* Accessed June 28, 2019. doi:10.1007/s11130-013-0360-2.

[117] Jung EunKyung. Younglun, Kim. Nami, Joo. 2013. "Physicochemical properties and antimicrobial activity ofRoselle (*Hibiscus sabdariffa L.*)." *J Sci Food Agric 2013, 10.1002/jsfa.6256.* Accessed June 28, 2019. doi:10.1002/jsfa.6256.

[118] Medina-Carrillo, Raquel, Enedina. María, Teresa, Sumaya-Martínez. María Luisa, Machuca-Sánchez. Leticia, Mónica, Sánchez-Herrera. Rosendo, Balois-Morales. Edgar, Iván, Jimenez-Ruiz. 2013 "Actividad antioxidante de extractos de cálices deshidratados de 64 variedades de Jamaica (*hibiscus sabdariffa l.)* en función de fenólicos y antocianinas totals." *Revista Ciencias Técnicas Agropecuarias, ISSN-1010-2760, RNPS-0111, Vol. 22, No. Esp. (diciembre, pp. 41-44), 2013.* Accessed June 28, 2019.

[119] Herranz-López, María. Salvador, Fernández-Arroyo. Almudena, Pérez-Sanchez. Enrique, Barrajón-Catalán. Raúl, Beltrán-Debón. Javier, Abel, Menéndez. Carlos, Alonso-Villaverde. Antonio, Segura-Carretero. Jorge, Joven. Vicente, Micol. 2012. "Synergism

of plant-derived polyphenols in adipogenesis: Perspectives and implications." *Phytomedicine 19 253–2.* Accessed June 29, 2019. doi:10.1016/j.phymed.2011.12.001.

[120] Alarcón-Aguilar, Francisco J. Alejandro, Zamilpa. Ma. Dolores, Perez-Garcia. Julio, C.Almanza-Perez. Eunice, Romero-Nuñez. Efrain, A.Campos-Sepulveda. Laura, I.Vazquez-Carrillo. Ruben, Roman-Ramos. 2007. "Effect of *Hibiscus sabdariffa* on obesity in MSG mice." *Journal of Ethnopharmacology 114. 66–71.* Accessed June 29, 2019. doi:10.1016/j.jep.2007.07.020.

[121] Carvajal-Zarrabal, Octavio. Patricia, M Hayward-Jones. Zaida, Orta-Flores. Cirilo, Nolasco-Hipolito. 2009, "Effect of *Hibiscus sabdariffa L.* Dried Calyx Ethanol Extract on Fat Absorption-Excretion, and BodyWeight Implication in Rat." *Journal of Biomedicine and Biotechnology Volume 2009, Article ID 394592, 5 pages* .Accessed June 29, 2019. doi: 10.1155/2009/394592.

[122] Huang, To-Wei, Chia-Ling Chang, Erl-Shyh Kao & Jenq-Horng Lin. 2015. "Effect of *Hibiscus sabdariffa* extract on high fat diet–induced obesity and liver damage in hamsters." *Food & Nutrition Research, 59:1, 29018.* Accessed June 29, 2019. doi:10.3402/fnr.v59.29018.

[123] Zanoli, Paola. Manuela, Zavatti. 2008. "Pharmacognostic and pharmacological profile of *Humulus lupulus L." Journal of Ethnopharmacology 116 (2008) 383–396.* Accessed June 26, 2019. doi:10.1016/j.jep.2008.01.011.

[124] Eri, Sanja. Bin, K. Khoo. Joseph, Lech. Thomas, G. Hartman. 2000. "Direct Thermal Desorption-Gas Chromatography and Gas Chromatography-Mass Spectrometry Profiling of Hop (*Humulus lupulus L.*) Essential Oils in Support of Varietal Characterization." *J. Agric. Food Chem, 48, 1140-1149* Accessed June 26, 2019. doi:10.1021/jf9911850.

[125] Bland, Jeffrey. Deanna, Minich. Robert, Lerman. Gary, Darland. Joseph, Lamb. Matthew, Tripp. Neile, Grayson. 2015 "Isohumulones from hops (*Humulus lupulus*) and their potential role

in medical nutrition therapy." *Pharma Nutrition 3 (2015) 46–52.* Accessed June 26, 2019. doi:10.1016/j.phanu.2015.03.001.

[126] Xin, Guang. Zeliang, Wei. Chengjie, Ji. Huajie, Zheng. Jun, Gu. Limei, Ma. Wenfang, Huang. Susan, Morris-Natschke. Jwu-Lai, Yeh. Rui, Zhang. Chaoyi, Qin. Li, Wen. Zhihua, Xing. Yu, Cao. Qing, Xia. Ke, Li.Hai, Niu. Kuo-Hsiung Lee. Wen, Huanga. 2017 "Xanthohumol isolated from *Humulus lupulus* prevents thrombosis without increased bleeding risk by inhibiting platelet activation and mtDNA release." *Free Radical Biology and Medicine 108 (2017) 247–257.* Accessed June 26, 2019. doi:10.1016/j.freeradbiomed. 2017.02.018.

[127] Nozawa, Hajime. 2005. "Xanthohumol, the chalcone from beer hops (*Humulus lupulus L.*), is the ligand for farnesoid X receptor and ameliorates lipid and glucose metabolism in KK-Ay mice." *Biochemicaland Biophysical Research Communications, Volume 336, Issue 3* Accessed June 26, 2019. doi:10.1016/j.bbrc. 2005.08.159.

[128] Shimura, Masako. Akiko, Hasumi. Toshiko, Minato. Mayu, Hosono. Yutaka, Miura. Satoru, Mizutani. Keiji, Kondo. Shinichi, Oikawa. Aruto, Yoshida. 2005. "Isohumulones modulate blood lipid status through the activation of PPARa" *Biochimica et Biophysica Acta 1736 51 – 60.* Accessed June 26, 2019. doi:10.1016/j.bbalip. 2005.06.008.

[129] Obara, Kunniaki. Mai, Mizutani. Yoshitaka, Hitomi. Hiroaki, Yajima. Keiji, Kondo. 2009. "Isohumulones, the bitter component of beer, improve hyperglycemia and decrease body fat in Japanese subjects with prediabetes." *Clinical Nutrition 28 (2009) 278–284.* Accessed June 26, 2019. doi:10.1016/j.clnu.2009.03.012.

[130] Lerman, Robert. Deanna M. Minich. Gary, Darland. Joseph, Lamb. Jyh-lurn, Chang. Alex, His. Jeffrey, S. Bland. Matthew, Tripp. 2010. "Subjects with elevated LDL cholesterol and metabolic syndrome benefit from supplementation with soy protein, phytosterols, hops rho iso-alpha acids, and *Acacia nilotica proanthocyanidins.*"

Journal of Clinical Lipidology 4, 59–68. Accessed June 26, 2019. doi:10.1016/j.jacl.2009.11.002.

[131] Borrelli, Francesca. Angelo, A. Izzo. Herb–Drug. 2009. "Interactions with St John's Wort (*Hypericum perforatum*): an Update on Clinical Observations.*" The AAPS Journal, Vol. 11, No. 4.* Accessed June 27, 2019. doi: 10.1208/s12248-009-9146-8.

[132] Guedes, Ana P. Manuel, Fernandes-Ferreira. 2012. "Hypericum sp: essential oil composition and biological Activities." *Phytochem Rev* 11:127–152 Accessed June 27, 2019. **doi:** 10.1007/s11101-012-9223-y.

[133] Velingkar, Vinay S., Girdharilal L. Gupta. Namita B. Hegde 2017. "A current update on phytochemistry, pharmacology and herb–drug interactions of *Hypericum perforatum. Phytochem Rev* 16:725–744 Accessed June 27, 2019. doi:10.1007/s11101-017-9503-7.

[134] Liu, Yan-Ru. Wei-Guang, Li. Li-Fang, Chen. Bing-Kun, Xiao. 2014. "ABTS+ scavenging potency of selected flavonols from *Hypericum perforatum L.* by HPLC-ESI/MS QQQ: Reaction observation, adduct characterization and scavenging activity determination.*" Food Research International 58, 47–58.* Accessed June 27, 2019. doi:10.1016/j.foodres.2014.01.051.

[135] Hernández-Saavedra, Diego. Hebert, L. Hernández-Montiel. Claudia, I. Gamboa-Gómez. Luis, M. Salgado. Rosalía, Reynoso-Camacho. 2013. "The effect of Mexican herbal infusions on diet-induced insulin resistance.*" Nutrafoods 12:55-63.* Accessed June 27, 2019. doi:10.1007/s13749-013-0015-9.

[136] Petenatti Marta. María, Gette. Gerardo, Camí. Mariana, Popovich. 2014. "Quantitative micrograph, HPLC and FTIR profiles of *Melissa officinalis* and *Nepeta cataria (Lamiaceae*) from Argentina Rev." *FCA UNCUYO. 46(2): 15-27.* Accessed June 27, 2019.

[137] Shakeri, Abolfazl. Amirhossein, Sahebkar. Behjat, Javadi. 2016. "A review of its traditional uses, phytochemistry and pharmacology, Journal of Ethnopharmacology." *Volume 188, Pages 204-228,* Accessed June 27, 2019. doi:10.1016/j.jep.2016.05.010.

[138] Rivera, Diego. Alonso, Verdeb. Jose, Fajardob. Concepcion, Obonc. Vicente, Consuegrab. Jose, Garcia-Botiab. Segundo, Riosd. Francisco, Alcaraza. Arturo, Valdesb. Alejandro, del Morale. Emilio, Laguna. 2019. "Ethnopharmacology in the Upper Guadiana River area (Castile-La Mancha, Spain)." *Journal of Ethnopharmacology 24. 111968.* Accessed June 27, 2019. doi:10.1016/j.jep.2019.111968.

[139] Bolkent, Sehnaz. Refiye, Yanardag. Omur, Karabulut-Bulan. 2005. "Protective role of *Melissa officinalis L.* extract on liver of hyperlipidemic rats: A morphological and biochemical study." *Journal of Ethnopharmacology 99. 391–398.* Accessed June 28, 2019. doi:10.1016/j.jep.2005.02.038.

[140] Zarei, Ali. Saeed, Changizi, Ashtiyani. Soheila, Taheri. Fateme, Rasekh. 2014. "Comparison between effects of different doses of *Melissa officinalis* and atorvastatin on the activity of liver enzymes in hypercholesterolemia rats." *Avicenna J Phytomed, 4 (1): 15-23.* Accessed June 28, 2019. doi:10.22038/AJP.2014.1073.

[141] Weidner, Christopher. Sylvia, Wowro. Anja, Freiwald.Vitam, Kodelja. Heba Abdel-Aziz. Olaf, Kelber. Sascha, Sauer. 2014. "Lemon balm extract causes potent antihyperglycemic and antihyperlipidemic effects in insulin-resistant obese mice." *Mol. Nutr. Food Res.58, 903-907.* Accessed June 28, 2019. doi:10.1002/mnfr.201300477.

[142] Kim, Jeongjun. Hyunghee, Lee. Jonghoon, Lim. Haerim, Lee. Seolah, Yoon. Soon, ShikShin. Michung, Yoon. 2017. "The lemon balm extract ALS-L1023 inhibits obesity and nonalcoholic fatty liver disease in female ovariectomized mice." *Food and Chemical Toxicology 106, 292-305.* Accessed June 28, 2019. doi:10.1016/j.fct.2017.05.059.

[143] Barbalho, Sandra.Débora, Damasceno. Ana Paula, Spada. Vanessa, da Silva. Karla Aparecida, Martuchi. Marie, Oshiiwa. Flávia, Machado. Claudemir, Mendez. 2011. "Metabolic Profile of Offspring from Diabetic Wistar Rats Treated with *Mentha piperita* (Peppermint)." *Evid Based Complement Alternat Med.*

2011;2011:430237. Accessed July 2, 2019. doi:10.1155/2011/430237.

[144] McKay Diane. Jeffrey, Blumberg. 2006. "A review of the bioactivity and potential health benefits of peppermint tea (*Mentha piperita L.*)." *Phytother Res. 2006 Aug;20(8):619-33.* Accessed July 2, 2019. doi:10.1002/ptr.1936.

[145] Clark, Menary. 1981. "Variations in composition of peppermint oil in relation to production areas." *Econ Bot 35: 59–69.* Accessed July 2, 2019.

[146] Maffei Massimo. Silvano, Scannerini. 1992. "Seasonal variations in fatty acids from non-polar lipids of developing peppermint leaves." *Phytochemistry 31: 479–484.* Accessed July 2, 2019.

[147] Ruiz del Castillo, Maria Luisa. Guillermo, Santa-Maria. Marta, Herraiz. Gracia, Blanch. 2003. "A comparative study of the ability of different techniques to extract menthol from *Mentha piperita.*" *J Chromatogr Sci 41: 385–389.* Accessed July 2, 2019. doi:10.1093/chromsci/41.7.385.

[148] Hoffmann Birgit. Livio, Lunder. 1984. "Flavonoids from *Mentha piperita leaves.*" *Planta Med 50: 361.* Accessed July 2, 2019. doi:10.1055/s-2007-969736.

[149] Gherman, Carmen. Monica, Culea. Onuc, Cozar. 2000. "Comparative analysis of some active principles of herb plants by GC/MS." *Talanta 53: 253–262.* Accessed July 2, 2019. doi:10.1016/s0039-9140(00)00458-6.

[150] Zheng, Wei. Shiow, Wang. 2001. "Antioxidant activity and phenolic compounds in selected herbs." *J Agric Food Chem 49: 5165– 5170.* Accessed July 2, 2019. doi:10.1021/jf010697n.

[151] Yang, Seun-Ah. Sang-Kyung, Jeon. Eun-Jung, Lee. Chang-Hyun, Shim. In-Seon, Lee. 2010. "Comparative study of the chemical composition and antioxidant activity of six essential oils and their components." *Natural Product Research, v. 24, n. 2, p. 140-51, 2010.* Accessed July 2, 2019. doi:10.1080/14786410802496598.

[152] Badal, Mani, Badal. Divya, Badal. Pourush, Ashish, Khare. Jyoti, Shrivastava. Vinod, Kumar. 2011. "Pharmacological Action of

Mentha piperita on Lipid Profile in Fructose-Fed Rats." *Iran J Pharm Res. 2011 Fall;10(4):843-8.* Accessed July 2, 2019. PMID:24250421.

[153] Barbalho, Sandra Maria. Ana Paula, Machado, Spada. Erick, Prado de Oliveira. Márcio, Emilio, Paiva-Filho. Karla, Aparecida, Martuchi. Náyra, Coelho, Leite. Renata, Maeda Deus. Vivianne, Sasaki. Lucas, Silva, Braganti. Marie, Oshiiwa. 2009. "*Mentha piperita* effects on wistar rats plasma lipids." *Braz. arch. biol. Technol., vol.52, n.5, pp.1137-1143.* Accessed July 2, 2019. doi: 10.1590/S1516-89132009000500011.

[154] Barbalho, Sandra Maria, Flávia Maria, Vasques. Farinazzi Machado. Marie, Oshiiwa. Marcio, Abreu. Ellen, Landgraf Guiger. Paschoal, Tomazela. Ricardo, Alvares Goulart. 2011. "Investigation of the effects of peppermint (*Mentha piperita*) on the biochemical and anthropometric profile of university students." *Ciênc. Tecnol. Aliment.* vol. 31 no.3 Campinas. Accessed July 2, 2019. doi: 10.1590/S0101-20612011000300006.

[155] Teixeira Bárbara. António, Marques. Cristina, Ramos. Carmo, Serrano. Olívia, Matos. Nuno, R Neng. José, M. F. Nogueira. Jorge, Alexandre, Saraiva. Maria, Leonor, Nunes. 2013. "Chemical composition and bioactivity of different oregano (*Origanum vulgare*) extracts and essential oil." *J Sci Food Agric.; 93(11):2707-14*. Accessed July 5, 2019. doi:10.1002/jsfa.6089.

[156] Gruenwald, Joerg, Thomas, Brendler. Christof, Jaenicke. 2000, "PDR for Herbal Medicines. Medical Economic Co." *Inc., Montvale, NJ (2000).* Accessed July 5, 2019.

[157] Pascual, María. Elena. Karla, Slowing. Emily, Carretero. Daniel, Sánchez Mata. 2001. "Lippia: Traditional uses, chemistry and pharmacology." *A review. J. Ethnopharmacol, 76, 201–214.* Accessed July 5, 2019. doi:10.1016/S0378-8741(01)00234-3.

[158] Atrea Ioannis, Aikaterini, Papavergou. Ioannis, Amvrosiadis. Ioannis, Savvaidis. 2009. "Combined effect of vacuum-packaging and oregano essential oil on the shelf-life of Mediterranean octopus

(*Octopus vulgaris*) from the Aegean Sea stored at 4°C." Accessed July 5, 2019. doiI:10.1016/j.fm.2008.10.005.

[159] Vujicic, Milica. Ivana, Nikolic. Vassiliki, Kontogianni. Tamara, Saksida. Pantelis, Charisiadis. Zorana, Orescanin-Dusic. Dusko, Blagojevic. Stanislava, Stosic-Grujicic. Andreas, Tzakos. Ivana, Stojanovic. 2015. "Methanolic extract of *Origanum vulgare* ameliorates type 1 diabetes through antioxidant, anti-inflammatory and anti-apoptotic activity." *Br J Nutr; 113(5):770-82.* Accessed July 5, 2019. doi:10.1017/S0007114514004048.

[160] Gutiérrez-Grijalva Erick. Manuel, A. Picos-Salas. Nayely, Leyva-López. Marilyn, Criollo-Mendoza. Gabriela, Vazquez-Olivo. Basilio, Heredia. 2017. "Flavonoids and Phenolic Acids from Oregano: Occurrence, Biological Activity and Health Benefits." *Plants (Basel).; 7(1).* Accessed July 5, 2019. doi:10.3390/plants7010002.

[161] Leyva-López, Nayely. Erick, Gutiérrez-Grijalva, Gabriela, Vazquez-Olivo. Basilio, Heredia. 2017. "Essential Oils of Oregano: Biological Activity beyond Their Antimicrobial Properties." *Molecules.; 22(6). pii: E989.* Accessed July 5, 2019. doi:10.3390/molecules22060989.

[162] Embuscado, Milda. 2015. "Spices and herbs: Natural sources of antioxidants—A mini review." *J. Funct. Foods, 18, 811–819.* Accessed July 5, 2019. doi:10.1016/j.jff.2015.03.005.

[163] Gonçalves, Sandra. Elsa, Moreira. Clara, Grosso. Paula, Andrade. Patrícia, Valentão. Anabela, Romano. 2017. "A. Phenolic profile, antioxidant activity and enzyme inhibitory activities of extracts from aromatic plants used in mediterranean diet." *J. Food Sci. Technol. 54, 219–227.* Accessed July 5, 2019. doi:10.1007/s13197-016-2453-z.

[164] Kogiannou, Dimitra. Nick, Kalogeropoulos. Panagiotis, Kefalas. Moschos, Polissiou. Adriana, Kaliora. 2013. "Herbal infusions; their phenolic profile, antioxidant and anti-inflammatory effects in HT29 and PC3 cells." *Food Chem. Toxicol. 2013, 61, 152–159.* Accessed July 5, 2019. doi:10.1016/j.fct.2013.05.027.

[165] Ocaña-Fuentes, Aurelio. Elena, Arranz. Francisco, Señorans. Guillermo, Reglero. 2010. "Supercritical fluid extraction of oregano (*Origanum vulgare*) essentials oils: Anti-inflammatory properties based on cytokine response on THP-1 macrophages." *Food Chem. Toxicol. 2010, 48, 1568–1575.* Accessed July 5, 2019. doi:10.1016/j.fct.2010.03.026.

[166] Han, Xuesheng. Tory, L. Parker. 2017. "Anti-inflammatory, tissue remodeling, immunomodulatory, and anticancer activities of oregano (*Origanum vulgare*) essential oil in a human skin disease model." *Biochim* 4, 73–77. Accessed July 5, 2019. doi:10.1016/j.biopen.2017.02.005.

[167] Ghadimi, Darya. Mohammad Taghi, Goodarzi. Nasrin Ziamajidi. Shirin, Moradkhani. Taiebeh, Ghyasvand. 2017, "The effect of aqueous extract of *Origanum Vulgare* leaves on Adiponectin and c-Cbl associated protein levels in adipose tissue of Streptozotocin-Nicotinamide induced diabetic rats," *Pharmacophore, 8(6S), e-1173374.* Accessed July 5, 2019.

[168] Tontonoz, Peter. Bruce, M. Spiegelman. 2008. "Fat and beyond: the diverse biology of PPARgamma." *Annu Rev Biochem. 77:289–312.* Accessed July 5, 2019. doi:10.1146/annurev.biochem.77.061307.091829.

[169] Dham, Shefali. Viral, Shah. Sondra, Hirsch. Mary Ann, Banerji. 2006. "The role of complementary and alternative medicine in diabetes." *Curr. Diab. Rep., 251–258.* Accessed July 5, 2019. PMID:16898580.

[170] Cazzola, Roberta. Benvenuto, Cestaro. 2014. "Antioxidant spices and herbs used in diabetes. In Diabetes: Oxidative Stress and Dietary Antioxidants"; *Preedy, V. R., Ed.; Academic Press: San Diego, CA, USA; pp. 89–97.* Accessed July 5, 2019. doi:10.1016/B978-0-12-405885-9.00009-7.

[171] Sarikurkcu, Cengiz. Gokhan, Zengin. Mustafa, Oskay. Sengul, Uysal. Ramazan, Ceylan. Abdurrahman, Aktumsek. 2015. "A. Composition, antioxidant, antimicrobial and enzyme inhibition activities of two *Origanum vulgare* subspecies (subsp. vulgare and

subsp. hirtum) essential oils." *Ind. Crop. Prod, 70, 178–184.* Accessed July 5, 2019. doi:10.1016/j.indcrop.2015.03.030.

[172] Quijano, Clara Elizabeth. Jorge, A. Pino. 2007. "Characterization of volatile compounds in guava (*Psidium guajava L.*) varieties from Colombia" *Revista CENIC Ciencias Químicas, Vol. 38, No. 3,* Accessed July 18, 2019.

[173] Valenzuela, Luz A. Blanca, L. Ortiz. Clara, E. Pérez. 2013. "Comparative study of the metabolic effect of arepas enriched with pectin extracted from guava (*Psidium guajava L.*) or commercial citrus pectin." *Rev. Colomb. Quim., 42 (3), 5-10.* Accessed July 18, 2019.

[174] Li, Wang. Hanchen, Zheng. Jacques, Bukuru. Norbert, De Kimpe. 2004. "Natural medicines used in the traditional Chinese medical system for therapy of diabetes mellitus." *Journal of Ethnopharmacology 92 1–21 Review.* Accessed July 18, 2019. doi: 10.1016/j.jep.2003.12.031.

[175] Sen, Shib Sankar, V. Sukumaran, Sib Sankar Giri, Se Chang Park. 2015 "Flavonoid fraction of guava leaf extract attenuates lipopolysaccharide induced inflammatory response via blocking of NF-kB signaling pathway in Labeo rohita macrophages." *Fish & Shellfish Immunology 47 85-92.* Accessed July 18, 2019. doi:10.1016/j.fsi.2015.08.031.

[176] Beidokhti, Maliheh Najari. Anna, K. Jäger. 2017. "Review of antidiabetic fruits, vegetables, beverages, oils and spices commonly consumed in the diet." *Journal of Ethnopharmacology 201 (2017) 26–41.* Accessed July 18, 2019. doi:10.1016/j.jep.2017.02.031.

[177] Ventosa, Marisabel. José Luis, Rodríguez. Olga Lidia, Zerqueira. 2018. "Determinación De Los Principales Carotenoides De La Guayaba (*Psidium Guajava*)." *Ciencia y Tecnología de Alimentos Vol. 18, No. 1,* Accessed July 18, 2019.

[178] Capote, Alina, Naivy Pérez-Alonso, Anabel Pérez, Dirk Wilken, André Gerth, Lutz Müller-Kuhrt, Elio Jiménez. 2009. "Efecto del cultivo in vitro en el perfil metabólico de *Psidium guajava L.*"

Biotecnología Vegetal Vol. 9, No. 1: 41 - 45, Accessed July 18, 2019.

[179] Díaz-de-Cerio, Elixabet. Alba, Rodríguez-Nogales. Francesca, Algieri. Miguel, Romero, Vito Verardo. Antonio, Segura-Carretero. Juan, Duarte. Julio, Galvez. 2017. "The hypoglycemic effects of guava leaf (*Psidium guajava L.*) extract are associated with improving endothelial dysfunction in mice with diet-induced obesity." *Food Research International 96 64–71*. Accessed July 18, 2019. doi:10.1016/j.foodres.2017.03.019.

[180] Guo, Xiangyu. Hisae, Yoshitomi. Ming, Gao. Lingling, Qin. Ying, Duan. Wen, Sun. Tunhai, Xu. Peifeng, Xie. Jingxin, Zhou. Liansha, Huang. Tonghua, Liu. 2013. "Guava leaf extracts promote glucose metabolism in SHRSP.Z-Leprfa/Izm rats by improving insulin resistance in skeletal muscle." *BMC Complementary and Alternative Medicine* 13:52 Accessed July 18, 2019. doi: 10.1186/1472-6882-13-52.

[181] Yoshitomi, Hisae. Xiangyu, Guo. Tonghua, Liu. Ming, Gao. 2012. "Guava leaf extracts alleviate fatty liver via expression of adiponectin receptors in SHRSP.ZLeprfa/ Izm rats." *Nutrition & Metabolism, 9:13*. Accessed July 18, 2019. doi:10.1186/1743-7075-9-13.

[182] Siani, Antonio. Mariana, Souza. Maria G, Henriques. Mónica, Ramos. 2013. "Anti-inflammatory activity of essential oils from *Syzygium cumini* and *Psidium guajava*." *Pharm. Biol. 51, 881–887.* Accessed July 18, 2019. doi:10.3109/13880209.2013.768675.

[183] Deguchi, Yoriko. Kouji, Miyazaki. 2010. "Anti-hyperglycemic and anti-hyperlipidemic effects of guava leaf extract." *Nutrition & Metabolism, 7:9*. Accessed July 18, 2019. doi: 10.1186/1743-7075-7-9.

[184] Juárez-Vázquez María del Carmen, Candy Carranza-Álvarez, Angel Josabad Alonso-Castro, Violeta F. González-Alcaraz, Eliseo Bravo-Acevedo, Felipe Jair Chamarro-Tinajero, EloySolano. 2013. "Ethnobotany of medicinal plants used in Xalpatlahuac, Guerrero,

México." *Journal of Ethnopharmacology 148 521–527.* Accessed July 18, 2019. doi: 10.1016/j.jep.2013.04.048.

[185] Bading Taika, Bayissi. Henri, Bouckandou. Alain, Souza. Henri Paul, Bourobou Bourobou. Louise, MacKenzie. Lisa A, Lione. 2018. "An overview of anti-diabetic plants used in Gabon: Pharmacology and Toxicology." *Journal of Ethnopharmacology 216 203-228.* Accessed July 19, 2019. doi: 10.1016/j.jep.2017.12.036.

[186] Hassani, Faezeh, Vahdati. Kobra, Shirani. Hossein, Hosseinzadeh. 2016. "Rosemary (*Rosmarinus officinalis*) as a potential therapeutic plant in metabolic syndrome: a review." *Naunyn Schmiedebergs Arch Pharmaco. 389(9):931-49.* Accessed July 12, 2019. doi:10.1007/s00210-016-1256-0.

[187] De Oliveira, Jonatas.Samira Esteves, Afonso, Camargo. Luciane, Dias, de Oliveira. 2019. "*Rosmarinus officinalis L.* (rosemary) as therapeutic and prophylactic agent." *J Biomed Sci; 26(1):5.* Accessed July 12, 2019. doi:10.1186/s12929-019-0499-8.

[188] Sánchez-Camargo, Andrea del Pilar. Miguel, Herrero. 2017. "Rosemary (*Rosmarinus officinalis*) as a functional ingredient: recent scientific evidence." *Current Opinion in Food Science, 14:13–19.* Accessed July 13, 2019. doi: 10.1016/j.cofs.2016.12.003.

[189] Sedighi, Rashin. Yantao, Zhao. Aaron, Yerke, Shengmin, Sang. 2015. "Preventive and protective properties of rosemary (*Rosmarinus officinalis L.)* in obesity and diabetes mellitus of metabolic disorders: a brief review." *Current Opinion in Food Science, 2:58–70.* Accessed July 13, 2019. doi:10.1016/j.cofs.2015.02.002.

[190] Yıldırım, Emel, Diraz. 2018. "The Effect of Seasonal Variation on *Rosmarinus officinalis (L.)"* Essential Oil Composition. *International Journal of Agriculture and Wildlife Science (IJAWS), 4(1): 33 – 38.* Accessed July 13, 2019. doi: 10.24180/ijaws.381564.

[191] Hegazy, Amany. Amal, Abdel-Azeem. Hala, Zeidan. Khadiga, Ibrahim. 2018. "Hypolipidemic and hepatoprotective activities of rosemary and thyme in gentamicin-treated rats." *Hum Exp Toxicol.;*

37:420–30. Accessed July 12, 2019. doi:10.1177/0960327117710534.

[192] Ibarra, Alvin. Julien, Cases. Antoine, Bily. Kan, He. Naisheng, Bai. Marc, Roller. Aurélie, Coussaert. Christophe, Ripoll. 2010. "Importance of extract standardization and in vitro/ex vivo assay selection for the evaluation of antioxidant activity of botanicals: a case study on three *Rosmarinus officinalis L. extracts*." *J Med Food 13, 1167–1175*. Accessed July 13, 2019. doi:10.1089/jmf.2009.0259.

[193] Arranz, Elena. Laura, Jaime. Monica, García-Risco. Tizina, Fornari. Guillermo, Reglero. Susana, Santoyo. 2015 "Anti-inflammatory activity of rosemary extracts obtained by supercritical carbon dioxide enriched in carnosic acid and carnosol." *Int J Food Sci Technol, 50:674-68*1. Accessed July 12, 2019. doi: 10.1111/jfs.12656.

[194] Tu, Zheng. Tijuana, Moss-Pierce. Paul, Ford. Alan, Jiang. 2013. "Rosemary (*Rosmarinus officinalis L.)* extract regulates glucose and lipid metabolism by activating AMPK and PPAR pathways in HepG2 cells." *J Agric Food Chem;61(11):2803-10.* Accessed July 13, 2019. doi:10.1021/jf400298c.

[195] Yun, Young Sook. Sachie Noda. Genta, Shigemori. Ryunosuke, Kuriyama. Shigeru, Takahashi. Mariko, Umemura. Yuji, Takahashi. Hideshi, Inoue. 2013. "Phenolic diterpenes from rosemary suppress cAMP responsiveness of gluconeogenic gene promoters." *Phytother Res 27:906–910.* Accessed July 13, 2019. doi:10.1002/ptr.4794.

[196] Cazzola, Roberta. Carla, Camerotto. Benvenuto, Cestaro. 2011. "Anti-oxidant, antiglycant, and inhibitory activity against alpha-amylase and alpha- glucosidase of selected spices and culinary herbs." *Int J Food Sci Nutr 62:175–184.* Accessed July 12, 2019. doi:10.3109/09637486.2010.529068.

[197] McCue, Patrick. Kalidas, Shetty. 2004. "Inhibitory effects of rosmarinic acid extracts on porcine pancreatic amylase in vitro." *Asia Pac J Clin Nutr 13:101–106*. Accessed July 13, 2019. PMID:15003922.

[198] Ibarra, Alvin. Julien, Cases. Marc, Roller. Amparo, Chiralt-Boix. Aurélie, Coussaert. Chrisophe, Ripoll. 2011. "Carnosic acid-rich rosemary (*Rosmarinus officinalis L.*) leaf extract limits weight gain and improves cholesterol levels and glycaemia in mice on a high-fat diet." *Br. J. Nutr. 2011, 106, 1182−9.* Accessed July 13, 2019. doi:10.1017/S0007114511001620.

[199] Harach, Taoufiq. Oliver, Aprikian. Irina, Monnard. 2010. "Rosemary *(Rosmarinus officinalis L.)* leaf extract limits weight gain and liver steatosis in mice fed a high-fat diet." *Planta Med 76,566–571.* Accessed July 12, 2019. doi:10.1055/s-0029-1240612.

[200] Babish, John. Linda M, Pacioretty. Jeffrey S, Bland. Deanna M, Minich. Jeffrey, Hu. Matthew L, Tripp. 2010. "Antidiabetic screening of commercial botanical products in 3T3-L1 adipocytes and db/db mice." *J. Med. Food, 13*, 535–547. Accessed July 12, 2019. doi:10.1089/jmf.2009.0110.

[201] Naimi, Madina. Theodoros, Tsakiridis. Theocharis, Stamatatos. Dimiris, Alexandropoulos. 2015. "Increased skeletal muscle glucose uptake by rosemary extract through AMPK activation." *Appl. Physiol. Nutr. Metab. 40*, 407–413. Accessed July 13, 2019. doi:10.1139/apnm-2014-0430.

[202] Gaya, Mauro. Victoria, Repetto. Judith, Toneatto. Claudia, Anesini. Graciela, Piwien-Pilipuk. Silvia, Moreno. 2013. "Antiadipogenic effect of carnosic acid, a natural compound present in *Rosmarinus officinalis*, is exerted through the C/EBPs and PPARgamma pathways at the onset of the differentiation program." *Biochim Biophys Acta 1830:3796–3806.* Accessed July 12, 2019. doi:10.1016/j.bbagen.2013.03.021.

[203] Apostolidis, Emmanouil., Young-In, Kwon. Kalidas, Shetty. 2006. "Potential of cranberry-based herbal synergies for diabetes and hypertension management." *Asia Pac J Clin Nutr 15:433–441.* Accessed July 12, 2019. PMID:16837438.

[204] Zhao, Yantao. Rashin, Sedighi. Pei, Wang. Huadong, Chen. Yingdong, Zhu. Shengmin, Sang. 2015. "Carnosic acid as a major bioactive component in rosemary extract ameliorates high-fat-diet-

induced obesity and metabolic syndrome in mice." *J Agric Food Chem 63:4843–4852.* Accessed July 13, 2019. doi:10.1021/acs.jafc. 5b01246.

[205] Al Sheyab, Fawzi, Mohammad. Nizar, Abuharfeil. Lina, Salloum. Rehan Bani, Hani. Dalal Saleh, Awad. 2011. "The effect of rosemary (*Rosmarinus officinalis.* L) plant extracts on the immune response and lipid profile in mice." *J. Biol. Life Sci., 3.* Accessed July 12, 2019. doi: 10.5296/jbls.v3i1.906.

[206] Romo Vaquero, María. María-Josefa, Yáñez-Gascón. Rocío, García Villalba. Mar, Larrosa. Emilie, Fromentin. Alvin, Ibarra. Marc, Roller. Francisco, Tomás-Barberán. Juan Carlos, Espín de Gea. María-Teresa, García-Conesa. 2012. "Inhibition of gastric lipase as a mechanism for body weight and plasma lipids reduction in Zucker rats fed a rosemary extract rich in carnosic acid." *PLoS One 2012, 7, e39773.* Accessed July 13, 2019. doi:10.1371/journal.pone.0039773.

[207] Romo-Vaquero, María. María-Victoria, Selma. Mar, Larrosa. María, Obiol. Rocío, García-Villalba. Rocío, González-Barrio. Nicolas, Issaly. John, Flanagan. Marc, Roller.Francisco, A, Tomás-Barberán.María-Teresa, García-conesa. 2014. "A Rosemary Extract Rich in Carnosic Acid Selectively Modulates Caecum Microbiota and Inhibits b-Glucosidase Activity, Altering Fiber and Short Chain Fatty Acids Fecal Excretion in Lean and Obese Female Rats." *PLoS ONE 9(4): e94687.* Accessed July 13, 2019. doi:10.1371/journal. pone.0094687.

[208] Koga, Kunimasa. Kyoji, Yoshino. Ken´ichi, Nomoto. 2006. "Effects of 50% ethanol extract from rosemary (*Rosmarinus officinalis*) on α-glucosidase inhibitory activity and the elevation of plasma glucose level in rats, and its active compound." *J Food Sci 71:S507–S512.* Accessed July 13, 2019. doi:10.1111/j.1750-3841.2006.00125.x.

[209] Afonso Milessa. Ana-Mara, de O Silva. Eliane, Carvalho. Diogo, Rivelli. Silvia, Barros. Marcelo, Rogero. Ana-María, Lottenberg. Rosângela, Torres. Jorge, Mancini-Filho. 2013. "Phenolic compounds from Rosemary (*Rosmarinus officinalis L.*) attenuate oxidative stress and reduce blood cholesterol concentrations in diet-

induced hypercholesterolemic rats." *Nutr Metab (Lond). 10(1):19.* Accessed July 13, 2019. doi: 10.1186/1743-7075-10-19.

[210] Wu, Ya-Ni. Jian, Huang. An-Lian, Zuo. Lei, Yao. 2011. "Research on the effects of rosemary (*Rosmarinus officinalis L.*) on the blood lipids and anti- lipid peroxidation in rats." *J Essential Oil Res 23:26–34.* Accessed July 13, 2019. doi:10.1080/10412905.2011.9700465.

[211] Vanithadevi, Balasubramanian. Carani-Venkataraman, Anuradha. 2008. "Effect of rosmarinic acid on insulin sensitivity, glyoxalase system and oxidative events in liver of fructose-fed mice." Int. *J. Diabetes Metab, 16, 35–44.* Accessed July 13, 2019.

[212] Hosseinzadeh, Saleh. Azizollah-Jafari, Kukhdan. Ahmadreza, Hosseini. Raham, Armand. 2015."The application of *Thymus vulgaris* in traditional and modern medicine: a review." *Global J. Pharmacol. 9 (3), 260–266.* Accessed July 28, 2019. doi: 10.5829/idosi.gjp.2015.9.3.94246.

[213] Stahl-Biskup, E., Venskutonis, R. P. Thyme. In: Peter, K. V. (Ed.), 2012. "Handbook of Herbs and Spices." *second ed.Woodhead Publishing, Cambridge, pp. 499–525, (Chapter 27).* Accessed July 28, 2019. doi:10.1201/9781439823002.

[214] Rustaiyan, Aabdolhossein. Shiva, Masoudi. Aazam, Monfared. Mohammad, Kamalinejad. Taraneh, Lajevardi. Soheila, Sedaghat. Mohammad, Yari. 2000. "Volatile constituents of three *Thymus* species grown wild in Iran." *Planta Med. 66, 197–198.* Accessed July 28, 2019. doi:10.1055/s-0029-1243136.

[215] Soliman, Kawther. Isamil, Badeaa. 2002. "Effect of oil extracted from some medicinal plants on different mycotoxigenic fungi." *Food Chem. Toxicol. 40, 1669–1675.* Accessed July 28, 2019. doi:10.1016/S0278-6915(02)00120-5.

[216] Kuete Victor. 2017. "Potential in the Management of Metabolic Syndrome. Chapter 12 in Medicinal Spices and Vegetables from Africa. Therapeutic Potential against Metabolic, Inflammatory, Infectious and Systemic Diseases." Academic Press, 2017, pp 315-327. Accessed July 28, 2019.

[217] Vila, Roser. 2002. "Flavonoids and further polyphenols in the genus *Thymus*," in *Thyme: The Genus Thymus. Medicinal and Aromatic Plants-Industrial Profiles*, eds. E. Stahl-Biskup and F. Saez (New York, NY: Taylor and Francis), 75. Accessed July 28, 2019.

[218] Nagoor Meeran Mohamed-Fizur.Hayate, Javed. Hasan, Al Taee. Sheikh, Azimullah. Shreesh, Ojha. 2017. "Pharmacological Properties and Molecular Mechanisms of Thymol: Prospects for Its Therapeutic Potential and Pharmaceutical Development." *Front Pharmacol. 8:380*. Accessed July 28, 2019. doi:10.3389/fphar. 2017.00380.

[219] Kulisi´c, Tea. Anita, Krisko. Verica, Dragovic´-Uzelac.Mladen, Milos. Greta, Pifat. 2007. "The effects of essential oils and aqueous tea infusions of oregano (*Origanum vulgare L. spp. hirtum*), thyme (*Thymus vulgaris L.*) and wild thyme (*Thymus serpyl- lum L.*) on the copper-induced oxidation of human low-density lipoproteins." *Int J Food Sci Nutr 2007; 58(2): 87–93*. Accessed July 28, 2019. doi: 10.1080/09637480601108307.

[220] Prasanth, Reddy. Ravi, Vital. Kandisa, Varsha. 2014. "Review on *Thymus vulgaris* Traditional Uses and Pharmacological Properties." *Medicinal and aromatic plants*. Accessed July 28, 2019. doi:10.4172/2167-0412.1000164.

[221] Hegazy Amany. Amal, Abdel-Azeem. Hala, Zeidan. Khadiga, Ibrahim. 2018. "Hypolipidemic and hepatoprotective activities of rosemary and thyme in gentamicin-treated rats." *Hum Exp Toxicol; 37:420–30*. Accessed July 28, 2019. doi: 10.1177/0960327 117710534.

[222] El-Houri, Rime. Dorota, Kotowska. Louise, Olsen. Sumangala, Bhattacharya. Lars, Christensen. Kai, Grevsen. Niels, Oksbjerg. Nils, Færgeman. Karsten, Kristiansen. Kathrine, Christensen. 2014. "Screening for bioactive metabolites in plant extracts modulating glucose uptake and fat accumulation." *Evid. Based Complement. Alternat. Med. 2014, 156398*. Accessed July 28, 2019. doi: 10.1155/2014/156398.

[223] Hotta, Mariko. Rieko, Nakata. Michiko, Katsukawa. Kazuyuki, Hori. Saori, Takahashi. Hiroyasu, Inoue. 2010. "Carvacrol, a component of thyme oil, activates PPARα and γ and suppresses COX-2 expression," *Journal of Lipid Research, vol. 51, no. 1, pp. 132–139.* Accessed July 28, 2019. doi:10.1194/jlr.M900255-JLR200.

[224] Christensen, Kathrine. Ariane, Minet. Henrik, Svenstrup. Kai, Grevsen. Hongbin, Zhang. Eva, Schrader. Gerald, Rimbach. Silvia, Wein. Siegfried, Wolffram. Karsten, Kristiansen. Lars, Christensen. 2009. "Identification of plant extracts with potential antidiabetic properties: effect on human peroxisome proliferator-activated receptor (PPAR), adipocyte differentiation and insulin-stimulated glucose uptake," *Phytotherapy Research*, vol. 23, no. 9, pp. 1316–1325. Accessed July 28, 2019. doi: 10.1002/ptr.2782.

[225] Kavalali, Gulsel. 2003. "Urtica: therapeutic and nutritional aspects of stinging nettles." *In Medicinal and Aromatic Plants—Industrial Profiles, Hardman R (ed.). Taylor & Francis: New York.* Accessed July 22, 2019.

[226] Mehri, Avin. Shirin, Hasani-Ranjbar. Bagher, Larijani. Mohammad, Abdollahi. 2011. "A systematic review of efficacy and safety of *Urtica dioica* in the treatment of diabetes." *Int. J. Pharmacology; 7(2):161–170.* Accessed July 22, 2019. doi:.10.3923/ijp.2011.161.170.

[227] Jakubczyk, Karolina. Katarzyna, Janda. Sylwia, Szkyrpan. Izabela, Gutowska. Jolanta, Wolska. "Stinging nettle (*Urtica dioica L.*) botanical characteristics, biochemical composition and health benefits." *Pomeranian J Life Sci. 2015;61(2):191-8.* Accessed July 22, 2019. PMID:27141606.

[228] Joshi, Bhuwan-Chandra. Minky, Mukhija. Ajudhia-Nath, Kalia. 2014. "Pharmacognostical review of *Urtica dioica L.*" *International Journal of Green Pharmacy. 8. 201-209.* Accessed July 22, 2019. doi: 10.4103/0973-8258.142669.

[229] Dhouibi, Raouia. Kamilia, Ksouda. Maryem, Ben-Salem. Zouher, Sahnoun. Serria, Hammami. Mounir, Zeghal. Hanen, Affes. 2019.

"Screening of pharmacological uses of *Urtica dioica* and others benefits. Prog Biophys." *Mol Biol. 2019 Jun 1. pii: S0079-6107(19)30098-7.* Accessed July 22, 2019. doi: 10.1016/j. pbiomolbio.2019.05.008.

[230] Jamison, Jennifer. 2003. "*Urtica dioica.*" Chapter 95 · In book: *Clinical Guide to Nutrition & Dietary Supplements in Disease Management,* pp.665-667, 2003. Accessed July 22, 2019. doi:10.1016/B978-0-443-07193-5.50099-6.

[231] Vajić, Una-Jovana. Jelica, Grujić-Milanović. Jelena, Ţivković. Katarina, Šavikin. Dejan, Gođevac. Zoran, Miloradović. Branko, Bugarski. Nevena, Mihailović-Stanojević. 2015. "Optimization of extraction of stinging nettle leaf phenolic compounds using response surface methodology." *Ind. Crops Prod. 74, 912–917.* Accessed July 22, 2019. doi:10.1016/j.indcrop.2015.06.032.

[232] Francišković, Marina. Raquel, Gonzalez-Pérez. Dejan, Orčić. Fermín, Sánchez de Medina. Olga, Martínez-Augustin. Emilija, Svirčev. Natasa, Simin. Neda, Mimica-Dukić. 2017. "Chemical Composition and Immuno-Modulatory Effects of *Urtica dioica L.* (Stinging Nettle) Extracts." *Phytother Res. 2017 Aug;31(8):1183-1191.* Accessed July 22, 2019. doi:10.1002/ptr.5836.

[233] Konić-Ristić, Aleksandra. Tatjana, Srdić-Rajić. Nevena, Kardum. Vesna, Aleksić-Veličković. Paul, Kroon. Wendy, Hollands. Paul, Needs. Nadiya, Boyko. Osman, Hayran. Mariam, Jorjadze. Maria, Glibetić. 2013. "Effects of bioactive-rich extracts of pomegranate, persimmon, nettle, dill, kale and Sideritis and isolated bioactives on arachidonic acid induced markers of platelet activation and aggregation." *J Sci Food Agric. 2013 Nov; 93(14):3581-7.* Accessed July 22, 2019. doi:10.1002/jsfa.6328.

[234] Barnes, Joanne. Linda, Anderson. David, Phillipson. 2002. "Herbal Medicines: a guide for healthcare professionals." - *2nd edition London: Pharmaceutical Press, 530 p.* Accessed July 22, 2019. doi: 10.1021/np0207320.

[235] Riehemann, Kristina. Klaus, Schulze-Osthoff. 1999. "Plant extracts from stinging nettle (*Urtica dioica*), an antirheumatic remedy,

inhibit the proinflammatory transcription factor NF-K~ FEBS." *lett. 442, 1, 89-94.* Accessed July 22, 2019. doi:10.1016/s0014-5793(98)01622-6.

[236] Obertreis, B., Giller, K., Teucher, T., Behnke, B., Schmitz, H., 1996. "Antiphylogistische Effekte von Extractum *Urtica dioica* foliorum im vergleich zu Kaffeoylapfel saure." *Arzneim. Forsch. Drug Res. 46, 1, 52-56.* Accessed July 22, 2019.

[237] Vajic, Una-Jovana. Jelica, Grujic-Milanovic. Zoran, Miloradovic. Djurdjica, Jovovic. Milan, Ivanov. Danijela, Karanovic. Katarina, Savikin. Branko, Bugarski. Nevena, Mihailovic-Stanojevic. 2018. "*Urtica dioica L.* leaf extract modulates blood pressure and oxidative stress in spontaneously hypertensive rats." *Phytomedicine. 2018 Jul 15;46:39-45.* Accessed July 22, 2019. doi:10.1016/j.phymed.2018.04.037.

[238] Obanda Diana. Peng, Zhao. Allison, Richard. David, Ribnicky. William, Cefalu. Jacqueline, Stephens. 2016. "Stinging Nettle *(Urtica dioica L.)* Attenuates FFA Induced Ceramide Accumulation in 3T3-L1 Adipocytes in an Adiponectin Dependent Manner." *PLoS One. 2016 Mar 3;11(3):e0150252.* Accessed July 22, 2019. doi:10.1371/journal.pone.0150252.

[239] Domola, Masoud-Shabani, Vivian, Vu .Christine, Robson-Doucette. Gary, Sweeney. Michael, Wheeler. 2010. "Insulin mimetics in Urtica dioica: structural and computational analyses of *Urtica dioica* extracts." *Phyto-ther Res. 2010;24:S175-82.* Accessed July 22, 2019. doi:10.1002/ptr.3062.

[240] Ozen, Tevfik. Halil, Korkmaz. 2003. "Modulatory effect of Urtica dioica L. (*Urticaceae*) leaf extract on biotransformation enzyme systems, antioxidant enzymes, lactate dehydrogenase and lipid peroxidation in mice." *Phytomedicine. 10 (5): 405-415.* Accessed July 22, 2019. doiI:10.1078/0944-7113-00275.

[241] Tahri, Abdelhafid. Sabah, Yamani. Abdelkhaleq, Legssyer, Mohammed, Aziz. Hassane, Mekhfi. Mohammed, Bnouham. Abderrahim, Ziyyat. 2000. "Acute diuretic, natriuretic and hypotensive effects of a continuous perfusion of aqueous extract of

Urtica dioica in the rat." *J Ethnopharmacol. 73 (1-2): 95-100.* Accessed July 22, 2019. doi:10.1016/S0378-8741(00)00270-1.

[242] Qayyum, Rahila. Hafiz-Misbah-ud-Din, Qamar., Shamim, Khan. Umme, Salma. Taous, Khan. Abdul-Jabbar, Shah. 2016. "Mechanisms underlying the antihypertensive properties of *Urtica dioica*." *J Transl Med.;14:254.* Accessed July 22, 2019. doi:10.1186/s12967-016-1017-3.

[243] Testai, Lara. Silvio, Chericoni. Vincenzo, Calderone. Giulia, Nencioni. Paola, Nieri. Ivana, Morelli. Enrica, Martinotti. 2002. "Cardiovascular effects of *Urtica dioica L.* (*Urticaceae*) roots extracts: in vitro and in vivo pharmacological studies." *J Ethnopharmacol. 2002 Jun;81(1):105-9.* Accessed July 22, 2019. doi10.1016/S0378-8741(02)00055-7.

[244] Farzami, Bijan., Vardasbi. Shokri, Khaghani. 2003. "Induction of insulin secretion by a component of *Urtica dioica leaf* extract in perifused Islets of Langerhans and it's in vivo effects in normal and streptozotocin diabetic rats." *J. Ethnopharmacology. 2003; 89(1):47–53.* Accessed July 22, 2019. doi: 10.1016/s0378-8741(03)00220-4.

[245] Bnouham, Mohamed. Fatima-Zahra, Merhfour. Abderrahim, Ziyyat. Hassane, Mekhfi. Mohammed, Aziz. Abdelkhaleq, Legssyer. 2003. "Antihyperglycemic activity of the aqueous extract of *Urtica dioica*." *Fitoterapia. 74 (7- 8):677-681.* Accessed July 22, 2019. doi: 10.1016/S0367-326X(03)00182-5.

[246] Rahimzadeh, Mahsa. Samaneh, Jahanshahi. Sohelia, Moein. Mahmood-Reza, Moein. 2014. "Evaluation of alpha-amylase inhibition by *Urtica dioica* and Juglans regia extracts." *Iran J Basic Med Sci*. (6):465-9. PMID: 25140210.

[247] Ahangarpour, Akram. Maryam, Mohammadian. Mahin, Dianat. 2012. "Antidiabetic effect of hydroalcholic *Urtica dioica* leaf extract in male rats with fructose-induced insulin resistance." *Iran J Med Sci. 2012 Sep;37(3):181-6.* Accessed July 22, 2019. PMID: 23267403.

[248] Daher, Costantine. Karmen, Baroody. George, Baroody. 2006. "Effect of *Urtica dioica* extract intake upon blood lipid profile in the rats." *Fitoterapia. 2006;77:183-8.* Accessed July 22, 2019. doi:10.1016/j.fitote.2006.01.010.

[249] Nassiri-Asl, Marjan. Farzaneh, Zamansoltani. Esmail, Abbasi. Mohammad-Mehdi. 2009. "Effects of *Urtica dioica* extract on lipid profile in hypercholesterolemic rats." *Zhong Xi Yi Jie He Xue Bao. 7 (5): 428-433.* Accessed July 22, 2019. doi:10.3736/jcim20090506.

[250] Avci, Gulcan. Esra, Kupeli. Abdullah, Eryavuz. Erdem, Yesilada. Ismail, Kucukkurt. 2006. "Antihypercholesterolaemic and antioxidant activity assessment of some plants used as remedy in Turkish folk medicine." *J Ethnopharmacol.;107(3):418-23.* Accessed July 22, 2019. doi: 10.1016/j.jep.2006.03.032.

[251] Johnson, Michelle. Anita, Lucius. Tessa, Meyer. Elvira, Gonzalez. 2011. "Cultivar evaluation and effect of fermentation on antioxidant capacity and in vitro inhibition of a-amylase and a-glucosidase by highbush blueberry (*Vaccinium corombosum*)." *J. Agric. Food Chem 59:8923–8930.* Accessed June 23, 2019. doi: 10.1021/jf201720z.

[252] Coban, Jale. Betul. Evran. Ferda, Özkan. Aydin, Çevik, Semra, Dogru. Mujdat, Uysal. 2013. "Effect of blueberry feeding on lipids and oxidative stress in the serum, liver and aorta of guinea pigs fed on a high-cholesterol diet." *Biosci. Biotechnol. Biochem. 77:389–391.* Accessed June 23, 2019. doi: 10.1271/bbb.120722.

[253] Gustinelli Graziele, Eliasson, Lovisa. Cecilia, Svelander., Marie, Alminger. Lilia, Ahrne. 2018. "Supercritical CO_2 extraction of bilberry (*Vaccinium myrtillus L.*) seed oil: Fatty acid composition and antioxidant activity." *The Journal of Supercritical Fluids 135: 91–97.* Accessed June 23, 2019. doi: DwdDDD10.1016/j.supflu.2018.01.002.

[254] Rohloff, Jens. Eivind, Uleberg. Arnfinn, Nes. Tore, Krogstad. Rolf, Nestby. Inger, Martinussen. 2015. "Nutritional composition of bilberries (*Vaccinium myrtillus*L.) from forest elds in Norway – Effects of geographic origin, climate, fertilization and soil

properties." *Journal of Applied Botany and Food Quality* 88, 274 – 287. Accessed July 22, 2019. doi:10.5073/JABFQ.2015.088.040.

[255] Naseri, Rozita. Fatemeh, Farzaei. Pouya, Haratipour. Seyed, Nabavi. Solomon, Habtemariam. Mohammad, Farzaei. Reza, Khodarahmi. Devesh, Tewari. Saeideh, Momtaz. 2018. "Anthocyanins in the Management of Metabolic Syndrome: A Pharmacological and Biopharmaceutical" *Review. Front. Pharmacol 9:1310.* Accessed June 23, 2019. doi: 10.3389/fphar.2018.01310.

[256] Suzuki, Rieko. Masami, Tanaka. Masakatsu, Takanashi. Aashiq, Hussain. Bo, Yuan. Hiro, Toyoda. 2011. "Anthocyanidins-enriched bilberry extracts inhibit 3T3-L1 adipocyte differentiation via the insulin pathway." *Nutrition & Metabolism 8:1–9.* Accessed June 23, 2019. doi: 10.1186/1743-7075-8-14.

[257] Tsuda, Takanori. Yuki, Ueno. Hitoshi, Kojo. Toshikawa, Yoshikawa, Toshihiko, Osawa. 2005. "Gene expression profile of isolated rat adipocytes treated with anthocyanins." *Biochimica et Biophysica Acta 1733:137–147.* Accessed June 24, 2019. doi: 10.1016/j.bbalip.2004.12.014.

[258] Defuria, Jason. Grace, Bennett. Katherine, Strissel. James, Perfield. Paul, Milbury. Andrew Greenberg. 2009. "Dietary blueberry attenuates whole-body insulin resistance in high fat-fed mice by reducing adipocyte death and its inflammatory sequelae." *J. Nutr. 139:1510–1516.* Accessed June 23, 2019. doi: 10.3945/jn.109. 105155.

[259] Prior, Ronald. Sam, Wilkes. Theo, Rogers. Ramesh, Khanal. Xianli, Wu. Tiffany, Hager. 2010. "Dietary black raspberry anthocyanins do not alter development of obesity in mice fed an obesogenic high-fat diet." *J. Agric. Food Chem 58:3977–3983.* Accessed June 23, 2019. doi: 10.1021/jf9030772.

[260] Liu, Yunen. Dehong, Tan. Lin, Shi. Xinwei, Liu. Yubiao, Zhang. Changci, Tong. 2015. "Blueberry anthocyanins-enriched extracts attenuate cyclophosphamide-induced cardiac injury." *PLoS ONE 10:e0127813.* Accessed June 23, 2019. doi: 10.1145/2818302.

[261] Shaughnessy, Kevin. Ian, Boswall. Adam, Scanlan. Katherine, Gottschall. Marva, Sweeney. 2009. "Diets containing blueberry extract lower blood pressure in spontaneously hypertensive stroke-prone rats." *Nutr Res 29(2):130–138.* Accessed June 24, 2019. doi: 10.1016/j.nutres.2009.01.001.

[262] Kalea, Anastasia. Kateryna, Clark. Dale, Schuschke. Aleksandra, Kristo. Dorothy, Klimis. 2010. "Dietary enrichment with wild blueberries (*Vaccinium angustifolium*) affects the vascular reactivity in the aorta of Young spontaneously hypertensive rats." *J Nutr Biochem 21(1):14–22.* Accessed June 24, 2019. doi: 10.1016/j.jnutbio.2008.09.005.

[263] Onyenekwe, Paul. Seiji, Hashimoto. 1999. "The composition of the essential oil of dried Nigerian ginger (*Zingiber officinale Roscoe*)" *Eur Food Res Technol (1999) 209: 407.* Accessed July 19, 2019. doi: 10.1007/s002170050517.

[264] Siew, Yinyin. Sogand, Zareisedehizadeh. Wei-Guang, Seetoh. Soek-Ying, Neo. Chay-Hoon, Tan. Hwee-Ling, Koh. 2014. "Ethnobotanical survey of usage of fresh medicinal plants in Singapore." *Journal of Ethnopharmacology 155. 1450–1466.* Accessed July 19, 2019. doi: 10.1016/j.jep.2014.07.024.

[265] Chen, Yan-Ping. Wenhui, Si. Jianhao, Zhang. Zhen-Yu, Chen. Hau, Yin Chung. 2018. "Antioxidant activities of ginger extract and its constituents toward lipids." *Food Chem.* 239: 1117–1125. Accessed July 19, 2019. doi:10.1016/j.foodchem.2017.07.055 2018/01/15/.

[266] Kou, Xingran. Yuqian, Kec. Xiaoqi, Wang. Ramim, Tanver Rahman. Yuzhang, Xie. Shangwei, Chen. Hongxin, Wang. 2018. "Simultaneous extraction of hydrophobic and hydrophilic bioactive compounds from ginger (*Zingiber officinale Roscoe*)." *Food Chemistry 257 223–229.* Accessed July 19, 2019. doi: 10.1016/j.foodchem.2018.02.125.

[267] Tasneem, Shumaila. Bin, Liu. Bin, Li. M. Iqbal, Choudhary. Wei, Wang. 2019. "Molecular pharmacology of inflammation: Medicinal plants as anti-inflammatory agents." Pharmacological

Research 139 126–140. Accessed July 19, 2019. doi:10.1016/j.phrs.2018.11.001.

[268] Han, Young-ah.. Chang-woo, Song. Woo-suk,. Koh. Gyu-hwan, Yon. Young-sup, Kim. Shi-yong, Ryu. Kyu-hong, Lee. 2013. "Antiinflammatory effects of the *Zingiber officinale roscoe* constituent 12-dehydrogingerdione in lipopolysaccharide-stimulated Raw 264.7 cells," *Phytotherapy research: PTR 27 (8) 1200–1205.* Accessed July 19, 2019. doi:10.1016/j.phrs.2018.11.001.

[269] Yiming, Li. Van H, Tran. Bhavani P, Kota. Srinivas, Nammi. Colin C, Duke. Basil D, Roufogalis. 2014. "Preventative Effect of *Zingiber officinale* on Insulin Resistance in a High-Fat High-Carbohydrate Diet-Fed Rat Model and its Mechanism of Action." *Basic & Clinical Pharmacology & Toxicology, 115, 209–215* doi: 10.1111/bcpt. 12196. Accessed July 19, 2019. doi: 10.1111/bcpt.12196.

[270] Lei, Lin. Yuwei, Liu. Xiaobo, Wang. Rui, Jiao. Ka-Ying, Ma. Yuk-Man, Li, . Lijun, Wang. Sun-Wa, Man. Shengmin, Sang. Yu, Huang. Zhen-Yu, Chen. 2014. "Plasma cholesterol-lowering activity of gingerol-and shogaol-enriched extract is mediated by increasing sterol excretion." *Journal of Agricultural and Food Chemistry, 62, 10515–10521.* Accessed July 19, 2019. doi:10.1021/jf5043344.

[271] Kim, Soon-Hee. Hee-Sook, Park. MoonJu, Hong. JiYoung, Yoo. Hoyoung, Lee. Ju Ah, Lee. Jinyoung, Hur. DaeYoung, Kwon. Myung-Sunny, Kim. 2016. "Tongqiaohuoxue decoction ameliorates obesity-induced inflammation and the prothrombotic state by regulating adiponectin and plasminogen activator inhibitor-1." *Journal of Ethnopharmacology 192.* Accessed July 19, 2019. doi: 10.1016/j.jep.2016.07.023.

[272] Atashak, Sirvan. Maghsoud, Peeri. Mohammad-Ali, Azarbayjani. Stephen, Stannard. 2014. "Effects of ginger (*Zingiber officinale Roscoe*) supplementation and resistance training on some blood oxidative stress markers in obese men." *Journal of Exercise Science & Fitness 12. 26-30.* Accessed July 19, 2019. doi: 0.1016/j.jesf.2014.01.002.

In: Phytochemicals from Mexican … ISBN: 978-1-53617-397-0
Editors: J. Basilio Heredia et al.

Chapter 5

TYPE-2 DIABETES MELLITUS

Hilda K. Saenz-Hidalgo[1], J. J. Buenrostro-Figueroa[1], A. Guevara-Aguilar[1], Hugo S. García[2] and R. Baeza-Jiménez[1,*]

[1]Centro de Investigación en Alimentación y Desarrollo, Delicias, Chihuahua, México

[2]UNIDA, Tecnológico Nacional de México, Instituto Tecnológico de Veracruz, Veracruz, Veracruz, México

ABSTRACT

Unfortunately, diabetes has increased rapidly. In 1980, 108 million people had diabetes, and in 2014, there were 422 million people. Such increment implies a severe public health situation with important effects on the finances of patients and their families, like for the economies of the countries affected. Diverse medication treatments are commercially available for diabetes; however, some of them are of limited efficacy or exert side effects, and then the use of plants is gaining importance. In the present chapter, a general description of diabetes (definition and treatment) is provided, and the use of Mexican medicinal plants with anti-diabetic potential is reviewed.

---

* Corresponding Author's Email: ramiro.baeza@ciad.mx.

Keywords: anti-diabetic potential, diabetes, medicinal plants, phytochemicals

INTRODUCTION

Mexico can be unexceptionally distinguished by its richness in plant species. Some of them have been utilized for food and medicinal applications by ancient civilizations and nowadays are still consumed and used in some communities. Some studies reported in technical literature have addressed that such crops are rich in bioactive compounds, not only in nutritional components. However, most people do not consume them because such plants are considered of low social stature.

On the other hand, in our country, there is not a real concern about what we are eating, which, combined with an elevated consumption of unhealthy foods and the lack of social environments for physical activity and public health legislation, has led to obesity and malnutrition, diabetes included.

Diabetes mellitus (Dm) is related to a complex metabolic disorder characterized by both type I- and type II-Dm, with a consequent reduction of life expectancy. Treatment implies hyperglycemic control and a healthy lifestyle.

In the present chapter, a general description of diabetes is provided, and the use of Mexican medicinal plants with anti-diabetic potential is reviewed.

DESCRIPTION OF THE DISEASE

Glucose uptake into myocytes and adipocytes is moderate by GLUT4 (glucose transporter). Such transporter is located within the cell in membrane vesicles, and when insulin interacts with its receptor (as a result of high blood glucose), those vesicles move to surface and fuse with the plasma membrane, increasing the number of glucose transporters in the plasma membrane. When insulin level decreases, glucose transporters undergo endocytosis, to be able to form small vesicles. The latter will be further fused with a larger endosome. When insulin levels rise again, the endosome (rich in glucose transporters) is divided into small vesicles, ready to return to the surface [1].

What Is Diabetes Mellitus?

Diabetes mellitus (Dm) is a severe and chronic disease that takes place when insulin is not produced in the amounts required in the pancreas or when our body does not appropriately utilize it (see Figure 1). The main effect related to Dm is hyperglycemia, and in the case of pregnant diabetic women, they are at risk of fetal death.

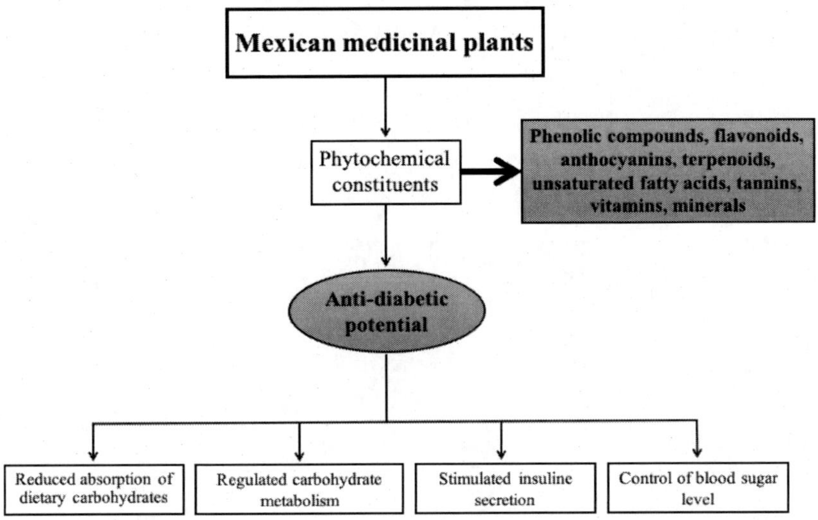

Figure 1. Anti-diabetic potential of Mexican medicinal plants.

There are two clinical types of Dm: type I- and type II-Dm. Type I (TIDm) is characterized by insulin-dependent, whereas type II (TIIDm) refers to insulin-resistant diabetes. In both types of diabetes the following symptoms can be observed: polyuria (excessive thirst and frequent urination), polydipsia (intake of large volumes of water), glucosuria (large amounts of glucose in the urine), ketonemia (elevated amounts of ketone bodies in blood) and ketonuria (large amounts of ketone bodies in urine) [1].

Dm in Mexico

According to statistics reported by the International Diabetes Federation, 425 million people have Dm in the world, and more than 46 million people live in the North America and Caribbean region, which is estimated to reach 62 million. Mexico is located in that region, and the incidence of Dm is very high: 12, 030, 100 cases of diabetes in adults were registered in 2017[2].

The diabetes country profiles reported by the World Health Organization in 2016, including for Mexico, the following data [3]:

- the number of Dm deaths in ages 30-69 was 23, 100 men and 22, 000 women; for ages 70+ was 17, 600 men and 24, 300 women
- proportional mortality (% of total deaths, all ages): 14% Dm
- the national response to Dm refers to policies, guidelines, and monitoring. Medicines in primary care facilities include: insulin, metformin and sulphonylurea; however, the following procedures are not generally available: retinal photocoagulation and renal replacement therapy by means of dialysis and transplantation.

Dm Treatment

Even when several efforts and assays have been conducted, TIDm cannot be prevented. However, those approaches have been useful for TIIDm. Therefore, it is worth noting that an early diagnosis (blood glucose testing) can improve the diabetic person´s life.

Table 1. Commercially available drugs for Dm treatment

Treatment	Description	Example
α-glucosidase inhibitors	Reduce the absorption of glucose in the intestine	Acarbose
Biguanides	Decrease the hepatic production of glucose	Metformin
Dipeptidyl peptidase-4 (DPP-4) inhibitors	DPP-4 inhibitors lower blood sugar by helping the body increase the level of the hormone insulin after meals	Sitagliptin
Glinides	Increase insulin secretion by β cell of pancreas	Prandin
Parenteral administration of glucagon-like peptide-1 (GLP-1) agonists	Stimulate post-prandial insulin secretion	
Sodium-glucose co-transporter-2 (SGLT2) inhibitors	SGLT2 inhibitors lower blood sugar by causing the kidneys to remove sugar from the body through the urine	Canagliflozin
Sulfonylureas	Increase insulin secretion by β cell of pancreas	Glibenclamide
Thiazolidinediones	which are agonists of the peroxisome proliferator-activated receptor-gamma (PPA-γ); increase the efficiency of insulin	Glitazones

Sources: Nazarian-Samani *et al.* [4]; Chinsembu [5]; FDA [6, 7]; Diabetes.co.uk [8, 9].

In Table 1 is indicated some medication used in Dm treatment. However, there are some drawbacks to such drugs: 1) limited efficacy, 2) narrow tolerability, 3) moderate to severe side effects, and 4) the need to take different drugs daily. As a result, a segment of the world population considers that using plants are less expensive and healthier than allopathic medicine.

Mexico is rich in plant diversity, and some of them are related to cultural and prehispanic values. Plants may contain chemical compounds with important bioactivities, anti-diabetic is included in such bioactivities and is described as follows.

MEXICAN MEDICINAL PLANTS WITH ANTI-DIABETIC POTENTIAL

Mexico is a country with a rich biodiversity of medicinal plants and a long history in the use of herbal medicine. Ancient civilizations used such plants, and the remaining communities that preserve their legacy also consume them fresh or preparing some type of extracts (teas, soups, infusions). Those plants contain chemical compounds with anti-diabetic potential. In Figure 2 is indicated the compounds responsible of such potential and the beneficial effects that they exert.

Some Mexican medicinal plants with anti-diabetic effects are described as follows, highlighting their phytochemical constituents as well as *in vitro* and *in vivo* studies carried out to demonstrate their potential.

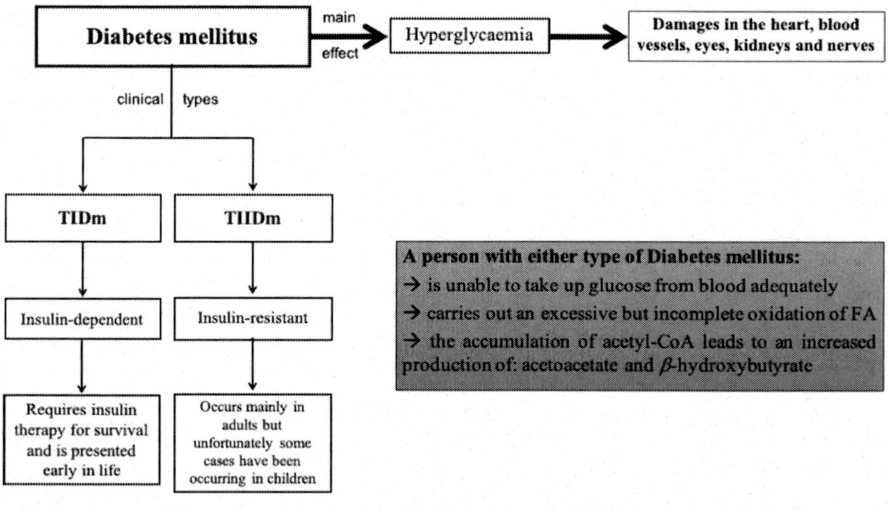

Figure 2. Diabetes mellitus: clinical types.

Agrimonia eupatoria L.

A. eupatoria L. is an indigenous plant from middle and northern Europe, temperate Asia and North America. The plant is a narrow, erect leaf perennial herb with pubescent stems (50 – 150 cm high). The whole plant yields a dye and a slightly bruised of the herb exhaling a peculiar but pleasant aromatic odor [10]. The phytochemical analysis of *A. eupatoria* revealed the presence of tannins, terpenoids, phenolic compounds (flavonoids), agrimony, oils, and many other bioactive chemical groups [11]. In another study, the phytochemical content of the *A. eupatoria* extract was analyzed by HPLC-MS, showing several phenolic compounds (up to 21), such as apigenin, kaempferol, and quercetin derivates, as well as catechin and oligomeric proanthocyanins [12]. The most abundant compounds are quercetin glycosides and proanthocyanins trimers.

On the other hand, the fresh herb contains palmitic and stearic acids, phytosterols and volatile oil. Oleic, linoleic, and linolenic acids are other fatty acid residues in *Agrimonia* seeds. The total polyphenol content has been estimated at around 8%, a significant source of such phytochemicals.

A. eupatoria is traditionally consumed as a tea to treat liver and gall bladder diseases, edemas, and pulmonary and gastrointestinal inflammatory diseases, and it is used in obesity and Dm. This plant exerts antibacterial, antiviral, antitumor, analgesic, antioxidant, immune-modulatory, antidiabetic, gastrointestinal, and hepatoprotective [11]. Biological effects are due to the presence of a high amount of tannins, flavonoids, and phenolic acids, which reduce the blood glucose level [13]. Dried leaves of *A. eupatoria* were used to evaluate the effect on glucose homeostasis in normal and streptozotocin (STZ)-diabetic mice. The authors reported a reduction in levels of hyperglycemia during the development of STZ-diabetes [14]. The water extracts of *Agrimonia* exhibit "insulin-like" activities, based on the stimulation of pancreatic β-cells and inhibition of α-glucosidase [15]. Kuczmannová *et al.* [16] evaluated the antidiabetic activities of *A. eupatoria* and *Cynara cardunculus* L. water infusions. Authors mentioned that both tested extracts exert similar antidiabetic activities, but a higher antioxidant activity was observed in *A. eupatoria*, which indicates a better clinical potential in the prevention or treatment of cardiovascular affections due to Dm and diseases associated with oxidative stress.

Allium sativum L.

Garlic (*A. sativum* L.) is considered among the oldest and the most cultivated plants, recognized for its medicinal uses related to cardiovascular diseases and Dm. It is used as a spice, food, and traditional medicine due to its several properties such as antibiotic, antiarthritic, hypoglycaemic effect, antitumor, antioxidant and antithrombotic [17].

In several studies have been reported the chemopreventive activity of garlic in diverse preparations, such as fresh garlic extract, aged garlic, and garlic oil. These activities have been attributed to the presence of volatile sulfur compounds, such as alliin, allicin, diallyl disulfide, diallyl trisulfide, diallyl sulfide, S-allyl cysteine, ajoene, and allyl mercaptan. S-allyl cysteine sulfoxide or alliin is an odorless compound released when a garlic bulb is cut. After that, the alliinase enzyme acts on alliin to produce diallythiosulfinate or allicin.

Allicin is an organosulfur compound, and plays a defensive role in the plant, with a unique odor easily detectable responsible for the characteristic smell and taste of the garlic [18]. Allicin has a high antioxidant activity, and it has been broadly reported for its effect on Dm. Allicin increases the levels of catalase and glutathione peroxidase, while allin stimulates insulin secretion in isolated β-cells in normal rats [19].

Hyperglycaemia is the main characteristic of a chronic metabolic disorder like Dm, and a consumption of 50 mg/kg of allicin has been reported to exert a hypoglycaemic effect [20]. Dm tends to increase the low-density lipoproteins-cholesterol (LDL-c) and decrease high-density lipoproteins-cholesterol (HDL-c) levels in the blood, promoting to coronary occlusions and blocks [21]. Oral administration of ethanol extract of *A. sativum* in diabetic mice reducing blood glucose level, accompanied by an increase in body weight [22].

Douaouya and Bouzerna [23] evaluated the effect of garlic on biochemical parameters of alloxan-induced diabetic mice. The oral administration of garlic extract decreased the blood glycosylated hemoglobin, serum glucose, total cholesterol (TC), triglycerides (TAG), total lipids and several enzymatic activities (glutamic oxalic transaminase, glutamic pyruvic transaminases, lactate dehydrogenase and phosphatase), while increases plasma insulin and glutathione content in liver, compared to an antidiabetic standard (glibenclamide) at a dose of 2.5 mg/kg. Other studies reported an important reduction in systolic and diastolic blood in hypertensive patients after the consumption of 200 mg of garlic powder three times per day [24]. Allicin contributes to reducing obesity, hypertension, and TIIDm by means of the stimulation of insulin secretion.

Cynara scolymus L.

C. scolymus L. or artichoke is one of the most commonly consumed foods around the world. Extracts of artichoke leaf have been reported to have important therapeutic activities, due to the presence of polyphenols, which are the active constituents responsible for the protective effects. Shütz et al. [25] identified and quantified the phenolic compounds present in artichoke heads, juice, and pomace by HPLC-DAD-ESI/MS. Caffeoylquinic acids (11) and flavonoids (8) were the main compounds detected. Among them, apigenin 7-*O*-glucuronide was the major component in all the samples. With respect to hydroxycinnamic acids, 1,5-Di-O-caffeoylquinic acid, and 5-O-caffeoylquinic acid (chlorogenic acid) were the main compounds present in both heads and pomace. Isomer 1-3-Di-O-caffeoylquinic acid (cynarine) was the predominant in juice, due to its processing. Loi et al. [26] proposed two active ingredients as underlying the hypoglycaemic effect of the artichoke: chlorogenic acid, that inhibits glucose 6-phosphate translocase (a component of the hepatic 6-phosphatase system regulating blood glucose); and di-caffeoylquinic acid, which modulates α-glucosidase activity and, in this sense, the carbohydrate metabolism.

Postprandial hyperglycemia is one of the most important pathophysiological states which leads to endothelial dysfunction, inflammatory reactions, as well as oxidative stress, and in consequence, a vascular failure can take place [27]. For that, Loi et al. [26] reported the use of a mixture of *Phaseolus vulgaris* and *C. scolymus* to control the appetite, food intake, and postprandial glycemia, highlighting the potential in the nutraceutical field. In other work, Fallah Huseini et al. [28] evaluated the effect of fiber-free artichoke leaf extract on patients with

TIIDm, found a reduction in blood levels of total cholesterol and LDL-c, but without effect on the blood levels of fasting glucose and postprandial glycemia. Authors concluded that artichoke extract may be used as an anti-hypercholesterolemic agent, suggesting the role of fiber in the antihyperglycemic effect.

Several studies have reported the potential of artichoke to reduce postprandial glycemia in rats [29] and clinical level [30]. The intake of artichoke leaf extract for 21 days reduced the levels of TC, TAG, LDL-c and VHDL-c, glucose levels, and plasma malondialdehyde level in STZ-induced diabetes rats [31].

One way to treat diabetes is the use of agents that decrease hyperglycemia levels by inhibiting the activity of carbohydrolases, such as α-glucosidase, avoiding the release of glucose from disaccharides and oligosaccharides. On the other hand, angiotensin-converting enzyme (ACE) plays an important role in the regulation of blood pressure. Villiger et al. [32] evaluated the commercial product Boldocynara®, a mixture of several plants, including artichoke. By itself, at the highest concentration used (100 μg/mL), artichoke showed low inhibition levels for α-glucosidase and moderate for ACE inhibition activities (15 and 31%, respectively).

Aloe vera L.

Aloe barbadensis is typically known as *Aloe vera*. This plant has been used for medicinal purposes since ancient times to treat skin irritation, burns, scalds, acne, dermatitis, digestive problems, asthma, immune system enhancement, peptic ulcers; due to it exerts several properties such as anti-inflammatory, antiviral, antitumor, anti-aging effect, hypoglycaemic, cardiovascular effect and anti-diabetes [33]. *A. vera* gel is into the plant leaf, composed of several bioactive compounds such as vitamins, enzymes, minerals, sugars, anthraquinones, sterols, dietary fiber, salicylic acid, and amino acids. A total of 13 phenolic compounds have been identified in the juice obtained from *A. vera* gel, being the main aloin and aloe-emodin (15-40%), and others such as aloenin and oleoresin [34]. It is not clearly defined which are the bioactive compounds responsible for the biological properties of *A. vera,* some studies suggest that is the polyphenol content (aloin and aloe-emodin), while others reported the role of some inorganic elements (Vanadium, manganese, copper), dietary fiber and especially polysaccharides (Acemannan) [35].

Pérez et al. [36] evaluated a polyphenol-rich extract (350 mg/kg) from *A. vera* on induced insulin resistance in mice. The extract was mainly composed of aloin (181-7 mg/g) and aloe-emodin (3.6 mg/g), and it was able to reduce body weight and blood glucose levels. The beneficial effect of *A. vera* was associated with its anti-inflammatory activity, due to there is a relation between a pro-inflammatory state and insulin resistance. For that, reducing the proinflammatory state help to reduces insulin resistance.

Several studies have indicated a decrease in fasting blood pressure, blood glucose, hemoglobin A1c levels in diabetic patients after oral administration of *A. vera* gel extracts. The use of *A. vera* has been successfully reported to treat patients with hyperlipidemia, due to it reduces LDL (18%), TC (15%), and TAG (30%) [37]. An administration of 300 mg/k of *A. vera* gel for 21 days in STZ-diabetic rats led to a reduced fasting blood glucose and in an improved insulin level in plasma [38]. Tanaka et al. [39] assayed five phytosterols from this plant (lophenol, 24-methyl-lophenol, 24-ethyl-lophenol, cycloartenol, and 24-methylene-cycloartanol) in diabetic mice with TIIDM for 28 days, reducing up to 64% the fasting blood glucose levels. A significant reduction in fasting blood glucose levels, TAG, TC, and LDL, as well as an increase in HDL-c was observed after oral administration of *A. vera* leaf extract during 21 days in alloxan-induced diabetic rabbits.

On the other hand, Yagi et al. [40] evaluated the hypoglycaemic effect of *A. vera* high molecular weight fractions on type 2 diabetic patients composed of a polysaccharide with acemannan and <10 ppm of barbaloin or aloin. Results showed a significant hypoglycaemic effect (glucose and triglycerides levels).

Vaccinium myrtillus L.

V. myrtillus (Bilberry) is one of the most medicinal plants widely used as an antioxidant, mucostimulant, hypoglycaemic, anti-inflammatory, vasoprotection and lipid-lowering agent. It is a potential natural source of anthocyanins andpolyphenolic compounds, which give color and high antioxidant content. Since ancient times, before to the availability of insulin, bilberry leaves were widely used to diabetes treatment, due to the presence of myrtillin, a gallic glycoside able to reduce glycosuria, postprandial hyperglycemia and insulin requirements [41, 42]. Hokkanen et al. [43] identified 31 different phenolic compounds present in bildberry leaves, including catechins, cinchonains, proanthocyanidins, phenolic acids, iridoids, and flavonols. Methanolic extracts of bilberry fruit inhibited the amylase and glucose enzyme activities, with high antioxidant activity in correlation with a high content of anthocyanins [44].

According to Edirisinghe and Burton-Freeman [45], mechanisms of anti-diabetic properties of berry polyphenols are divided into insulin-dependent and independent. For insulin-dependent, there are enhancing pancreatic β-cell function by modification of oxidative stress responses, increasing insulin production, reducing apoptosis and promoting β -cell proliferation, or enhancing peripheral tissue insulin sensitivity through modification of inflammation/oxidative stress responses in peripheral tissues. In the case of insulin-independent mechanism, may occur inhibition of digestives enzymes involved in carbohydrate breakdown, reduced glucose absorption in the gastrointestinal, as well as modifications in metabolism or energy status.

Supplementation with bilberry reduced glucose levels compared to the control (glibenclamide). Bilberry increases the insulin levels, reduced TC, LDL-c, VLDL-c, TG, and prevent the decline in HDL-C, profile not observed when glibenclamide was used [46]. Sidorova et al. [47] evaluated the hypoglycaemic and hypolipidemic effect of *V. myrtillus* leaf extract in Winstar rats with STZ and high-fructose, diet-induced diabetes. Significant reduction in blood glucose level was observed (50% reduction of hyperglycemia), normalizing the parameters of lipid metabolism with a high antioxidant effect on diene compounds in the blood serum of rats, also reduced hyperglycemia.

Preclinical studies are promising, but human data are limited and scarcely quality [48]. On the other hand, and well said by Paracelsus, "The dose makes the poison," Madić et al. [49] evaluated the cytotoxicity of several "antidiabetic" herbal plants, including bilberry by *A. cepa* test system. The author reported that a concentration of 400 μg/mL had a strongly mitodepressive effect on cell division, with a high prophase accumulation, showing the cytotoxicity effect of the bilberry extract. For that, the use of bilberry leaf preparation is not recommended [50]. Therefore, it is important to evaluate the toxicity level of all plants used as a natural remedy for any disease.

Calendula officinalis L.

C. Officinalis L. belongs to the Asteraceae family, and it is commonly known as marigold or mejorana. It is an herb used in traditional medicine, and antioxidant potential, as well as several pharmacological activities, such as anti-inflammatory, antigenotoxic effect, analgesic, and antidiabetic, have been distinguished [51, 52]. Some studies reported phytochemicals as the main responsible for the several functions related to the antidiabetic activity.

Olennikov and Kashchenko [53] evaluated an ethanolic extract and their chloroform, ethyl acetate and butanol fractions on amylase activity. Authors found that ethanolic extract and its ethyl acetate fractions had the highest inhibition values of amylase activity, with an IC50 of 38.02 and 24.52 μg/mL. Inhibition activities were attributed to the presence of phenolic compounds, such as caffeic acid, aesculetin, quercetin, and isorhamnetin, present in insignificant concentrations, but with activity at low concentration (1.02-2.64 μg/mL), compared to the control acarbose (9.54 μg/mL). The other 18 dominant compounds showed anti-amylase activity, identified as isoquercetin, isorhamnetin-3-*O*-β-D-glucopyranoside, 3-5, di-*O*-caffeoylquinic acid and quercetin-3-*O*-(6´-acetyl)- β-D-glucopyranoside). In another study, a new compound showed amylase and α-glucosidase inhibition, isolated from *C. officinalis* flowers and identified as esculetin-7-*O*-neohesperiodoside, a coumarin glucoside [54].

A dose of 100 mg/kg of hydroalcoholic extract of *C. officinalis* reduced body weight, blood glucose, and urine sugar, restoring the levels of serum lipid in alloxan diabetic rats, as insulin does. An *in vivo* study was conducted in patients with diabetic foot ulcers, which were treated with hydroglycolic extract of the plant and it was observed a 78% of reduction in the amount of exudate, fibrin slough, and necrotic tissue, indicating a potential use of these extracts for the treatment of foot ulcers [55]. Recently, lutein isolated from marigold flowers was successfully used to reduced blood glucose levels in mice, as well as reduce the malondialdehyde levels (used as oxidative stress marker) [56].

Other studies have been shown no effect on antidiabetic activity. Diabetic nephropathy is a type of micro and macrovascular complications caused by chronic hyperglycemia. In a study to evaluate the effect of hydroalcoholic extract of marigold (300 mg/kg) histopathological changes and biochemical blood profiles and histopathological changes in kidney of STZ-induced diabetic rats, they reported that the extracts decreased blood urea nitrogen and creatinine, total oxidant status and malondialdehyde, while it increased the total antioxidant capacity. However, no effect on blood glucose was observed [57].

Psidium guajava

Psidium guajava is plant is commonly known as Guayaba, and it is typical of hot climates. It belongs to the Myrtaceae family, and its fruit is rich in vitamins B1, B2, B6, and mainly in Vitamin C. The leaves of the plant are rich in flavonoids, particularly quercetin. Its antibacterial and antidiarrheal effects have been demonstrated, as well as its ability to relax the smooth intestinal muscle and inhibit bowel contractions [58].

Different parts of the plant have been studied in order to obtain a clinical treatment of Dm due to their better insulin sensitivity [59]. The bioactive constituents that characterize this plant for its hypoglycaemic effect in TIIDm are strictinin, isostrictinin and pedunculagin [60]. The active phytoconstituent Vescalagin, isolated from *P. guajava*, exerted beneficial effects on the insulin-resistant mouse hepatocytes [61].

A study conducted by Huang et al. [62] showed the antihyperglycemic effect in STZ-induced diabetic rats when they were fed with this plant. In other work carried out by Oh *et al.* [63], butanol-soluble extracts from *P. guajava* were used in young and adult diabetic mice.

Hypericum perforatum L.

Commonly known as St. John's wort, this plant belongs to the Hypericaceae family, order Malpighiales. This plant has been used for a long time in the treatment of rheumatism, hemorrhoids, neuralgia [64], and mostly as an antidepressant [65]. The phytochemicals with antidiabetic properties reported for this plant include phloroglucinols (hyperforins), xanthones, and flavonoids (procyanidins) [66].

Husain et al. [67] carried out a study where hyperglycemia-induced rats were treated with extracts of the plant. From their findings, the authors noted a relationship between blood glucose levels and the dose administered, meaning that the extracts were effective against metabolic disturbances in type-2 diabetic animals. In another study conducted by Can et al. [68], improved blood glucose levels were measured in rats with diabetic disorders after the treatment with ethanol esplanates of the plant.

Hibiscus sabdariffa L.

This plant is found in several parts of the world, but in Latin American countries is known as Flor de Jamaica. This plant belonging to the Malvaceae family and is widely used for constipation, heart ailment, high blood pressure, urinary tract infections, cancer, diabetes [69]. With respect to the phytochemical's components of this plant, malic acid, anthocyanins, ascorbic acid, and minerals (mainly Ca and Fe) have been reported [70].

In an *in vitro* study conducted by Addedmlui and Ohbo [71], it was investigated that extracts of *H. sabdariffa* inhibited both a-amylase and a-glucosidase, then this plant has the potential for its anti-diabetic properties.

On the other hand, Wisetmuen et al. [72] observed that the ethanolic extract of dried *H. sabdariffa* causes a significant decrease in fasting blood glucose of chronic diabetic rats and an increment in the basal insulin levels of normal rats and chronic diabetic rats. In a similar study, Peng et al. [73], mentioned that extracts of *H. sabdariffa* led to lower blood glucose levels and ameliorating insulin resistance in adult rats.

Zingiber officinale L.

It is commonly known as ginger, and it belongs to the Zingiberaceae family. This plant has been employed as a condiment, and it confers multiple benefits for human health. Ginger rhizomes are characterized by a high content of antioxidant compounds, which exhibit potential properties for cancer prevention [74], as well as anti-inflammatory [75], and antinausea [76] benefits.

In a research carried out by Iranloye et al. [77], aqueous extracts of raw ginger were administered in adult rats for 1 month, and from their results, authors noted a reduction in blood glucose fasting and a hypoglycaemic effect. Similar findings were reported by Al-Amin et al. [78], they mentioned a significant effect on lowering serum glucose, cholesterol and TAG levels in diabetic rats when using raw ginger. In another study, Andallu et al. [79] assayed a low dose of ginger for 30 days in a diabetic patient, and their results showed a significant reduction in blood glucose, TAG and TC, after the treatment.

Humulus lupulus L.

This plant belongs to the Cannabinaceae family, being humulon and lupulon the active compounds that can be distinguished. Due to the flavor and aroma that these components confer, they have been widely applied in the manufacture of beer. After being used in the manufacturing process, they are converted into isohumulones (by means of isomerization), which are potent angiogenic inhibitors [80] and exert antibacterial property [81].

Other benefits related to isohumulones are an improvement in glucose tolerance and reduced insulin resistance in diabetic and obese mice, as well as reduced blood glucose levels [82]. Studies conducted by Bland et al. [83] demonstrated that isohumulones do not only reduce insulin resistance but also reduces metabolic inflammation.

Melissa officinalis L.

This plant is well known as lemon balm and belongs to the Lamiaceae family. For a long time, this plant was used as a vegetable and to add flavor. Essential oil can be obtained, and it is used for topical applications due to antibacterial and antifungal properties [84]. The most important phytochemical constituents include triterpenes, phenolic acids and flavonoids [85].

Yen et al. [86] conducted an *in vitro* assay, where it was noted that the essential oil of this plant was capable of regulating glucose content. On the other hand, an *in vivo* study showed that the essential oil of *M. officinalis* was able to reduce plasma glucose in adult rats [87]. The findings of Weidner et al. [88] demonstrated that ethanolic extracts of this plant led to a reduction in insulin resistance in adult rats.

Mentha piperita L.

M. Piperita L. belongs to the Lamiaceae family. It has been shown that polyphenolic compounds are responsible for their antimicrobial effect and antioxidant capacity. Alexa et al. [89] concluded that the extracts of this plant are a valuable source of antioxidants due to the presence of ferulic and rosmarinic acids. Essential oil is obtained from this plant, and its chemical composition includes menthol (49.89%) and menthone (20.84%), whereas isomenthone (7.25%), 1.8-cineole (6.73%) and ciscarane (4.99%) are the minor constituents. This oil is suitable to prevent induced diseases related to oxidative stress [90].

It has been reported that this plant can play a role in diabetic complications, targeting aging, and oxidative stress-related diseases due to studies that showed a positive glycation inhibitory and antioxidative activities [91]. On the other hand, in a research conducted by Narendhirakannan et al. [92], it was mentioned that the administration of the leaves of this plant in rats, has exerted a positive effect on glycemia by means of the inhibition of proteolysis caused by lack of insulin and decreasing the activity of glycogen phosphorylase, thus re-establishing the hepatic glycogen.

Barbalho et al. [93] studied the effect of the juice of *M. Piperita* on the offspring from diabetic rats, by the evaluation of lipid profile and glycemia levels. It was found that the juice had the potential to prevent Dm, since its consumption in rats reduced the levels of cholesterol, glucose, LDL-c, and TAG, with a significant increase in HDL-c levels.

Origanum vulgare L.

O. vulgare belongs to the Lamiaceae family. From this plant, the essential oil is extracted, which is mainly composed of oxygenated monoterpenes (carvacrol, thymol, ß-fenchyl alcohol, and δ-terpineol) and monoterpene hydrocarbons (γ-terpinene and α-terpinene). Additionally, 1-methyl-3-(1-methylethyl)-benzene (6.8%) also represented a substantial fraction of the oil [94].

The use of this plant as an antidiabetic has been studied. It was observed that oregano has a positive effect on the inhibition of α-glucosidase and α-amylase [95]. In order to help in the treatment of hyperglycemia, it has been studied the effect of phenolic antioxidants on the inhibition of amylase, and it can be mentioned the potential inhibitory effect (57%) of oregano on porcine pancreatic amylase in an *in vitro* study carried out by McCue et al. [96].

In vivo study, it has been demonstrated that aqueous extracts of this plant decreased glucose levels in STZ diabetic rats without modifying insulin secretion, so it has been concluded that its use has an anti-hyperglycaemic effect [97].

Citrus aurantium

C. aurantium is known for its antioxidant capacity, due to the presence of several phenolic acids (caffeic, cinnamic, benzoic acids) and flavonoids (catechin, epicatechin gallate, esculetin and quercetin, coumaric acid, hesperidin, quercitrin, kaempferol, and rhamnetin) [98].

In the treatment of TIIDm, glucagon-like peptide-1 (GLP-1) is used to regulate the secretion of insulin in pancreatic cells. For this reason, the stimulation of secretion in NCI-H716 cells was evaluated with hexane fractions of this plant, and it was observed that the extracts increased almost ten-fold compared to the negative control [99]. Flavonoids have also been studied in Dm. Among them, hesperidin (a by-product of citrus crops) was reported by Akiyama et al. [100] when in a study with type 2 diabetes Goto-Kakizaki (GK) rats, it led to normalized glucose metabolism by altering the activities of glucose-regulating enzymes.

On the other hand, Jia et al. [101] investigated the possible hypoglycaemic and hypolipidemic effects of neohesperidin derived from this plant, on KK-Ay mice. The treatment with neohesperidin decreased glycosylated serum protein, serum glucose, and fasting glucose.

Urtica dioica

U. dioica belongs to the family Urticaceae. The presence of different phytochemicals has been detected, namely, tannins, flavonoids, scopoletin, fatty acids, sterols, isolectins, polysaccharides and sterols [102], as well as alkenes, aromatic rings, aliphatic flour, ethers, alcohols, carboxylic acids, nitro compounds, esters, phenols hydrogen and bonded alcohols. Methanolic extracts of this plant has been analysed by GC-MS and the following compounds were identified: Oxime- methoxy-phenyl, 2, 6,-Nonadienal, triethyl (2-phenylethoxy), 2,3,-dihydro, 2,5,5,8a-Tetramethyl-1,2,3,5,6,7,8, 8a-octahydronaphthalen-1-ol, 2H-Indeno[1,2-b]furan-2-one, 3,3a, 4,5,6,7,8, 8b-octahydro-8,8-dimet, 4, 4a,5,6,7,8-hexahydro-1-methoxy, diethyl mercaptal, pentaacetate, Silane, [1,1-Bicyclopropyl-2-octanoic acid 2 hexyl-methyl ester, Estra-1,3,5(10)- trien-17B-ol, Benzofuran, 1- Dodecanamine, N, N-dimethyl, 2(3H)-Naphthalenone, 1-Hydroxy-2-(2,3,4,6- tetra-O-acetyl-beta-d-glucopyranosyl)-9H-xanthe, 3, 7-dimethyl, 1, 2, 3- Butanetriol, Cyclopropaneoctanoic acid, 2-[2-pentylcycloproyl)methyl]-methyl, and Ethyl iso- allochlate [103].

In one *in vitro* research, it was demonstrated an increase in insulin secretion on Islets of Langerhans induced by an active component of *U. dioica* and it was correlated with the decrease in glucose levels [104]. On the other hand, Bnouham et al. [105] studied peripheral glucose consumption in rat diaphragm, applying intraperitoneal doses of aqueous extracts of *U. dioica* and it was observed a decrease in blood glucose levels, which suggests an enhanced glucose utilization if the extracts are combined with insulin. In other study, it was demonstrated the inhibitory activity of α-amylase and the inhibition effect (IC50 = 1.89 mg/mL) of *U. dioica* leaves using hydroethanolic extracts [106].

With respect to *in vivo* studies, Bnouham et al. [107] employed aqueous extracts of this plant on neonatal STZ induced-diabetic rats. After

5 weeks of supplying the extracts, a reduction in plasma glucose was noted compared to untreated diabetic controls. On the other hand, Farzami et al. [104] cited the effect of *U. dioica* on the production of insulin in normal and STZ diabetic rats, mentioning an increase in the level of insulin and a reduction in the blood glucose level. In another study, Ahangarpour et al. [108] investigated the antidiabetic effect of the hydroalcoholic extract of *U. dioica* on male rats with insulin resistance, and it was noted a decrease in serum glucose and in the index of insulin resistance, which indicates a potential application in TIIDm.

Other studies conducted by Golalipour et al. [109, 110] referred that hydroalcoholic extracts of this plant not only have a positive effect on blood glucose level but also exert a protective effect of pancreatic β-cells and hepatocytes morphometric changes in STZ diabetic rats. On the other hand, Namjou et al. [111] observed that the use of hydroalcoholic extracts in non-ovarized female rats has a hyperglycaemic effect; however, in ovarized female rats, it caused a significant decrease in glucose levels.

Thymus vulgare

T. vulgare belongs to the Lamiaceae family and it is recognized by the oil obtained, which is rich in thymol, carvacrol, flavonoids and rosmarinic acid [112] and the following bioactive compounds: Erythritol, γ-Terpinene, Octan-2-one, 3,6-dimethyl-, DL-Leucine, N-glycyl-, Trans-2-Caren-4-ol, Dithiocarbamate, S-methyl-,N-(2-methyl-3-oxobutyl)-, 5-Caranol, trans, trans-(+)-, Phenol,2-methyl-5-(1-methylethyl)-, Phenol,2-methyl-5-(1-methylethyl)-, acetate, (5β)Pregnane-3,20β-diol, 14α,18α-[4-methyl-3-oxo(1-oxa-4-, β-Bisabolene, 1-Chloro-1-n-decyloxy-1-silacyclopentane, 7-epi-trans-sesquisabinene hydrate, (-)-Spathulenol, Caryophyllene oxide, 1-Heptatriacotanol, Ppropiolic acid, 3-(1-hydroxy-2-isopropyl-5-methylcyclohexyl), Naphthalene,1,2,3,4,4a,5,6,7-octahydro-4a-methyl, Paromomycin, Gibberellic acid, Phytol, 2,7-Diphenyl-1,6-dioxopyridazino[4,5:2,'3']pyrrolo[4,'5'-d] pyrida, Androst-4,6-dien-3,11,17-trione, 9-mercapto-, Ricinoleic acid, 4a,10a-Methanophenenthren-9β-ol, 11-syn-bromo-1,2,3,4,4a and (+)-γ- Tocopherol, O-methyl [113]. Due to its complex chemical composition, the volatile fraction of *T.*

vulgare has been shown to possess anti-diabetic activity related to the inhibition of both α-glucosidase and α-amylase enzymes [113]. This beneficial effect has been related to its antioxidant capacity and to the inhibition of glycation [87].

In vivo studies with *T. Vulgare* extracts demonstrated the therapeutic effects in the prevention of injuries in diabetic male rats induced by certain diseases, against histopathological changes [114], as well as a positive effect on the control of diabetes, by decreasing glucose levels in rats [115, 116].

Rosmarinus officinalis

R. officinalis also belongs to the Lamiaceae family. It is recognized as an important source of flavonoids, phenolic acids, and di- and tri-terpenoids. The antioxidant compounds found in this plant are rosmarinic acid, carnasol and carnosic acid [117], whereas its oil is rich in α-pinene, 1,8-cineole, camphene and verbenone [118].

The effect of the essential oils of *R. officinalis* on the activity of glycation was studied by Mahomoodally et al. [119], and a moderate inhibition was noted. Kwon et al. [95], investigated the activity of aqueous extracts *R. officinalis* on the inhibition of α-glucosidase, and inhibition of 70% was attained. Koga et al. [120] evaluated ethanolic extracts of *R. officinalis* and it was confirmed the inhibitory activity of α-glucosidase (IC50 of 290 mg/mL in maltose substrate and 150 mg/mL in sucrose substrate).

On the other hand, Bakirel et al. [121] explored the capacity of ethanolic extracts of this plant on the antidiabetic activity in rabbits, and an antidiabetogenic effect was observed in both normal and diabetic rabbits, decreasing blood glucose levels and increasing serum insulin levels. n-butanol and diethyl ether extracts of this plant were found to be rich in flavonoids and showed a reduction of 50% and 60% of glucose level in STZ-induced diabetes mice, respectively [122]. Aqueous extracts of *R. officinalis* have also been observed to reduce the production of serum alkaline phosphatase, aspartate aminotransferase, and alanine transaminase, which are indicative enzymes of hepatic dysfunction caused by Dm [123]. In other research, Selmi et al. [124] demonstrated that the essential oil of *R. officinalis* reduces glucose levels in alloxan-induced diabetes rats and that it also exerted a protective effect against liver oxidative stress.

Conclusion

The use of medicinal plants for the treatment of several affections is gaining importance. Plants are known to contain diverse chemical compounds that exert different bioactivities, anti-diabetic included. As it has been mentioned, from traditional plants have been extracted a-glucosidase and a-amylase inhibitors, phytochemicals with antioxidant activity, as well as some important minerals for the regulation of insulin and the control of blood sugar levels. However, there is still a lot of studies to do. On one hand, more research is needed to recover new and novel compounds that could exert anti-diabetic potential, and then the incorporation of such constituents into food in order to prepare functional foods or new medication. Furthermore, on the other hand, public health policies are also required in order to promote healthier diets and physical activity.

References

[1] Nelson, David L, Albert L Lehninger, and Michael M Cox. 2008. *Lehninger Principles of Biochemistry*: Macmillan. Reprint.

[2] International Diabeter Federation, IDF. 2019. *Idf North America and Caribbean Members.* https://idf.org/our-network/regions-members/north-america-and-caribbean/members/66-mexico.html (accessed July 2019).

[3] World Health Organization, WHO 2016. *"World Health Organization Diabetes in México."* https://www.who.int/diabetes/country-profiles/mex_en.pdf (accessed July 2019).

[4] Nazarian-Samani, Zeinab, Robert DE Sewell, Zahra Lorigooini, and Mahmoud Rafieian-Kopaei. 2018. "Medicinal Plants with Multiple Effects on Diabetes Mellitus and Its Complications: A Systematic Review." *Current Diabetes Reports* 18,(10): 72.

[5] Chinsembu, Kazhila C. 2019. "Diabetes Mellitus and Nature's Pharmacy of Putative Antidiabetic Plants." *Journal of Herbal Medicine* 15: 100230.

[6] Food and Drug Administration, FDA. 2016. *Information on Dipeptidyl Peptidase-4 (Dpp-4) Inhibitors*. https://www.fda.gov/drugs/information-drug-class/information-dipeptidyl-peptidase-4-dpp-4-inhibitors (accessed July 2019).

[7] Food and Drug Administration, FDA. 2018. *Sodium-Glucose Cotransporter-2 (Sglt2) Inhibitors*. https://www.fda.gov/drugs/postmarket-drug-safety-information-patients-and-providers/sodium-glucose-cotransporter-2-sglt2-inhibitors (accessed July 2019).

[8] Diabetes.co.uk. 2019. *Prandial Glucose Regulators (Glinides).* https://www.diabetes.co.uk/diabetes-medication/prandial-glucose-regulators.html (accessed July 2019).

[9] Diabetes.co.uk. 2019. *Sglt2 Inhibitors (Gliflozins).* https://www.diabetes.co.uk/diabetes-medication/sglt2-inhibitors.html (accessed August 2019).

[10] Rahimi, Maryam. 2015. "A Review: Anti Diabetic Medicinal Plants Used for Diabetes Mellitus." *Bulletin of Environment, Pharmacology and Life Sciences* 4: 163-80.

[11] Al-Snafi, Ali Esmail 2015. "The Pharmacological and Therapeutic Importance of Agrimonia Eupatoria-a Review." *Asian Journal of Pharmaceutical Science Technology* 5,(2): 112-17.

[12] Kuczmannová, Anika, Peter Gál, Lenka Varinská, Jakub Treml, Ivan Kováč, Martin Novotný, Tomáš Vasilenko, Stefano Dall'Acqua, Milan Nagy, and Pavel Mučaji. 2015. "*Agrimonia Eupatoria* L. and *Cynara Cardunculus* L. Water Infusions: Phenolic

Profile and Comparison of Antioxidant Activities." *Molecules* 20,(11): 20538-50.

[13] Ivanova, Diana, Deyana Vankova, and Milka Nashar. 2013. "*Agrimonia Eupatoria* Tea Consumption in Relation to Markers of Inflammation, Oxidative Status and Lipid Metabolism in Healthy Subjects." *Archives of Physiology and Biochemistry* 119,(1): 32-37.

[14] Swanston-Flatt, SK, C Day, CJ Bailey, and PR Flatt. 1990. "Traditional Plant Treatments for Diabetes. Studies in Normal and Streptozotocin Diabetic Mice." *Diabetologia* 33,(8): 462-64.

[15] Gray, Alison M, and Peter R Flatt. 1998. "Actions of the Traditional Anti-Diabetic Plant, *Agrimony Eupatoria* (Agrimony): Effects on Hyperglycaemia, Cellular Glucose Metabolism and Insulin Secretion." *British Journal of Nutrition* 80,(1): 109-14.

[16] Kuczmannová, Anika, Andrea Balažová, Eva Račanská, Miroslava Kameníková, Silvia Fialová, Jaroslav Majerník, Milan Nagy, Peter Gál, and Pavel Mučaji. 2016. "*Agrimonia Eupatoria* L. And *Cynara Cardunculus* L. Water Infusions: Comparison of Anti-Diabetic Activities." *Molecules* 21,(5): 564.

[17] Thomson, Martha, and Muslim Ali. 2003. "Garlic [*Allium Sativum*]: A Review of Its Potential Use as an Anti-Cancer Agent." *Current Cancer Drug Targets* 3,(1): 67-81.

[18] Padiya, Raju, and Sanjay K Banerjee. 2013. "Garlic as an Anti-Diabetic Agent: Recent Progress and Patent Reviews." *Recent Patents on Food, Nutrition Agriculture* 5,(2): 105-27.

[19] Thomson, Martha, Zainab M Al-Amin, Khaled K Al-Qattan, Lemia H Shaban, and Muslim Ali. 2007. "Anti-Diabetic and Hypolipidaemic Properties of Garlic (*Allium sativum*) in Streptozotocin-Induced Diabetic Rats." *International Journal of Diabetes & Metabolism* 15: 108-15.

[20] Salehi, Bahare, Paolo Zucca, Ilkay Erdogan Orhan, Elena Azzini, Charles Oluwaseun Adetunji, Soheb Anwar Mohammed, Sanjay K Banerjee, Farukh Sharopov, Daniela Rigano, and Javad Sharifi-Rad. 2019. "Allicin and Health: A Comprehensive Review." *Journal Trends in Food Science Technology*.

[21] Mamun, MA, N Hasan, F Shirin, MH Belal, MAJ Khan, MN Tasnin, MD Islam, A Islam, T Ara, and MRU Karim. 2017. "Antihyperglycemic and Antihyperlipidemic Activity of Ethanol Extract of Garlic (*Allium sativum*) in Streptozotocin-Induced Diabetic Mice." *International Journal of Medical and Health Research* 3,(2): 63-66.

[22] Shariatzadeh, SMA, M Soleimani Mehranjani, M Mahmoodi, MH Abnosi, HR Momeni, AR Dezfulian, and M Noori. 2008. "Effects of Garlic (*Allium sativum*) on Blood Sugar and Nephropathy in Diabetic Rats." *Journal of Biological Sciences* 8,(8): 1316-21.

[23] Douaouya, Lilia, and Noureddine Bouzerna. 2016. "Effect of Garlic (*Allium sativum* L) on Biochemical Parameters and Histopathology of Pancreas of Alloxaninduced Diabetic Rats." *International Journal of Pharmacy and Pharmaceutical Sciences* 8,(6): 202-06.

[24] Stabler, Sarah N, Aaron M Tejani, Fong Huynh, and Claire Fowkes. 2012. "Garlic for the Prevention of Cardiovascular Morbidity and Mortality in Hypertensive Patients." *Cochrane Database of Systematic Reviews*,(8): 1-21.

[25] Schütz, Katrin, Dietmar Kammerer, Reinhold Carle, and Andreas Schieber. 2004. "Identification and Quantification of Caffeoylquinic Acids and Flavonoids from Artichoke (*Cynara scolymus* L.) Heads, Juice, and Pomace by Hplc-Dad-Esi/Ms N." *Journal of Agricultural Food Chemistry* 52,(13): 4090-96.

[26] Loi, Barbara, Noemi Fantini, Giancarlo Colombo, Gian Luigi Gessa, Antonella Riva, Ezio Bombardelli, Paolo Morazzoni, and Mauro Carai. 2013. "Reducing Effect of a Combination of *Phaseolus vulgaris* and *Cynara scolymus* Extracts on Food Intake and Glycemia in Rats." *Phytotherapy Research* 27,(2): 258-63.

[27] Node, Koichi, and Teruo Inoue. 2009. "Postprandial Hyperglycemia as an Etiological Factor in Vascular Failure." *Cardiovascular Diabetology* 8,(1): 23.

[28] Fallah Huseini, H, S Kianbakht, and R Heshmat. 2012. "*Cynara scolymus* L. In Treatment of Hypercholesterolemic Type 2 Diabetic

Patients: A Randomized Double-Blind Placebo-Controlled Clinical Trial." *Journal of Medicinal Plants* 1,(41): 85-65.

[29] Fantini, Noemi, Giancarlo Colombo, Andrea Giori, Antonella Riva, Paolo Morazzoni, Ezio Bombardelli, and Mauro AM Carai. 2011. "Evidence of Glycemia-Lowering Effect by a *Cynara scolymus* L. Extract in Normal and Obese Rats." *Phytotherapy Research* 25,(3): 463-66.

[30] Nazni, P, T Poongodi Vijayakumar, P Alagianambi, and M Amirthaveni. 2006. "Hypoglycemic and Hypolipidemic Effect of *Cynara scolymus* among Selected Type 2 Diabetic Individuals." *Pakistan Journal of Nutrition* 5,(2): 147-51.

[31] Heidarian, Esfandiar, and Yadollah Soofiniya. 2011. "Hypolipidemic and Hypoglycemic Effects of Aerial Part of *Cynara scolymus* in Streptozotocin-Induced Diabetic Rats." *Journal of Medicinal Plants Research* 5,(13): 2717-23.

[32] Villiger, Angela, Filippo Sala, Andy Suter, and Veronika Butterweck. 2015. "In Vitro Inhibitory Potential of *Cynara scolymus*, *Silybum marianum*, *Taraxacum officinale*, and *Peumus boldus* on Key Enzymes Relevant to Metabolic Syndrome." *Phytomedicine* 22,(1): 138-44.

[33] Chatterjee, Poulomi, Bodhisattwa Chakraborty, and Subhangkar Nandy. 2013. "*Aloe vera* Plant: Review with Significant Pharmacological Activities." *Mintage Journal of Pharmaceutical & Medical Sciences* 2,(3): 21-24.

[34] Rodríguez, Elena Rodríguez, Jacinto Darias Martín, and Carlos Díaz Romero. 2010. "*Aloe vera* as a Functional Ingredient in Foods." *Critical Reviews in Food Science Nutrition* 50,(4): 305-26.

[35] Christaki, Efterpi V, and Panagiota C Florou-Paneri. 2010. "*Aloe vera*: A Plant for Many Uses." *Journal of Food, Agriculture and Environment* 8,(2): 245-49.

[36] Pérez, Yolanda Y, Enrique Jiménez-Ferrer, Alejandro Zamilpa, Marcelino Hernández-Valencia, Francisco J Alarcón-Aguilar, Jaime Tortoriello, and Rubén Román-Ramos. 2007. "Effect of a Polyphenol-Rich Extract from *Aloe vera* Gel on Experimentally

Induced Insulin Resistance in Mice." *The American Journal of Chinese Medicine* 35,(06): 1037-46.

[37] Shapiro, Karen, and William C Gong. 2002. "Natural Products Used for Diabetes." *Journal of the American Pharmaceutical Association* 42,(2): 217-26.

[38] Rajasekaran, Subbiah, Kasiappan Ravi, Karuran Sivagnanam, and Sorimuthu Subramanian. 2006. "Beneficial Effects of *Aloe vera* Leaf Gel Extract on Lipid Profile Status in Rats with Streptozotocin Diabetes." *Clinical Experimental Pharmacology and Physiology* 33,(3): 232-37.

[39] Tanaka, Miyuki, Eriko Misawa, Yousuke Ito, Noriko Habara, Kouji Nomaguchi, Muneo Yamada, Tomohiro Toida, Hirotoshi Hayasawa, Mitunori Takase, and Masanori Inagaki. 2006. "Identification of Five Phytosterols from *Aloe vera* Gel as Anti-Diabetic Compounds." *Biological and Pharmaceutical Bulletin* 29,(7): 1418-22.

[40] Yagi, Akira, Sahar Hegazy, Amal Kabbash, and Engy Abd-El Wahab. 2009. "Possible Hypoglycemic Effect of *Aloe vera* L. High Molecular Weight Fractions on Type 2 Diabetic Patients." *Saudi Pharmaceutical Journal* 17,(3): 209-15.

[41] Allen, Frederick M 1927. "Blueberry Leaf Extract: Physiologic and Clinical Properties in Relation to Carbohydrate Metabolism." *Journal of the American Medical Association* 89,(19): 1577-81.

[42] Cignarella, Andrea, Milena Nastasi, Elena Cavalli, and Lina Puglisi. 1996. "Novel Lipid-Lowering Properties of *Vaccinium myrtillus* L. Leaves, a Traditional Antidiabetic Treatment, in Several Models of Rat Dyslipidaemia: A Comparison with Ciprofibrate." *Thrombosis Research* 84,(5): 311-22.

[43] Hokkanen, Juho, Sampo Mattila, Laura Jaakola, Anna Maria Pirttilä, and Ari Tolonen. 2009. "Identification of Phenolic Compounds from Lingonberry (*Vaccinium vitis-idaea* L.), Bilberry (*Vaccinium myrtillus* L.) and Hybrid Bilberry (*Vaccinium X intermedium ruthe* L.) Leaves." *Journal of Agricultural and Food Chemistry* 57,(20): 9437-47.

[44] Güder, A, M Gür, and MS Engin. 2018. "Antidiabetic and Antioxidant Properties of Bilberry *(Vaccinium myrtillus* Linn.) Fruit and Their Chemical Composition." *Journal of Agricultural Science and Technology*.

[45] Edirisinghe, Indika, and Britt Burton-Freeman. 2016. "Anti-Diabetic Actions of Berry Polyphenols–Review on Proposed Mechanisms of Action." *Journal of Berry Research* 6,(2): 237-50.

[46] Asgary, Sedigheh, Mahmood RafieianKopaei, Amirhossein Sahebkar, Fatemeh Shamsi, and Najmeh Goli-malekabadi. 2016. "Anti-Hyperglycemic and Anti-Hyperlipidemic Effects of *Vaccinium myrtillus* Fruit in Experimentally Induced Diabetes (Antidiabetic Effect of *Vaccinium myrtillus* Fruit)." *Journal of the Science of Food and Agriculture* 96,(3): 764-68.

[47] Sidorova, Yulia, Vladimir Shipelin, Vladimir Mazo, Sergey Zorin, Nikita Petrov, and Alla Kochetkova. 2017. "Hypoglycemic and Hypolipidemic Effect of *Vaccinium myrtillus* L. Leaf and *Phaseolus vulgaris* L. Seed Coat Extracts in Diabetic Rats." *Nutrition* 41: 107-12.

[48] Ulbricht, Catherine, Ethan Basch, Samuel Basch, Steve Bent, Heather Boon, Dilys Burke, Dawn Costa, Carla Falkson, Nicole Giese, and Michael Goble. 2009. "An Evidence-Based Systematic Review of Bilberry (*Vaccinium myrtillus*) by the Natural Standard Research Collaboration." *Journal of Dietary Supplements* 6,(2): 162-200.

[49] Madić, Višnja, Jelena Jovanović, Aleksandar Stojilković, Marina Jušković, and Perica Vasiljević. 2018. "Evaluation of Cytotoxicity of 'Anti-Diabetic'herbal Preparation and Five Medicinal Plants: An Allium Cepa Assay." *Biologica Nyssana* 8,(2): 151-58.

[50] Helmstädter, A, and N Schuster. 2010. "*Vaccinium myrtillus* as an Antidiabetic Medicinal Plant–Research through the Ages." *Die Pharmazie-An International Journal of Pharmaceutical Sciences* 65,(5): 315-21.

[51] Preethi, KC, Girija Kuttan, and Ramadasan Kuttan. 2006. "Antioxidant Potential of an Extract of *Calendula officinalis*.

Flowers in Vitro. And in Vivo." *Pharmaceutical Biology* 44,(9): 691-97.

[52] Singh, Mukesh Kr, Pankaj Sahu, K Nagori, D Dewangan, T Kumar, A Alexander, H Badwaik, and DK Tripathi. 2011. "Organoleptic Properties in-Vitro and in-Vivo Pharmacological Activities of *Calendula officinalis* Linn: An over Review." *Journal of Chemical and Pharmaceutical Research* 3,(4): 655-63.

[53] Olennikov, Daniil N, and Nina I Kashchenko. 2014. "Componential Profile and Amylase Inhibiting Activity of Phenolic Compounds from *Calendula officinalis* L. Leaves." *The Scientific World Journal* 2014.

[54] Olennikov, Daniil N, Nina I Kashchenko, and Cecile Vennos. 2017. "A New Esculetin Glycoside from *Calendula officinalis* (Asteraceae) and Its Bioactivity." *Farmacia* 65,(5): 698-702.

[55] Buzzi, Marcelo, Franciele de Freitas, and Marcos Winter. 2016. "A Prospective, Descriptive Study to Assess the Clinical Benefits of Using *Calendula officinalis* Hydroglycolic Extract for the Topical Treatment of Diabetic Foot Ulcers." *Ostomy Wound Manage* 62,(3): 8-24.

[56] Kusmiati, Wiobet Caesarianto, Fifi Afiati, and Rahmi Hutabarat. 2019. "Effect Lutein of Marigold Flower (*Tagetes erecta* L.) on Decreasing Glucose and Malondialdehyde Levels in Alloxan-Induced Blood Mice." Paper presented at the AIP Conference Proceedings, 03 july 2019.

[57] Salehi, Iraj, Shirin Moradkhani, Amaneh Mohammadi Roushandeh, Hamideh Nazari, and Alireza Pouyandeh Ravan. 2015. "Controversial Effects of *Calendula officinalis* L. On Biochemical and Pathological Factors of Nephropathy in Diabetic Wistar Rats." *Avicenna Journal of Medical Biochemistry* 3,(2).

[58] Begum, Sabira, Syed Imran Hassan, Syed Nawazish Ali, and Bina S Siddiqui. 2004. "Chemical Constituents from the Leaves of *Psidium guajava*." *Natural Product Research* 18,(2): 135-40.

[59] Fatima, Amreen, Parul Agrawal, and Prem Prakash Singh. 2012. "Herbal Option for Diabetes: An Overview." *Asian Pacific Journal of Tropical Disease* 2: S536-S44.

[60] Goel, Richa, Daksh Bhatia, Sadaf Jamal Gilani, and Deepti Katiyar. 2012. "Medicinal Plants as Anti-Diabetics: A Review." *International Bulletin of Drug Research* 1,(2): 100-07.

[61] Chang, Cicero LT, Yenshou Lin, Arlene P Bartolome, Yi-Ching Chen, Shao-Chih Chiu, and Wen-Chin Yang. 2013. "Herbal Therapies for Type 2 Diabetes Mellitus: Chemistry, Biology, and Potential Application of Selected Plants and Compounds." *Evidence-Based Complementary and Alternative Medicine* 2013.

[62] Huang, Chin-Shiu, Mei-Chin Yin, and Lan-Chi Chiu. 2011. "Antihyperglycemic and Antioxidative Potential of *Psidium guajava* Fruit in Streptozotocin-Induced Diabetic Rats." *Food and Chemical Toxicology* 49,(9): 2189-95.

[63] Oh, Won Keun, Chul Ho Lee, Myung Sun Lee, Eun Young Bae, Cheon Bae Sohn, Hyuncheol Oh, Bo Yeon Kim, and Jong Seog Ahn. 2005. "Antidiabetic Effects of Extracts from *Psidium guajava*." *Journal of Ethnopharmacology* 96,(3): 411-15.

[64] Upton, R, A Graff, E Williamson, D Bunting, DM Gatherum, EB Walker, V Butterweck, U Lieflünder, A Nahrstedt, and H Winterhoff. 1997. "St. John's Wort Monograph In: American Herbal Pharmacopoeia and Therapeutic Compendium." *HerbalGram* 32,(40).

[65] Müller, WE, A Singer, and M Wonnemann. 2001. "Hyperforin-Antidepressant Activity by a Novel Mechanism of Action." *Pharmacopsychiatry* 34,(Sup. 1): 98-102.

[66] Ghosal, Shibnath. 2001. "St. John's Wort Composition U.S. Patent No. 6,224,906."Patent and Trademark Office. Washington, DC.

[67] Husain, Gulam Mohammed, Paras Nath Singh, and Vikas Kumar. 2009. "Beneficial Effects of a Standardized Hypericum Perforatum Extract in Rats with Experimentally Induced Hyperglycemia." *Drug Discoveries and Therapeutics* 3,(5).

[68] Can, Özgür Devrim, Yusuf Öztürk, Nilgün Öztürk, Gianni Sagratini, Massimo Ricciutelli, Sauro Vittori, and Filippo Maggi. 2011. "Effects of Treatment with St. John's Wort on Blood Glucose Levels and Pain Perceptions of Streptozotocin-Diabetic Rats." *Fitoterapia* 82,(4): 576-84.

[69] Patel, Seema 2014. "*Hibiscus sabdariffa*: An Ideal yet under-Exploited Candidate for Nutraceutical Applications." *Biomedicine and Preventive Nutrition* 4,(1): 23-27.

[70] Jung, EunKyung, YoungJun Kim, and Nami Joo. 2013. "Physicochemical Properties and Antimicrobial Activity of Roselle (*Hibiscus sabdariffa* L.)." *Journal of the Science of Food and Agriculture* 93,(15): 3769-76.

[71] Ademiluyi, Adedayo O, and Ganiyu Oboh. 2013. "Aqueous Extracts of Roselle (*Hibiscus sabdariffa* Linn.) Varieties Inhibit A-Amylase and A-Glucosidase Activities in Vitro." *Journal of Medicinal Food* 16,(1): 88-93.

[72] Wisetmuen, Eamruthai, Patchareewan Pannangpetch, Bunkerd Kongyingyoes, Upa Kukongviriyapan, Wiboonchai Yutanawiboonchai, and Arunporn Itharat. 2013. "Insulin Secretion Enhancing Activity of Roselle Calyx Extract in Normal and Streptozotocin-Induced Diabetic Rats." *Pharmacognosy Research* 5,(2): 65.

[73] Peng, Chiung-Huei, Charng-Cherng Chyau, Kuei-Chuan Chan, Tsung-Hsien Chan, Chau-Jong Wang, and Chien-Ning Huang. 2011. "*Hibiscus sabdariffa* Polyphenolic Extract Inhibits Hyperglycemia, Hyperlipidemia, and Glycation-Oxidative Stress While Improving Insulin Resistance." *Journal of Agricultural and Food Chemistry* 59,(18): 9901-09.

[74] Shukla, Yogeshwer, and Madhulika Singh. 2007. "Cancer Preventive Properties of Ginger: A Brief Review." *Food and Chemical Toxicology* 45,(5): 683-90.

[75] Afzal, M, D Al-Hadidi, M Menon, J Pesek, and MSI Dhami. 2001. "Ginger: An Ethnomedical, Chemical and Pharmacological Review." *Drug Metabolism and Drug Interactions* 18,(3-4): 159-90.

[76] Panahi, Yunes, Alireza Saadat, Amirhossein Sahebkar, Farshad Hashemian, Mojgan Taghikhani, and Ehsan Abolhasani. 2012. "Effect of Ginger on Acute and Delayed Chemotherapy-Induced Nausea and Vomiting: A Pilot, Randomized, Open-Label Clinical Trial." *Integrative cancer Therapies* 11,(3): 204-11.

[77] Iranloye, BO, AP Arikawe, G Rotimi, and AO Sogbade. 2011. "Anti-Diabetic and Anti-Oxidant Effects of *Zingiber officinale* on Alloxan-Induced and Insulin-Resistant Diabetic Male Rats." *Nigerian Journal of Physiological Sciences* 26,(1).

[78] Al-Amin, Zainab M, Martha Thomson, Khaled K Al-Qattan, Riitta Peltonen-Shalaby, and Muslim Ali. 2006. "Anti-Diabetic and Hypolipidaemic Properties of Ginger (*Zingiber officinale*) in Streptozotocin-Induced Diabetic Rats." *British Journal of Nutrition* 96,(4): 660-66.

[79] Andallu, B, B Radhika, and V Suryakantham. 2003. "Effect of Aswagandha, Ginger and Mulberry on Hyperglycemia and Hyperlipidemia." *Plant Foods for Human Nutrition* 58,(3): 1-7.

[80] Shimamura, Mariko, Tadahiko Hazato, Hiromi Ashino, Yukio Yamamoto, Eri Iwasaki, Hiroyasu Tobe, Kei Yamamoto, and Shozo Yamamoto. 2001. "Inhibition of Angiogenesis by Humulone, a Bitter Acid from Beer Hop." *Biochemical and Biophysical Research Communications* 289,(1): 220-24.

[81] Simpson, WJ, and ARW Smith. 1992. "Factors Affecting Antibacterial Activity of Hop Compounds and Their Derivatives." *Journal of Applied Bacteriology* 72,(4): 327-34.

[82] Yajima, Hiroaki, Emiko Ikeshima, Maho Shiraki, Tomoka Kanaya, Daisuke Fujiwara, Hideharu Odai, Nobuyo Tsuboyama-Kasaoka, Osamu Ezaki, Shinichi Oikawa, and Keiji Kondo. 2004. "Isohumulones, Bitter Acids Derived from Hops, Activate Both Peroxisome Proliferator-Activated Receptor Α and Γ and Reduce Insulin Resistance." *Journal of Biological Chemistry* 279,(32): 33456-62.

[83] Bland, Jeffrey S, Deanna Minich, Robert Lerman, Gary Darland, Joseph Lamb, Matthew Tripp, and Neile Grayson. 2015.

"Isohumulones from Hops (*Humulus lupulus*) and Their Potential Role in Medical Nutrition Therapy." *PharmaNutrition* 3,(2): 46-52.

[84] Mimica-Dukic, Neda, Biljana Bozin, Marina Sokovic, and Natasa Simin. 2004. "Antimicrobial and Antioxidant Activities of *Melissa officinalis* L. (Lamiaceae) Essential Oil." *Journal of Agricultural and Food Chemistry* 52,(9): 2485-89.

[85] Mencherini, Teresa, Patrizia Picerno, Carla Scesa, and Rita Aquino. 2007. "Triterpene, Antioxidant, and Antimicrobial Compounds from *Melissa officinalis*." *Journal of natural products* 70,(12): 1889-94.

[86] Yen, Hsiu-Fang, Chi-Ting Hsieh, Tusty-Jiuan Hsieh, Fang-Rong Chang, and Chin-Kun Wang. 2015. "In Vitro Anti-Diabetic Effect and Chemical Component Analysis of 29 Essential Oils Products." *Journal of Food and Drug Analysis* 23,(1): 124-29.

[87] Hasanein, Parisa, and Hassan Riahi. 2015. "Antinociceptive and Antihyperglycemic Effects of *Melissa officinalis* Essential Oil in an Experimental Model of Diabetes." *Medical Principles and Practice* 24,(1): 47-52.

[88] Weidner, Christopher, Sylvia J Wowro, Anja Freiwald, Vitam Kodelja, Heba Abdel-Aziz, Olaf Kelber, and Sascha Sauer. 2014. "Lemon Balm Extract Causes Potent Antihyperglycemic and Antihyperlipidemic Effects in Insulin-Resistant Obese Mice." *Molecular Nutrition and Food Research* 58,(4): 903-07.

[89] Alexa, Ersilia, Corina Danciu, Isidora Radulov, Diana Obistioiu, Renata Maria Sumalan, Adriana Morar, and Cristina Adriana Dehelean. 2018. "Phytochemical Screening and Biological Activity of *Mentha× piperita* L. And *Lavandula angustifolia* Mill. Extracts." *Analytical Cellular Pathology* 2018: 1-7.

[90] Benabdallah, A, M Boumendjel, O Aissi, C Rahmoune, M Boussaid, and C Messaoud. 2018. "Chemical Composition, Antioxidant Activity and Acetylcholinesterase Inhibitory of Wild Mentha Species from Northeastern Algeria." *South African Journal of Botany* 116: 131-39.

[91] Ramkissoon, JS, MF Mahomoodally, N Ahmed, and AH Subratty. 2013. "Antioxidant and Anti–Glycation Activities Correlates with

Phenolic Composition of Tropical Medicinal Herbs." *Asian Pacific Journal of Tropical Medicine* 6,(7): 561-69.

[92] Narendhirakannan, RT, S Subramanian, and M Kandaswamy. 2006. "Biochemical Evaluation of Antidiabetogenic Properties of Some Commonly Used Indian Plants on Streptozotocin-Induced Diabetes in Experimental Rats." *Clinical and Experimental Pharmacology and Physiology* 33,(12): 1150-57.

[93] Barbalho, Sandra M, Débora C Damasceno, Ana Paula Machado Spada, Vanessa Sellis da Silva, Karla Aparecida Martuchi, Marie Oshiiwa, Flávia Machado, MV Farinazzi, and Claudemir Gregório Mendes. 2011. "Metabolic Profile of Offspring from Diabetic Wistar Rats Treated with *Mentha piperita* (Peppermint)." *Evidence-Based Complementary and Alternative Medicine* 2011.

[94] Teixeira, Barbara, António Marques, Cristina Ramos, Carmo Serrano, Olívia Matos, Nuno R Neng, José MF Nogueira, Jorge Alexandre Saraiva, and Maria Leonor Nunes. 2013. "Chemical Composition and Bioactivity of Different Oregano (*Origanum vulgare*) Extracts and Essential Oil." *Journal of the Science of Food and Agriculture* 93,(11): 2707-14.

[95] Kwon, Young-In I, Dhiraj A Vattem, and Kalidas Shetty. 2006. "Evaluation of Clonal Herbs of Lamiaceae Species for Management of Diabetes and Hypertension." *Asia Pacific Journal of Clinical Nutrition* 15,(1): 107-18.

[96] McCue, Patrick, Dhiraj Vattem, and Kalidas Shetty. 2004. "Inhibitory Effect of Clonal Oregano Extracts against Porcine Pancreatic Amylase in Vitro." *Asia Pacific Journal of Clinical Nutrition* 13,(4): 401-08.

[97] Lemhadri, A, N-A Zeggwagh, M Maghrani, H Jouad, and M Eddouks. 2004. "Anti-Hyperglycaemic Activity of the Aqueous Extract of *Origanum vulgare* Growing Wild in Tafilalet Region." *Journal of Ethnopharmacology* 92,(2-3): 251-56.

[98] Nadia, Zeghad, and Merghem Rachid. 2013. "Antioxidant and Antibacterial Activities of *Thymus vulgaris* L." *Medicinal and Aromatic Plant Research* 1,(1): 5-11.

[99] Choi, Eun-Kyeong, Ki-Suk Kim, Hea Jung Yang, Min-Hee Shin, Hyo-Weon Suh, Ki-Beom Lee, Kwang Seok Ahn, Jae Young Um, Seok-Geun Lee, and Byung-Cheol Lee. 2012. "Hexane Fraction of *Citrus aurantium* L. Stimulates Glucagon-Like Peptide-1 (Glp-1) Secretion Via Membrane Depolarization in Nci-H716 Cells." *BioChip Journal* 6,(1): 41-47.

[100] Akiyama, Satoko, Shin-ichi Katsumata, Kazuharu Suzuki, Yumi Nakaya, Yoshiko Ishimi, and Mariko Uehara. 2009. "Hypoglycemic and Hypolipidemic Effects of Hesperidin and Cyclodextrin-Clathrated Hesperetin in Goto-Kakizaki Rats with Type 2 Diabetes." *Bioscience, Biotechnology and Biochemistry* 73,(12): 2779-82.

[101] Jia, Sheng, Ying Hu, Wenna Zhang, Xiaoyong Zhao, Yanhong Chen, Chongde Sun, Xian Li, and Kunsong Chen. 2015. "Hypoglycemic and Hypolipidemic Effects of Neohesperidin Derived from *Citrus aurantium* L. In Diabetic Kk-a Y Mice." *Food and Function* 6,(3): 878-86.

[102] Asgarpanah, Jinous, and Razieh Mohajerani. 2012. "Phytochemistry and Pharmacologic Properties of *Urtica dioica* L." *Journal of Medicinal Plants Research* 6,(46): 5714-19.

[103] Al-Tameme, Huda Jasim, Mohammed Yahya Hadi, and Imad Hadi Hameed. 2015. "Phytochemical Analysis of *Urtica dioica* Leaves by Fourier-Transform Infrared Spectroscopy and Gas Chromatography-Mass Spectrometry." *Journal of Pharmacognosy and Phytotherapy Research* 7,(10): 238-52.

[104] Farzami, Bijan, D Ahmadvand, S Vardasbi, FJ Majin, and SH Khaghani. 2003. "Induction of Insulin Secretion by a Component of *Urtica dioica* Leave Extract in Perifused Islets of Langerhans and Its in Vivo Effects in Normal and Streptozotocin Diabetic Rats." *Journal of Ethnopharmacology* 89,(1): 47-53.

[105] Bnouham, Mohamed, Wafaa Benalla, Said Bellahcen, Zineb Hakkou, Abderrahim Ziyyat, Hassane Mekhfi, Mohammed Aziz, and Abdelkhaleq Legssyer. 2012. "Antidiabetic and Antihypertensive Effect of a Polyphenol-Rich Fraction of *Thymelaea hirsuta* L. In a Model of Neonatal Streptozotocin-

Diabetic and Ng-Nitro-L-Arginine Methyl Ester-Hypertensive Rats." *Journal of Diabetes* 4,(3): 307-13.

[106] Nickavar, Bahman, and Nasibeh Yousefian. 2011. "Evaluation of A-Amylase Inhibitory Activities of Selected Antidiabetic Medicinal Plants." *Journal Für Verbraucherschutz und Lebensmittelsicherheit* 6,(2): 191-95.

[107] Bnouham, Mohamed, Fatima Zahra Merhfour, Abderrahim Ziyyat, Mohamed Aziz, Abdelkhaleq Legssyer, and Hassane Mekhfi. 2010. "Antidiabetic Effect of Some Medicinal Plants of Oriental Morocco in Neonatal Non-Insulin-Dependent Diabetes Mellitus Rats." *Human and Experimental Toxicology* 29,(10): 865-71.

[108] Ahangarpour, Akram, Maryam Mohammadian, and Mahin Dianat. 2012. "Antidiabetic Effect of Hydroalcholic *Urtica dioica* Leaf Extract in Male Rats with Fructose-Induced Insulin Resistance." *Iranian Journal of Medical Sciences* 37,(3): 181.

[109] Golalipour, Mohammad Jafar, and Vahid Khori. 2007. "The Protective Activity of *Urtica dioica* Leaves on Blood Glucose Concentration and Beta-Cells in Streptozotocin-Diabetic Rats." *Pakistan Journal of Biology Sciences* 10,(8): 1200-4.

[110] Golalipour, Mohammad Jafar, Soraya Ghafari, and Mohammad Afshar. 2010. "Protective Role of *Urtica dioica* L.(Urticaceae) Extract on Hepatocytes Morphometric Changes in Stz Diabetic Wistar Rats." *Turkish Journal of Gastroenterology* 21,(3): 262-69.

[111] Namjou, Abdolrasoul, Esfandiar Heidarian, and Mahmoud Rafieian-Kopaei. 2018. "Effects of *Urtica dioica* Hydro-Alcoholic Extract on Blood Serum Glucose and Lipid Profiles of Female Wistar Rats with Long-Term Estrogen Deficiency." Paper presented at the *Veterinary Research Forum*, 15 december 2018.

[112] Vladimir-Knežević, Sanda, Biljana Blažeković, Marija Kindl, Jelena Vladić, Agnieszka Lower-Nedza, and Adelheid Brantner. 2014. "Acetylcholinesterase Inhibitory, Antioxidant and Phytochemical Properties of Selected Medicinal Plants of the Lamiaceae Family." *Molecules* 19,(1): 767-82.

[113] Aljarah, Ali Kadhim, and Imad Hadi Hameed. 2018. "In Vitro Anti-Diabetic Properties of Methanolic Extract of *Thymus vulgaris* Using A-Glucosidase and A-Amylase Inhibition Assay and Determination of Its Bioactive Chemical Compounds." *Indian Journal of Public Health Research and Development* 9,(3): 388-92.

[114] Koohi-Hosseinabadi, Omid, Maryam Moini, Alireza Safarpoor, Amin Derakhshanfar, and Masood Sepehrimanesh. 2015. "Effects of Dietary *Thymus vulgaris* Extract Alone or with Atorvastatin on the Liver, Kidney, Heart, and Brain Histopathological Features in Diabetic and Hyperlipidemic Male Rats." *Comparative Clinical Pathology* 24,(6): 1311-15.

[115] Ekoh, Solomon Nwanze, Emmanuel Iroha Akubugwo, Victor Chibueze Ude, and Nzubechukwu Edwin. 2014. "Anti-Hyperglycemic and Anti-Hyperlipidemic Effect of Spices (*Thymus vulgaris, Murraya Koenigii, Ocimum Gratissimum* and *Piper Guineense*) in Alloxan-Induced Diabetic Rats." *International Journal of Bioscience* 4,(2): 179-87.

[116] Akan, Zafer, Melek Dikilidal, Hulya Ozdemir, Gokhan Oto, and Ahmet Yilmaz. 2014. "Effects of *Thymus vulgaris* L. And *Thymbra spicata* L. On Diabetes Mellitus Associated Cognitive Impairment and Neuropathy: *Thymus vulgaris* and Cognitive Function Improvements." *Medical Science and Discovery* 1,(1): 16-21.

[117] Kontogianni, Vassiliki G, Goran Tomic, Ivana Nikolic, Alexandra A Nerantzaki, Nisar Sayyad, Stanislava Stosic-Grujicic, Ivana Stojanovic, Ioannis P Gerothanassis, and Andreas G Tzakos. 2013. "Phytochemical Profile of *Rosmarinus officinalis* and *Salvia officinalis* Extracts and Correlation to Their Antioxidant and Anti-Proliferative Activity." *Food Chemistry* 136,(1): 120-29.

[118] Atti-Santos, Ana Cristina, Marcelo Rossato, Gabriel Fernandes Pauletti, Luciana Duarte Rota, Juarez Ciro Rech, Marcia Regina Pansera, Fabiana Agostini, Luciana Atti Serafini, and Patrick Moyna. 2005. "Physico-Chemical Evaluation of *Rosmarinus officinalis* L. Essential Oils." *Brazilian Archives of Biology and Technology* 48,(6): 1035-39.

[119] Mahomoodally, F, Z Aumeeruddy-Elalfi, Katharigatta N Venugopala, and M Hosenally. 2018. "Antiglycation, Comparative Antioxidant Potential, Phenolic Content and Yield Variation of Essential Oils from 19 Exotic and Endemic Medicinal Plants." *Saudi Journal of Biological Sciences*.

[120] Koga, K, H Shibata, K Yoshino, and K Nomoto. 2006. "Effects of 50% Ethanol Extract from Rosemary (*Rosmarinus officinalis*) on A-Glucosidase Inhibitory Activity and the Elevation of Plasma Glucose Level in Rats, and Its Active Compound." *Journal of Food Science* 71,(7): S507-S12.

[121] Bakırel, Tülay, Utku Bakırel, Oya Üstüner Keleş, Sinem Güneş Ülgen, and Hasret Yardibi. 2008. "In Vivo Assessment of Antidiabetic and Antioxidant Activities of Rosemary (*Rosmarinus officinalis*) in Alloxan-Diabetic Rabbits." *Journal of ethnopharmacology* 116,(1): 64-73.

[122] Belmouhoub, M, I Chebout, and M Iguer-ouada. 2018. "Antidiabetic and Anti-Hypercholesterolemic Effects of Flavonoid-Rich Fractions of *Rosmarinus officinalis* in Streptozotocin-Induced Diabetes in Mice." *Phytothérapie*: 1-7.

[123] Ramadan, Kholoud S, Olfat A Khalil, Enas N Danial, Hanan S Alnahdi, and Najla O Ayaz. 2013. "Hypoglycemic and Hepatoprotective Activity of *Rosmarinus officinalis* Extract in Diabetic Rats." *Journal of Physiology and Biochemistry* 69,(4): 779-83.

[124] Selmi, Slimen, Kais Rtibi, Dhekra Grami, Hichem Sebai, and Lamjed Marzouki. 2017. "Rosemary (*Rosmarinus officinalis*) Essential Oil Components Exhibit Anti-Hyperglycemic, Anti-Hyperlipidemic and Antioxidant Effects in Experimental Diabetes." *Pathophysiology* 24,(4): 297-303.

In: Phytochemicals from Mexican … ISBN: 978-1-53617-397-0
Editors: J. Basilio Heredia et al.

Chapter 6

MEXICAN MEDICINAL PLANTS WITH ANTI-CANCER PROPERTIES

Erick Paul Gutiérrez-Grijalva[1,*], Marilyn Shomara Criollo-Mendoza[2], Luis Angel Cabanillas-Bojórquez[2] and J. Basilio Heredia[2]

[1]Cátedras CONACYT-Centro de Investigación en Alimentación y Desarrollo, A.C., Culiacán, México
[2]Centro de Investigación en Alimentación y Desarrollo, A.C., Culiacán, México

ABSTRACT

Cancer is a major public health problem worldwide, and the different types of cancer represent the second leading cause of death in the world. The incidence of cancer is increasing rapidly due to a myriad of factors such as genetic predisposition, unhealthy lifestyles, bad nutrition patterns, pollution, etc. The conventional treatment of cancer is by drugs such as chemotherapeutic agents that affect cancer cells through several

* Corresponding Author's Email: erick.gutierrez@ciad.mx.

mechanisms. These treatments are costly and can not be easily accessed by low- and middle- income populations. Medicinal plants have been traditionally used to treat different types of cancer, and contemporary science has focused on the systematic study of plants, identifying their active agents and engaging *in vitro* and *in vivo* strategies for their potential use as anti-cancer agents. *In vitro* studies have allowed the elucidation of the mode of action of some plants; however, *in vivo* studies are very limited. This chapter will focus only on the compounds that have already been identified in plants used against cancer and the available *in vivo* studies.

Keywords: cancer, phytochemicals, natural products, anticancer, medicinal plants

1. Introduction

Cancer is a major public health problem worldwide because it is the second leading cause of morbidity and mortality among not transmissible diseases. The incidence and mortality of the different types of cancer are increasing rapidly as a reflection of aging as well as population growth, in addition to changes in the distribution of the main risk factors as well as changes in the prevalence of the disease [1]. Cancer is considered a group of diseases with similar characteristics that include: preventing growth-inhibitory signals, evading apoptotic cell death, unlimited cell proliferation, capacitive formation of new blood vessels and metastases to other body tissues, as well as genomic instability and tumor inflammation of the promoter; this may originate from the transformation of a single cell, which may be due to inherited genetic factors or exposition to external carcinogenic agents. Besides, it is important to mention that there are differences in the molecular mechanisms involved in the carcinogenesis within each cell type as well as in the pattern of their propagation; as a result, there are currently more than 100 different types of cancer. Also, the stroma associated with the tumor (normal cells) plays a role important in cancer since they contribute to the development and expression of certain distinctive capacities of the tumor [2, 3].

Among the risk factors associated with the development of cancer are: biological infections such as hepatitis B, HIV, and human papillomavirus, exposure to synthetic chemicals, excessive exposure to sunlight and UV radiation, as well as overweight, tobacco use, alcohol and high fat consumption, these factors are associated with the development of 70-95% incidence of cancers. On the other hand, aging and family history represent unavoidable risk factors, examples of the latter are the genetic mutations BRCA1 and BRCA2 in specific types of breast cancer and HPC1 and TLR in prostate cancer while with the increase in age, the body begins to reduce stress resistance and the ability to repair cellular and DNA damage [4-6].

Early diagnosis of cancer provides a greater opportunity for patient care, as well as research on the cause, prevention, diagnosis, and treatment thereof. According to Salto-Tellez and Cree [7], the WHO Blue Books' have been the reference for the classification of cancer worldwide, covering all the sites of organs in 10 volumes, in which the characteristics of each type of cancer are explained, as well as the diagnostic criteria, the pathological characteristics and the associated molecular alterations to provide international standards for cancer diagnosis and research [8]. One of the important general classifications used in the diagnosis of this disease is related to the type of tissue in which it originated (histological type), this classification is composed of 6 groups of categories, among which are: carcinoma, which It is a type of cancer of epithelial origin and accounts for 80 to 90% of all cases of cancer diagnosed; sarcoma, which originates in supportive and connective tissues (bones, tendons, and cartilage); myeloma, originated in the bone marrow plasma cells; leukemia, also known as blood cancers, is native to the bone marrow, and are associated with an overproduction of immature white blood cells; lymphoma appears in lymph nodes or glands and organs such as the spleen, tonsils, and thymus. On the other hand, within the most common types of cancer, we can find colon and rectal cancer, endometrial, liver, leukemia, breast, prostate, and kidney cancer, which are the cause of high mortality rates around the world [9].

Globally, in 2018, lung and breast cancer were the highest incidence, followed by colorectal and prostate cancer (data shown in Figure 1); about 18 078 957 cases were diagnosed, of which the vast majority (9 456 418) were diagnosed in males, being in these the most prevalent types of cancer: lung, prostate, colorectal and stomach, while in women the type of cancer with the highest incidences was breast, lung, colorectal and cervical cancer. These data coincide with the occurrence of cases reported in Mexico, however, as the first place in the group of cancers causing a higher number of deaths, the colorectal type is found, and breast and prostate cancer are also included in the list, as well as liver cancer [10].

To combat this disease, several types of treatments have been designed, such as surgery, radiotherapy, chemotherapy, immunotherapy, hormonal therapy, and stem cell transplantation; which are used according to the type and grade of cancer that the patient presents, and can be used as a joint for greater effectiveness [11]. Currently, the new cancer treatments are focused on specific molecular alterations of each tumor type, resulting in precise, effective toxic therapies and with minor side effects in patients [12].

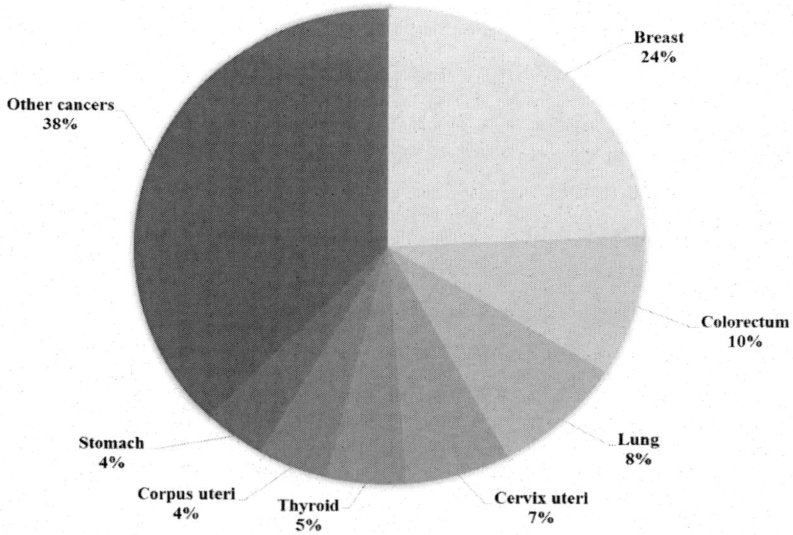

Figure 1. Cancer incidence worldwide during 2018. Adapted from data obtained from WHO [10].

In this sense, the therapeutic targeting of the main hallmarks of cancer are the following: EGFR inhibitors for the inhibition of sustaining proliferative signaling, cyclin-dependent kinase inhibitors for the evading growth suppressors, immune-activating anti-CTLA4 mAb for de avoiding immune destruction, telomerase inhibitors for the enabling replicative immortality, selective anti-inflammatory drugs for tumor-promoting inflammation, inhibitors of HGF/c-Met for the activating invasion and metastasis. Inhibitors of VEGF signaling for the induction of angiogenesis, PARP inhibitors for genome instability and mutation, proapoptotic BH3 mimetics for the resisting cell death, and aerobic glycolysis inhibitors for the deregulating cellular energetics [3].

Medicinal plants have been a source of molecules with therapeutic potential. For the development of drugs derived from these natural products, it is necessary to consider the availability, the profile of bioactive compounds, and the difficulty in collecting wild samples [13]. The use of these for the treatment of cancer has a long history, in this context, they are found within the clinically proven medicines for the treatment of cancer, to the vinca alkaloids of Madagascar: vinblastine and vincristine, isolated from *Catharanthus roseus*, as well as two active agents derived from semisynthetic of etoposide and teniposide of the natural product epipodophyllotoxin (extracted from the root of *Podophyllum peltatum*). Another important anti-cancer contribution derived from plants is paclitaxel (Taxol®) obtained from the leaves of several *Taxus* species. While within the most recent discoveries, we find three clinically active agents: topotecan, irinotecan (CPT-11), and belotecan, obtained semisynthetically from camptothecin, isolated from the Chinese ornamental tree *Camptotheca acuminata* [14].

On the basis of the therapeutic objectives mentioned above, research is continuing to elucidate new compounds with potential treatment or as adjuvants in the treatment of this disease, mainly from secondary metabolites such as: terpenes, carotenoids, flavonoids and alkaloids, which can be found in some plants listed in the Mexican herbal pharmacopoeia, whose effects *in vivo* in various cancers are described below.

2. Research Methodology

Relevant information regarding medicinal plants and their anti-cancer potential was searched based on recent scientific literature (2010-2019) from the Web of Science and Scopus databases. The keywords used for literature search included the terms: medicinal plants, cancer, *Allium sativum*, *Cynara scolymus*, *Aloe vera*, *Vaccinium myrtillus*, *Cinnamomum verum*, *Elettaria cardamomum*, *Aesculus hippocastanum*, *Echinacea pallida*, *Hypericum perforatum*, *Zingiber officinale*, *Humulus lupulus*, *Citrus aurantium*, *Origanum vulgare*, *Rosmarinus officinalis*, *Thymus vulgaris*, polyphenols, terpenes, alkaloids. For this chapter, we only searched for information regarding *in vivo* studies of plants listed in the Herbal Mexican Pharmacopoeia [15].

3. Phytochemical Constituents in Medicinal Plants

Cancer is a major worldwide health problem due to its high incidence and mortality, cancer is the second cause of death in the world, so it has been studied extensively. This disease is linked to various factors, such as dietary, genetic factors, and the environment. In recent years, it has been reported that a based diet on fruits and vegetables could reduce cancer risk. Therefore, studies have focused on its phytochemical characterization [16, 17]. In this sense, natural compounds are potential anti-cancer sources due to their phytochemical composition. A large number of plants have been studied to prove the anti-cancer effect on cell cancer; therefore, it has been identified that plants with different compounds as phenolic compounds, terpenoids, alkaloids inhibit cancer proliferation and invasion (Figure 2) [18, 16]. Because multiple reports indicate that secondary plant metabolites have anticancer effects, and the increase in drug resistance of some cancer cells, the study of new plants with anticancer properties is currently being carried out [19].

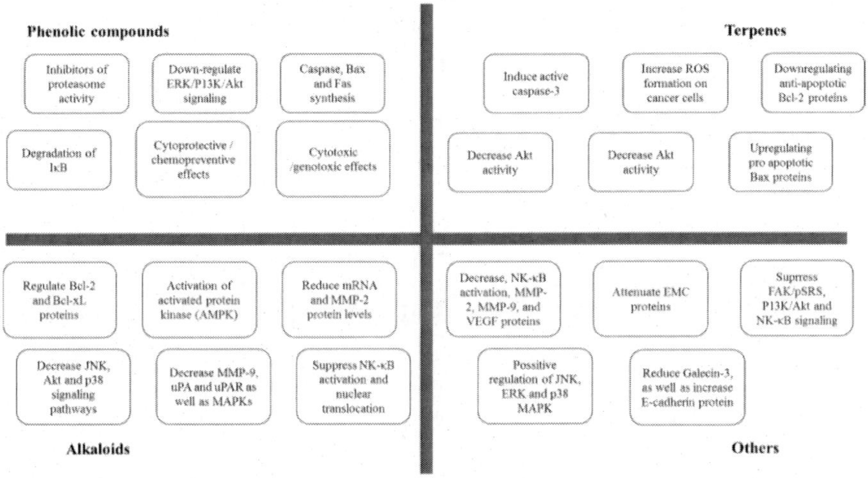

Figure 2. Summarization of the anti-cancer mechanism of action of phenolic compounds, terpenes, alkaloids, and other compounds.

3.1. Phenolic Compounds

Phenolic compounds are secondary metabolites found in plants, and their basic structure is a benzene ring with an -OH group attached, to which phenolic's antioxidant capacity is mainly attributed and linked to potential health benefits [20]. Polyphenols as flavonoids and their complexes have shown potential antitumor and anticancer activity by inhibiting the proteasome due to the degradation of IκB, which negatively regulates the transcription of angiogenesis, proliferation, migration, and inhibition of apoptosis [20-22].

It has been reported that polyphenols could have antitumor effects by two pathways, cytoprotective/chemopreventive due to their antioxidant character or cytotoxic/genotoxic effects by pro-oxidative properties of these compounds which could produce toxicity [20, 23]. Likewise, phenolic compounds could inhibit cancer proliferation due to its possible interaction with the cell cycle regulators and transcription factors, DNA, and other protein as well as affect the intracellular reactions [20, 24-26]. In this sense, some phenolic compounds may induce apoptosis to cancer cells directly or indirectly affecting the stability of proteins as well as inhibiting proteasome activity, nevertheless, the mechanism is not fully understood, but, this could be due to the electrophilic aromatic ketone group that interacts with cancer cells [20, 27, 28].

On the other hand, flavonoids are an important phytochemical group that has high antioxidant activity, which is linked to preventing different diseases like cancer, likewise, it has been studied the anti-cancer effects of flavonoids both *in vitro* and *in vivo*, in this sense, multiple works have been published, it is essential to highlight that flavonoids are a large group, therefore not all compounds have been studied as well as the different types of cancer. According to Rodriguez-Garcia, Sanchez-Quesada, and Gaforio [17], some flavonoids could reduce the risk of suffering some cancers. Some compounds that have been studied are wagonin, hesperetin, nobiletin, and kaempferol. Although it has been mentioned that flavonoids could reduce the risk of cancer, there is still controversy; therefore, studies are still being carried out to elucidate it. In this sense, these flavonoids could regulate the protein synthesis of MMP-2, MMP-9, and uPA, as well as p-ERK1/2 [29]. Also, flavonoids down-regulate some signaling cascade as ERK/P13K/Akt and reduced the expression of TIMP-2, p-Akt, and Bcl-2, which leads to an increase in caspase synthesis, Bax, and Fas. Therefore, some flavonoids could induce apoptosis and reduce cell proliferation, inhibit the cell cycle, as well as the angiogenesis of different cancer cells [16, 30]. Table 1 summarizes some of the reported phenolic with anticancer potential.

3.2. Terpenes

Terpenes, also called terpenoids, are a large group of natural compounds, which consist of isoprene units. Terpenes can be classified as monoterpenes, sesquiterpenes, sesterterpenes, polyterpenes, and diterpenes. These compounds have been studied to evaluate their health effects in various diseases such as cancer [16, 39]. Regarding the study of the effect of sesquiterpenes on cancer, Flores-Guzmán et al. [40] reported the antiproliferative and low toxicity activity of one sesquiterpene (5β-hydroxy-hirsutinolide) in the tumor (MDA-MB-231, CaSki, and SKLu-1 cells) and non-tumor cells. The authors suggest that this could be due to a different type of cell death than necrosis, such as induction of active caspase-3, which is considered one of the hallmarks during the apoptotic process; however, this has not yet been elucidated.

On the other hand, it has been reported that limonene (one of the most studied terpenoids) inhibits the *in vitro* proliferation of cancers. In this sense, Vieira et al. [41] reported that this monoterpene, both alone or combined with other compounds, promoted apoptosis to cancer cells. Also, the authors mention that the mechanism by which apoptosis is induced in cancer cells depends on the type of cells studied because, in some cases, cell death may be caused by an increase of ROS formation and caspase-3 expression. Likewise, limonene could diminish Akt activity as well as upregulation of pro-apoptotic Bax protein, an increase of cytosol cytochrome c level, and downregulation of anti-apoptotic Bcl-2; therefore, limonene could be involved in the activation of the mitochondrial apoptotic signaling pathway [42-44]. Table 2 summarizes some of the terpenes that have been studied for their anti-cancer potential.

Table 1. Some phenolic compounds that have been studied for their anti-proliferative potential against cancer cell lines

Phenolic compounds	Cancer Cell Lines	Reference
Wagonin	OC2 human oral cancer and MDA-MB-231 human breast cancer cells	[16]
Kaempferol	HCCC9810 and QBC939 cholangiocarcinoma, MIA PaCa-2, PANC-1 and SNU-213 human pancreatic cancer cells	[31]
Nobiletin	HepG2 liver cancer, MDA-MB-231 breast cancer, and AGS gastric cancer cells	[16]
Curcumin	CL1-5 Lung adenocarcinoma and CaOV3 human Ovarian cancer cells	[32, 33]
Magnolol	MDA-MB-231 breast cancer and PC-3 human prostate carcinoma cells	[34, 35]
Ferulic acid	Mia PaCa-2 pancreatic cancer, TT medullary thyroid cancer and A549 human lung adenocarcinoma cells	[36-38]

Table 2. Summarization of some terpenes that have been studied for their anti-proliferative potential against cancer cell lines

Terpenoid	Cancer Cell Lines	References
Sesquiterpene: Aromatic turmerone	HL-60 leukemia, P388D1 lymphoblasts, U937 histiocytic lymphoma, L-1210, HCC, and MDA-MB-231 breast cancer cells	[45, 46, 16, 47]
Triterpene: Maslinic acid	DU145 prostate cancer, OE33 human esophageal squamous cancer and SGC-7901 gastric cancer cells	[48, 49]

3.3. Alkaloids and Others

Alkaloids are a group of phytochemicals that contain a heterocyclic ring with nitrogen. It has been reported that a wide variety of plants from different botanic families contain a great diversity of alkaloids, such as Leguminosae, Ranunculaceae, Menispermaceae, Papaveraceae, and Loganiaceae families. Likewise, it has been reported that these compounds have potential anti-cancer and anti-metastatic effects [50, 51, 16]. There are several reports on the antiproliferative activity alkaloids on different cancer cell lines; however, this effect could be due to different routes, in this sense, it has been reported that alkaloids could regulate the proteins Bcl-2 and Bcl-xL, which are antiapoptotic proteins. On the other hand, it has been reported that alkaloids can induce cell death by activation of activated protein kinase (AMPK) [52].

Similarly, evodiamine (a quinone alkaloid) has been reported with antimetastatic and anti-invasive activity to colon, lung and breast cancers [16], this could be attributed to the reduction in the levels of mRNA and MMP-2 proteins, also, the decrease in MMP-2 by evodiamine is related to a decrease in the translocation of NF-κBp65. In addition, the anti-invasive activity of this compound could be due to the decrease in JNK, Akt, p38 signaling pathways, as well as ERK1/2 expression (p-ERK1/2). As for the antimetastatic effect, this could be due to the decrease in MMP-9 and the urokinase plasminogen activator (uPA) and uPA receptor (uPAR).

Therefore, the decrease in MAPKs such as JNK, p38 MAPK, and ERK are important proteins related to the antimetastatic as well as the anti-invasive effect of phytochemical compounds [53, 54, 16, 55].

In this sense, matrine (sophora alkaloid) has anti-invasive and antimetastatic activity in malignant tumors by a decrease in the expression of the EGFR/Akt/MMP-9 pathway [56]. Besides, it has been reported that other alkaloids as hirsutine, glaucine, sanguinarine decrease the invasion and metastasis of different types of cancer cell lines, effect attributed to the decreased expression of the tissue-type plasminogen activator (TPA) of MMP-9, COX-2 and prostaglandin E2 (PGE2), as well as the suppression of the constitutive expression of MMP-2 and MMP-9 and the NF-κB activation. Matrine also reduces the stimulated degradation of the NF-κBα inhibitor and NF-κB nuclear translocation [57, 58, 16, 59], therefore, makes them compounds of interest for different types of cancer treatment.

On the other hand, other phytochemical compounds such as quinones (isolated from plants as well as marine sources), isothiocyanates (isolated from vegetables), and cyclopeptide glucosides (isolated from plants) have been found effective against several cancer cell lines [16]. In this sense, quinones like emodin, rhein, and shikonin have been reported to inhibit cell migration and invasion of cancer cells through decreased activation of NF-κB, as well as the MMP-2, MMP-9 and VEGF proteins, as well as regulation of caspase-3 [60]. Similarly, it has been reported that rhein decreased the activation of NF-κB, Akt, and attenuated ECM proteins [61]. Also, it has been found that shikonin inhibited the expression of MMP-2 and MMP-9, as well as the positive regulation of JNK, ERK, and p38 MAPK and reduced the phosphorylation of Akt and mTOR [62].

According to Zhang et al. [63], the compound sulforaphane (an isothiocyanate) has anticarcinogenic activity; this could be due to the protein reduction of MMP-2, galectin-3, and MMP-9 as well as an increasing of E-cadherin which causes cell invasion inhibition. Similarly, the compound RA-XII, a cyclopeptide glucoside that has been found to effective against several cancer cell lines, in this sense, RA-XII has been reported to suppress MAPK, FAK/pSRC, PI3K/Atk and NF-kB signaling molecules as well as the reduction of MMP-9, uPA and ECM-associated proteinases [16]. Table 3 summarizes some of the alkaloids and other compounds that have been studied for their anti-cancer potential.

Table 3. Some alkaloids and other compounds that have been studied for their anti-proliferative potential against cancer cell lines

	Compounds	Cancer Cell Line	References
Alkaloids	Evodiamine	Nasopharyngeal carcinoma HONE1 and CNE1, MDA-MB-231 human breast cancer, Lewis lung carcinoma, B16-F10 melanoma, and 26-L5 colon carcinoma cells	[54, 16, 64]
	Matrine	NCI-H1299 nonsmall cell lung and Huh-7 hepatocellular carcinoma cancer cells	[56, 65]
	Hirsutine	Breast 4T1 cancer cells	[58]
	Glaucine	Human breast MCF-7 and MDA-MB-231 cancer cells	[57]
Others	Emodin	Leukemia K562 cell lines	[66-68]
	Rhein	Pancreatic LTC-14 cells stellate cells, PANC-1 (human adenocarcinoma line), SW480 and SW620 (human colorectal carcinoma lines)	[58, 61]
	Shikonin	U87 and U251 human glioblastoma, PC-3 and DU145 prostate and MDA-MB-231 breast cancer cells	[69, 62, 70]
	Sulforaphane	Human glioblastoma (T98G, U87, LN229, U373, and U251), human prostate cancer DU145s, U251MG, 87MG, and U373MG glioblastoma cell lines	[63, 71-73]
	RA-XII	4T1 mouse mammary carcinoma cells	[74, 16]

4. MEDICINAL PLANTS WITH ANTICANCER POTENTIAL

The Mexican population has traditionally used medicinal plants as remedies against different types of cancer. Today, they are still used due to their ethnobotanical importance in herbal medicine. Recent studies have focused on these plants as a means to record their ethnobotanical use, their scientific bioactive properties, and elucidate their constituents. Nowadays, medicinal plants are recognized as the source of many compounds with anti-cancer activity, such as alkaloids, terpenes, and phenolic compounds [75, 76].

Nowadays, there is a limited amount of *in vivo* studies assessing the anticancer properties of phytochemicals from Mexican medicinal plants mainly due to economic or infrastructure limitations. Nonetheless, *in vivo* studies are essential in the present and future studies before phytochemicals can be considered as anticancer agents. This section will focus only on *in vivo* pre-clinical and clinical studies.

4.1. *Allium sativum*

Allium sativum, commonly known as garlic, is a member of the Amaryllidaceae family, used worldwide as spice and remedy for various ailments and diseases. Garlic is originally from central Asia and is distributed in Chine, the Near East, the Mediterranean regions, Southern Europe, northern Africa, and Mexico [77, 78]. Garlic has been studied for its antiatherosclerotic, antithrombotic, antimicrobial, anti-inflammatory, antioxidant, and anticancer potential [78]. Here we present *in vivo* studies conducted on the biological potential of garlic as an anticancer agent.

A case-study by Desai et al. [79] suggests that there is a protective effect of high garlic (*Allium sativum*) consumption against breast cancer. The study was performed following primary breast cancer cases of women aged 30 – 79, and with women with no history of breast cancer. The authors mention that this may be to the reported content of flavonols and organo-sulfur compounds in garlic.

A population-based cohort study by Shi et al. [80] reported that people who consumed garlic occasionally had a lower risk for mortality than those who did not. Their results were related to the report by Kim and Kwon [81], who mentioned that the relation between garlic consumption and the incidence of cancer is not conclusive.

Moreover, a report by Jin et al. [82] evaluated the relation between tobacco smoking-alcohol drinking and garlic consumption on esophageal cancer in Chinese citizens. Interestingly, the authors found that a high intake of raw garlic may reduce esophageal cancer risk and might be a potential dietary recommendation for daily smokers and drinkers in the prevention of this type of cancer.

Alkhatib, Binsiddiq, and Backer [83] evaluated the antitumor activity of docetaxel loaded in a nanoemulsion formulated with garlic oil in Ehrlich ascites carcinoma-bearing mice. The authors found that nano emulsification of docetaxel with garlic oil showed a cardioprotective property through its antioxidant activity, besides exerting antitumor activity.

Li et al. [84] evaluated S-allylmercapto-L-cysteine, a compound derived from garlic that has been studied for its anti-cancer potential, in combination with rapamycin, a mTor inhibitor. Interestingly, the authors reported an enhanced tumor-suppressing ability in BALB/c nude mice, which is hypothetically induced by the upregulation of the Bax/Bcl-2 ratio.

Similarly, diallyl trisulfide, other compounds isolated from garlic, has been evaluated for its anticancer potential. Jiang et al. [85] tested diallyl trisulfide on gastric cancer and reported a suppression effect of cancer cell proliferation and induction of cell cycle arrest, with concomitant increased expression of cyclin A2 and cyclin B1. Diallyl trisulfide inhibited tumor migration and invasion in a xenograft model by modulating MMP9 and E-cadherin protein expressions.

Garlic has also been evaluated against benign prostatic hyperplasia, for instance, Chung et al. [86] reported that garlic administration decreased factors related to prostate cancer, such as prostate weight ratio, suppressed mRNA expression level of androgen receptor, dihydrotestosterone serum levels, and the growth of prostatic tissue in benign prostatic hyperplasia-induced rats. They also showed that garlic administration decreased the levels of inflammatory proteins, iNOS, and COX-2 in prostatic tissue, which was related to the antioxidant properties of garlic and its constituents.

A group of researchers reported the Shandong Intervention Trial which aimed to reduce the prevalence of advanced precancerous gastric lesions, the intervention trial consisted of a randomized trial with a total follow-up of 14.7 years after antibiotic treatment ended and 7.3 years after garlic and vitamin treatments ended, intended to evaluate the long-term effect of these interventions on gastric cancer incidence. Long-term administration of garlic and vitamin supplements was not statistically associated with a reduction in gastric cancer incidence and mortality [87].

4.2. *Aloe vera*

Aloe vera is a plant that belongs to the Aloeaceae botanical family. *Aloe vera* is a perennial plant known for its chemical properties and ethnobotanical use. There are more than 300 species of Aloe, and the most common are *A. barbadensis* and *A. arborescens. Aloe vera* grows in southern and eastern Africa, North Africa, Mediterranean countries, and Mexico [88, 89].

There is a reduced number of *in vivo* studies regarding Aloe and its anti-cancer properties. Briefly, Kocik et al. [90] reported that they fed mice with aloe vera for three days after tumor cell grafting with 150 μL daily dose of Aloe vera gel significantly diminished the number of newly-formed blood vessels, which may suggest an anti-proliferative effect. The authors suggest that Aloe vera has the potential for further studies to be used concomitantly with chemotherapy to increase tumor regression and survival time.

Another study by Tomasin, de Andrade, and Gomes-Marcondes [91] evaluate the immunomodulatory, anticancer, and antioxidant properties of *Aloe vera* and honey in tumor progression was analyzed in Walker 256 carcinoma-bearing rats. The authors found that A. vera and honey solution increased oxidative stress in tumor cells by modulating antioxidant and proteolysis enzymes, modulation of the immune system, and alteration of the proinflammatory status of cancer cells in the subjects of study. This effect is associated with the phytochemical content found in *A. vera* and honey, namely aloin, aloe-emodin, acemannan, and some flavonoids.

Furthermore, Pandiri et al. [92] reported that when F344 rats with intestinal tumors were exposed to chronic doses of Aloe vera extract in drinking water, the pathogenesis of human colorectal cancer like MPAK, WNT, and TGF-β was altered in large intestinal tumors.

4.3. *Vaccinium myrtillus*

Vaccinium myrtillus, also known as bilberries, belong to the genus Vaccinium that contains around 200 species. Bilberries are native to Europe and can also be found distributed in North America. Bilberries are considered to be an important source of phytochemicals such as phenolic compounds, for instance, anthocyanins, flavonoids, and phenolic acids, which are related with antioxidant, anti-inflammatory and anticancer potential benefits [93, 94]. The phytochemical content of bilberries has led some studies on their anticancer properties, however *in vivo* studies are scarce, and some are mentioned below.

A study by Choi et al. [95] evaluated the antioxidant potential of bilberry (*Vaccinium myrtillus*) against doxorubicin-induced toxicity in rats and mice. The authors report a decreased cardiotoxicity through protection against myocardial damage during doxorubicin-induced cardiotoxicity. Bilberry extracts, at a concentration of 500 mg/kg, enhanced CPK activity, and retention of cardiac glutathione. Furthermore, *V. myrtillus* supplementation increased cardiac creatine phosphokinase activity and total glutathione levels and managed to reduce serum lipid peroxidation.

Similarly, bilberry has been tested for its capacity to reduce doxorubicin-induced cardiotoxicity by Ashour et al. [94]. The authors reported that bilberry extract at a concentration of 100 mg/kg administered every day during ten consecutive days significantly inhibited doxorubicin-induced elevated levels of lactate dehydrogenase, creatine phosphokinase, creatine kinase isoenzyme-MB, and troponin I levels, which was reflected with less histological changes in treated rats.

4.4. *Cinnamomum verum*

Cinnamomum verum, commonly known as cinnamon, belongs to the Lauraceae family and comprises around 250 aromatic trees, mostly distributed in Asia [96]. Cinnamon bark (*Cinnamomum verum*) is one of the most popular spices used in traditional cooking and as herbal medicine [96, 97]. Scientific *in vivo* studies using cinnamon as a potential therapeutic agent is scarce.

An *in vivo* anti-inflammatory study using eight-week-old male BALB/c mice was performed by Hong et al. [98], they administered cinnamon water ad libitum, cinnamon water was prepared at a final concentration of 3.62% of yield. The authors evaluated the effect of cinnamon water extract on lipopolysaccharide-induced tumor necrosis factor-α (TNF-α). Interestingly, the authors found that cinnamon water decreased serum levels of TNF-α and IL-6, which are pro-inflammatory factors. Furthermore, an *in vitro* study showed that cinnamon treatment decreases TNF-α and IL-6 by decreasing the expression of the mRNA.

Once a relationship has been established between a particular plant species and a bioactive constituent, this compound isolated and further tests are performed. In this sense, the effect of 2-methoxycinnamaldehyde, a compound identified and extracted from cinnamon bark, was evaluated in male nude mice by an *in vivo* tumor xenograft study by Perng et al. [96]. The study showed that 2-methoxycinnamaldehyde could induce apoptosis and suppress tumor cell growth in human hepatocellular carcinoma SK-Hep-1. The compound isolated from cinnamon also shrank the tumor size, which was associated with suppression of expression of nuclear transcription factors NF-κB, cyclooxygenase-2, prostaglandin E_2, and topoisomerase I and II. These results may be of therapeutic significance.

Another study by Perng et al. [96] tested the effect of 2-methoxycinnamaldehyde in hepatocellular carcinoma Hep 3B cells. 2-methoxycinnamaldehyde showed a significant effect on the growth of Hep 3B xenografts in nude mice, and the compound reduced about 65% of tumor size. The authors also reported that 2-methoxycinnamaldehyde could induce apoptosis, suppress tumor cell growth, and decreased transcription factor NF-κB, topoisomerase I and II, and showed an anti-inflammatory effect through the inhibition of COX-2 and prostaglandin E2.

2-methoxycinnamaldehyde was also tested against human lung adenocarcinoma A549 cells by Wong et al. [99]. *In vivo* xenograft of A549 cells in nude mice showed that 2-methoxycinnamaldehyde significantly suppressed tumor growth at concentrations of 10 and 20 mg/kg/day with an approximate reduction of 58% in tumor size. Cinnamon 2-methoxycinnamaldehyde downregulated transcription factor NF-κB binding activity and induced apoptosis and topoisomerase I and II activities.

Another study by Zhang et al. [100] evaluated the tumor growth inhibition capacity of cinnamon extracts in SCID mice. Mice were inoculated intraperitoneally with SKOV3 human ovarian cancer cells and treated with the cinnamon suspension at 0.3 mg/g for 4 weeks. The authors observed an anti-angiogenesis and anti-tumor effect of the cinnamon extract through reducing tumor burden and ascites volume, and the cinnamon extract also decreased vascular endothelial growth factor mRNA and protein expression. Furthermore, cinnamon also decreased the blood vessel count in the tumor. It is important to mention that the assays in this report were performed in a small number of individuals, only 5 rats, so more studies are needed to assess their results.

4.5. *Zingiber officinale*

Zingiber officinale or ginger belongs to the Zingiberaceae family and is originally from South-East Asia, and over the years, it has spread to Nigeria, Sierra Leone, Indonesia, Nepal, and Mexico. Ginger has been cultivated for thousands of years, and ethnobotanically used both as a spice and as a medicinal plant [101]. Some studies have reported that phytochemicals from ginger are related to its chemopreventive effects. Most of the anticancer reports of ginger are related to phytochemicals like gingerol, paradols, and shogaol [101, 102].

A study by Zick et al. [103], evaluated the anticancer potential of ginger. Briefly, the authors used ginger powder at a concentration of 2g/d, the study was conducted using 30 subjects exposed to 2 g/d of ginger or placebo during 28 days, and biopsies of the colon were obtained to evaluate the effects. Furthermore, the authors report a decreased content of prostaglandin E2, 12-HETE, and 15-HETE. These results suggest the potential of ginger to decrease eicosanoid levels, which is related to a downregulation of the inflammatory process involved in the early stages of colorectal cancer [104].

Guahk et al. [105] studied the anti-inflammatory effect of water ginger extract containing gingerol and shogaol, in C57BL/6 mice irradiated with UV-B light. The authors reported that ginger treatment mitigated the inflammation caused by UV-B light, by decreasing UV-B induced hyperplasia, infiltration of leukocytes, and dilation of blood vessels mice skin. This effect was attributed to the reduced levels of IL-1β and IL-6 in C57BL/6 mice after 24 h the UV-B exposure.

Conclusion

As can be evidenced by the low amount of published *in vivo* studies, there is still work to do to evaluate the clinical potential of phytochemicals from medicinal plants as anticancer agents. Since cancer is one of the leading causes of death worldwide, only second to cardiovascular diseases, further studies are needed, so these phytochemicals are beginning to being considered as complementary medicine in the treatment of different types of cancer.

References

[1] Bray, F., Ferlay, J., Soerjomataram, I., Siegel, R. L., Torre, L. A. & Jemal, A. (2018). Global cancer statistics 2018: GLOBOCAN

estimates of incidence and mortality worldwide for 36 cancers in 185 countries. *CA Cancer J Clin*, *68* (6), 394-424. doi: 10.3322/caac.21492.

[2] Pecorino, L. (2012). *Molecular Biology of Cancer—Mechanisms, Targets and Therapeutics*. 3rd ed. Oxford, UK.

[3] Hanahan, D. & Weinberg, R. A. (2011). Hallmarks of cancer: the next generation. *Cell*, *144* (5), 646-74. doi: 10.1016/ j.cell.2011.02.013.

[4] Colditz, G. A. & Wei, E. K. (2012). Preventability of cancer: the relative contributions of biologic and social and physical environmental determinants of cancer mortality. *Annu Rev Public Health*, *33*, 137-56. doi: 10.1146/annurev-publhealth-031811-124627.

[5] Wu, S., Powers, S., Zhu, W. & Hannun, Y. A. (2015). Substantial contribution of extrinsic risk factors to cancer development. *Nature 529*, 43. doi: 10.1038/nature16166.

[6] Vineis, P. & Wild, C. P. (2014). Global cancer patterns: causes and prevention. *Lancet*, *383* (9916), 549-57. doi: 10.1016/s0140-6736(13)62224-2.

[7] Salto-Tellez, M. & Cree, I. A. (2019). Cancer taxonomy: pathology beyond pathology. *European Journal of Cancer*, *115*, 57-60. doi: https://doi.org/10.1016/j.ejca.2019.03.026.

[8] WHO. (2019). *Classification of Tumours*. World Health Organization, accessed July 2019. https://whobluebooks.iarc.fr/.

[9] NIH. (2019). *Cancer Classification*. National Cancer Institute, accessed july 2019. https://training.seer.cancer.gov/disease/categories/classification.html.

[10] WHO. (2019). *Cancer Today*. World Health Organization, accessed July 2019. http://gco.iarc.fr/today/online-analysis-pie?v=2018&mode=cancer&mode_population=continents&population=900&populations=900&key=total&sex=2&cancer=39&type=0&statistic=5&prevalence=0&population_group=0&ages_group%5B%5D=0&ages_group%5B%5D=17&nb_items=7&group_cancer=1&include_nmsc

=1&include_nmsc_other=1&half_pie=0&donut=0&population_group_globocan_id=.

[11] NIH. (2019). *Cancer. Types of treatment*. National Cancer Institute, accessed July 2019. https://www.cancer.gov/espanol/ cancer/ tratamiento/tipos.

[12] Zugazagoitia, J., Guedes, C., Ponce, S., Ferrer, I., Molina-Pinelo, S. & Paz-Ares, L. (2016). Current Challenges in Cancer Treatment. *Clin Ther*, *38* (7), 1551-66. doi: 10.1016/j.clinthera.2016.03.026.

[13] Atanasov, A. G., Waltenberger, B., Pferschy-Wenzig, E. M., Linder, T., Wawrosch, C., Uhrin, P., Temml, V., Wang, L., Schwaiger, S., Heiss, E. H., Rollinger, J. M., Schuster, D., Breuss, J. M., Bochkov, V., Mihovilovic, M. D., Kopp, B., Bauer, R., Dirsch, V. M. & Stuppner, H. (2015). Discovery and resupply of pharmacologically active plant-derived natural products: A review. *Biotechnol Adv*, *33* (8), 1582-1614. doi: 10.1016/j.biotechadv.2015.08.001.

[14] Cragg, G. M. & Newman, D. J. (2013). Natural products: a continuing source of novel drug leads. *Biochim Biophys Acta*, *1830* (6), 3670-95. doi: 10.1016/j.bbagen.2013.02.008.

[15] Secretaría, de Salud. (2013). *Farmacopea Herbolaria de los Estados Unidos Mexicanos*. 2nd ed. D.F., México: Secretaría de Salud.

[16] Md. Shahid, S., Hong-Jie, Z. & Siu Wai, T. (2018). Perspectives of Plant Natural Products in Inhibition of Cancer Invasion and Metastasis by Regulating Multiple Signaling Pathways. *Current Medicinal Chemistry*, *25* (38), 5057-5087. doi: http://dx.doi.org/ 10.2174/0929867324666170918123413.

[17] Rodriguez-Garcia, C., Sanchez-Quesada, C. & Gaforio, J. J. (2019). Dietary Flavonoids as Cancer Chemopreventive Agents: An Updated Review of Human Studies. *Antioxidants*, *8* (5), 23. doi: 10.3390/ antiox8050137.

[18] Jiang, Y. L. & Liu, Z. P. (2011). Natural Products as Anti-Invasive and Anti-Metastatic Agents. *Current Medicinal Chemistry*, *18* (6), 808-829. doi: 10.2174/092986711794927711.

[19] Alper, M. & Gunes, H. (2019). The Anticancer and Anti-inflammatory Effects of *Centaurea solstitialis* Extract on Human

Cancer Cell Lines. *Turkish Journal of Pharmaceutical Sciences*, *16* (3), 273-281. doi: 10.4274/tjps.galenos.2018.27146.

[20] Golonko, A., Pienkowski, T., Swislocka, R., Lazny, R., Roszko, M. & Lewandowski, W. (2019). Another look at phenolic compounds in cancer therapy the effect of polyphenols on ubiquitin-proteasome system. *European Journal of Medicinal Chemistry*, *167*, 291-311. doi: https://doi.org/10.1016/j.ejmech.2019.01.044.

[21] Németh, Z. H., Wong, H. R., Odoms, K., Deitch, E. A., Szabó, C., Vizi, E. S. & Haskó, G. (2004). *Proteasome Inhibitors Induce Inhibitory κB (IκB) Kinase Activation, IκBα Degradation, and Nuclear Factor κB Activation in HT-29 Cells. 65* (2), 342-349. doi: 10.1124/mol.65.2.342 %J Molecular Pharmacology.

[22] Rastogi, N. & Mishra, D. P. (2012). Therapeutic targeting of cancer cell cycle using proteasome inhibitors. *Cell Division*, *7* (1), 26. doi: 10.1186/1747-1028-7-26.

[23] Mileo, A. M. & Miccadei, S. (2016). Polyphenols as Modulator of Oxidative Stress in Cancer Disease: New Therapeutic Strategies *Oxidative Medicine and Cellular Longevity*, *17*, doi: 10.1155/ 2016/6475624.

[24] Ma, Q., Kinneer, K., Ye, J. & Chen, B. J. (2003). Inhibition of Nuclear Factor κB by Phenolic Antioxidants: Interplay between Antioxidant Signaling and Inflammatory Cytokine Expression. *Molecular Pharmacology*, *64* (2), 211-219. doi: 10.1124/ mol.64. 2.211.

[25] Ruiz, P. A. & Haller, D. (2006). Functional Diversity of Flavonoids in the Inhibition of the Proinflammatory NF-κB, IRF, and Akt Signaling Pathways in Murine Intestinal Epithelial Cells. *The Journal of Nutrition*, *136* (3), 664-671. doi: 10.1093/jn/136.3.664 %J The Journal of Nutrition.

[26] Shin, S. Y., Woo, Y., Hyun, J., Yong, Y., Koh, D., Lee, Y. H. & Lim, Y. (2011). Relationship between the structures of flavonoids and their NF-κB-dependent transcriptional activities. *Bioorganic & Medicinal Chemistry Letters*, *21* (20), 6036-6041. doi: https://doi. org/10.1016/j.bmcl.2011.08.077.

[27] Smith, D. M., Daniel, K. G., Wang, Z., Guida, W. C., Chan, T. H. & Dou, Q. P. (2004). Docking studies and model development of tea polyphenol proteasome inhibitors: Applications to rational drug design. *Proteins*, *54* (1), 58-70. doi: 10.1002/prot.10504.

[28] Yang, H., Landis-Piwowar, K. R., Chen, D., Milacic, V. & Dou, Q. P. (2008). Natural Compounds with Proteasome Inhibitory Activity for Cancer Prevention and Treatment. *Current Protein and Peptide Science*, *9* (3), 227-239. doi: 10.2174/138920308784533998.

[29] Dong, P., Zhang, Y., Gu, J., Wu, W., Li, M., Yang, J., Zhang, L., Lu, J., Mu, J., Chen, L., Li, S., Wang, J. & Liu, Y. (2011). Wogonin, an active ingredient of Chinese herb medicine Scutellaria baicalensis, inhibits the mobility and invasion of human gallbladder carcinoma GBC-SD cells by inducing the expression of maspin. *Journal of Ethnopharmacology*, *137* (3), 1373-1380. doi: https://doi.org/ 10.1016/j.jep.2011.08.005.

[30] Qin, Y., Cui, W., Yang, X. & Tong, B. (2016). Kaempferol inhibits the growth and metastasis of cholangiocarcinoma *in vitro* and *in vivo*. *Acta Biochimica et Biophysica Sinica*, *48* (3), 238-245. doi: 10.1093/ abbs/gmv133 %J Acta Biochimica et Biophysica Sinica.

[31] Lee, J. & Kim, J. H. (2016). Kaempferol Inhibits Pancreatic Cancer Cell Growth and Migration through the Blockade of EGFR-Related Pathway *In Vitro*. *PLOS ONE*, *11* (5), e0155264. doi: 10.1371/ journal.pone.0155264.

[32] Chen, H. W., Yu, S. L., Chen, J. J. W., Li, H. N., Lin, Y. C., Yao, P. L., Chou, H. Y., Chien, C. T., Chen, W. J., Lee, Y. T. & Yang, P. C. (2004). Anti-Invasive Gene Expression Profile of Curcumin in Lung Adenocarcinoma Based on a High Throughput Microarray Analysis. *Molecular Pharmacology*, *65* (1), 99. doi: 10.1124/mol.65.1.99.

[33] Ji, C., Cao, C., Lu, S., Kivlin, R., Amaral, A., Kouttab, N., Yang, H., Chu, W., Bi, Z., Di, W., Wan, Y. J. C. C. & Pharmacology. (2008). Curcumin attenuates EGF-induced AQP3 up-regulation and cell migration in human ovarian cancer cells. *Cancer Chemotherapy and Pharmacology*, *62* (5), 857-865. doi: 10.1007/s00280-007-0674-6.

[34] Hwang, E. S. & Park, K. K. (2010). Magnolol Suppresses Metastasis via Inhibition of Invasion, Migration, and Matrix Metalloproteinase-2/-9 Activities in PC-3 Human Prostate Carcinoma Cells. *Bioscience, Biotechnology, and Biochemistry*, *74* (5), 961-967. doi: 10.1271/bbb.90785.

[35] Liu, Y., Cao, W., Zhang, B., Liu, Y. q., Wang, Z. y., Wu, Y. p., Yu, X. j., Zhang, X. d., Ming, P. h., Zhou, G. b. & Huang, L. (2013). The natural compound magnolol inhibits invasion and exhibits potential in human breast cancer therapy. *Scientific Reports*, *3*, 3098. doi: 10.1038/srep03098 https://www.nature.com/articles/ srep03098# supplementary-information.

[36] Dodurga, Y., Eroğlu, C., Seçme, M., Elmas, L., Avcı, Ç. B. & Şatıroğlu-Tufan, N. L. (2016). Anti-proliferative and anti-invasive effects of ferulic acid in TT medullary thyroid cancer cells interacting with URG4/URGCP. *Tumor Biology*, *37* (2), 1933-1940. doi: 10.1007/s13277-015-3984-z.

[37] Fahrioğlu, U., Dodurga, Y., Elmas, L. & Seçme, M. (2016). Ferulic acid decreases cell viability and colony formation while inhibiting migration of MIA PaCa-2 human pancreatic cancer cells *in vitro*. *Gene*, *576* (1, Part 3):476-482. doi: https://doi.org/10.1016/ j.gene.2015.10.061.

[38] Tsai, C. M., Yen, G. C., Sun, F. M., Yang, S. F. & Weng, C. J. (2013). Assessment of the Anti-invasion Potential and Mechanism of Select Cinnamic Acid Derivatives on Human Lung Adenocarcinoma Cells. *Molecular Pharmaceutics*, *10* (5), 1890-1900. doi: 10.1021/ mp3006648.

[39] Silva, G. N. S., Rezende, L. C. D., Emery, F. S., Gosmann, G. & Gnoatto, S. C. B. (2015). Natural and Semi synthetic Antimalarial Compounds: Emphasis on the Terpene Class. *Mini Reviews in Medicinal Chemistry*, *15* (10), 809-836.

[40] Flores-Guzmán, F., Alvarado-Sansininea, J. J., López-Muñoz, H., Escobar, M. L., Espinosa-Trejo, M., Tavera-Hernandez, R., Jiménez-Estrada, M. & Sánchez-Sánchez, L. (2019). Antiproliferative, cytotoxic and apoptotic activity of the bentonite

transformation of sesquiterpene lactone glaucolide B to 5β-hydroxy-hirsutinolide on tumor cell lines. *European Journal of Pharmacology*, *856*, 172406. doi: https://doi.org/10.1016/j.ejphar.2019.172406.

[41] Vieira, A. J., Beserra, F. P., Souza, M. C., Totti, B. M. & Rozza, A. L. (2018). Limonene: Aroma of innovation in health and disease. *Chemico-Biological Interactions*, *283*, 97-106. doi: https://doi.org/10.1016/j.cbi.2018.02.007.

[42] Berliocchi, L., Ciociaro, A., Russo, R., Cassiano, M. G. V., Blandini, F., Rotiroti, D., Morrone, L. A. & Corasaniti, M. T. (2011). Toxic profile of bergamot essential oil on survival and proliferation of SH-SY5Y neuroblastoma cells. *Food and Chemical Toxicology*, *49* (11), 2780-2792. doi: https://doi.org/10.1016/j.fct.2011.08.017.

[43] Jia, S. S., Xi, G. P., Zhang, M., Chen, Y. B., Lei, B., Dong, X. S. & Yang, Y. M. (2013). Induction of apoptosis by D-limonene is mediated by inactivation of Akt in LS174T human colon cancer cells. *Oncology reports*, *29* (1), 349-354. doi: https://doi.org/10.3892/or.2012.2093.

[44] Xiu-Zhen, Z., Ling, W., Dong-Wu, L., Guang-Yan, T. & Hong-Yu, Z. (2014). Synergistic Inhibitory Effect of Berberine and d-Limonene on Human Gastric Carcinoma Cell Line MGC803. *Journal of Medicinal Food*, *17* (9), 955-962. doi: 10.1089/jmf.2013.2967.

[45] Aratanechemuge, Y., Komiya, T., Moteki, H., Katsuzaki, H., Imai, K. & Hibasami, H. (2002). Selective induction of apoptosis by ar-turmerone isolated from turmeric (*Curcuma longa* L.) in two human leukemia cell lines, but not in human stomach cancer cell line. *International Journal of Molecular Medicine*, *9* (5), 481-484. doi: 10.3892/ijmm.9.5.481.

[46] Kim, D., Suh, Y., Lee, H. & Lee, Y. (2013). Immune activation and antitumor response of ar-turmerone on P388D1 lymphoblast cell implanted tumors. *International Journal of Molecular Medicine*, *31* (2), 386-392. doi: 10.3892/ijmm.2012.1196.

[47] Park, S. Y., Kim, Y. H., Kim, Y. & Lee, S. J. (2012). Aromatic-turmerone attenuates invasion and expression of MMP-9 and COX-2 through inhibition of NF-κB activation in TPA-induced breast cancer cells. *Journal of Cellular Biochemistry*, *113* (12), 3653-3662. doi: 10.1002/jcb.24238.

[48] Lin, C. c., Yan, S. l. & Yin, M. c. (2014). Inhibitory effects of maslinic acid upon human esophagus, stomach and pancreatic cancer cells. *Journal of Functional Foods*, *11*, 581-588. doi: https://doi.org/ 10.1016/j.jff.2014.08.020.

[49] Park, S. Y., Nho, C. W., Kwon, D. Y., Kang, Y. H., Lee, K. W. & Park, J. H. Y. (2013). Maslinic acid inhibits the metastatic capacity of DU145 human prostate cancer cells: possible mediation via hypoxia-inducible factor-1α signalling. *British Journal of Nutrition*, *109* (2), 210-222. doi: 10.1017/S0007114512000967.

[50] Gupta, A. P., Pandotra, P., Kushwaha, M., Khan, S., Sharma, R. & Gupta, S. (2015). "Chapter 9 - Alkaloids: A Source of Anticancer Agents from Nature." In *Studies in Natural Products Chemistry*, edited by Rahman Atta ur, 341-445. Elsevier.

[51] Lu, J. J., Bao, J. L., Chen, X. P., Huang, M. & Wang, Y. T. (2012). Alkaloids Isolated from Natural Herbs as the Anticancer Agents %J Evidence-Based Complementary and Alternative Medicine. *Evidence-Based Complementary and Alternative Medicine*, *12*. doi: 10.1155/2012/485042.

[52] Kumar, A. & Jaitak, V. (2019). Natural products as multidrug resistance modulators in cancer. *European Journal of Medicinal Chemistry*, *176*, 268-291. doi: https://doi.org/10.1016/j.ejmech. 2019.05.027.

[53] Dhillon, A. S., Hagan, S., Rath, O. & Kolch, W. (2007). MAP kinase signalling pathways in cancer. *Oncogene*, *26*, 3279. doi: 10.1038/ sj.onc.1210421.

[54] Du, J., Wang, X. F., Zhou, Q. M., Zhang, T. L., Lu, Y. Y., Zhang, H. & Su, S. B. (2013). Evodiamine induces apoptosis and inhibits metastasis in MDA-MB-231 human breast cancer cells *in vitro* and

in vivo. *Oncology Reports*, *30* (2), 685-694. doi: 10.3892/or.2013.2498.

[55] Wu, G. S. (2007). Role of mitogen-activated protein kinase phosphatases (MKPs) in cancer. *Cancer Metastasis Reviews*, *26* (3), 579-585. doi: 10.1007/s10555-007-9079-6.

[56] Liao, H., Zhao, X., Qu, J., Zhang, J. & Cai, H. (2015). Matrine suppresses invasion and metastasis of NCI-H1299 cells by enhancing microRNA-133a expression. *International journal of clinical and experimental medicine*, *8* (7), 10714-10722.

[57] Kang, H., Jang, S. W., Pak, J. H. & Shim, S. (2015). Glaucine inhibits breast cancer cell migration and invasion by inhibiting MMP-9 gene expression through the suppression of NF-κB activation. *Molecular Cellular Biochemistry*, *403* (1), 85-94. doi: 10.1007/s11010-015-2339-9.

[58] Lou, C., Takahashi, K., Irimura, T., Saiki, I. & Hayakawa, Y. (2014). Identification of Hirsutine as an anti-metastatic phytochemical by targeting NF-κB activation. *International journal of oncology*, *45* (5), 2085-2091. doi: 10.3892/ijo.2014.2624.

[59] Park, S. Y., Jin, M. L., Kim, Y. H., Lee, S. J. & Park, G. (2014). Sanguinarine inhibits invasiveness and the MMP-9 and COX-2 expression in TPA-induced breast cancer cells by inducing HO-1 expression. *Oncology Reports*, *31* (1), 497-504. doi: 10.3892/or.2013.2843.

[60] Liu, A., Chen, H., Wei, W. T., Ye, S., Liao, W., Gong, J. M., Jiang, Z. C., Wang, L. & Lin, S. Z. (2011). Antiproliferative and antimetastatic effects of emodin on human pancreatic cancer. *Oncology Reports*, *26* (1), 81-89. doi: 10.3892/or.2011.1257.

[61] Tsang, S. W. & Bian, Z. X. (2015). Anti-fibrotic and Anti-tumorigenic Effects of Rhein, a Natural Anthraquinone Derivative, in Mammalian Stellate and Carcinoma Cells. *Phytotherapy Research*, *29* (3), 407-414. doi: 10.1002/ptr.5266.

[62] Chen, Y., Zheng, L., Liu, J., Zhou, Z., Cao, X., Lv, X. & Chen, F. (2014). Shikonin inhibits prostate cancer cells metastasis by reducing matrix metalloproteinase-2/-9 expression via AKT/mTOR

and ROS/ERK1/2 pathways. *International Immunopharmacology*, *21* (2), 447-455. doi: https://doi.org/10.1016/j.intimp.2014.05.026.

[63] Zhang, Z., Li, C., Shang, L., Zhang, Y., Zou, R., Zhan, Y. & Bi, B. (2016). Sulforaphane induces apoptosis and inhibits invasion in U251MG glioblastoma cells. *SpringerPlus*, *5* (1), 235. doi: 10.1186/s40064-016-1910-5.

[64] Peng, X., Zhang, Q., Zeng, Y., Li, J., Wang, L. & Ai, P. (2015). Evodiamine inhibits the migration and invasion of nasopharyngeal carcinoma cells *in vitro* via repressing MMP-2 expression. *Cancer Chemotherapy Pharmacology*, *76* (6), 1173-1184. doi: 10.1007/s00280-015-2902-9.

[65] Qian, L., Liu, Y., Xu, Y., Ji, W., Wu, Q., Liu, Y., Gao, Q. & Su, C. (2015). Matrine derivative WM130 inhibits hepatocellular carcinoma by suppressing EGFR/ERK/MMP-2 and PTEN/AKT signaling pathways. *Cancer Letters*, *368* (1), 126-134. doi: https://doi.org/ 10.1016/j.canlet.2015.07.035.

[66] Chun-Guang, W., Jun-Qing, Y., Bei-Zhong, L., Dan-Ting, J., Chong, W., Liang, Z., Dan, Z. & Yan, W. (2010). Anti-tumor activity of emodin against human chronic myelocytic leukemia K562 cell lines *in vitro* and *in vivo*. *European Journal of Pharmacology*, *627* (1), 33-41. doi: https://doi.org/10.1016/j.ejphar.2009.10.035.

[67] Chen, H., Wei, W., Guo, Y., Liu, A., Tong, H., Wang, Z., Tan, W., Liu, J. & Lin, S. (2011). Enhanced effect of gemcitabine by emodin against pancreatic cancer *in vivo* via cytochrome C-regulated apoptosis. *Oncology Reports*, *25* (5), 1253-1261. doi: 10.3892/or.2011.1174.

[68] Ko, J. C., Su, Y. J., Lin, S. T., Jhan, J. Y., Ciou, S. C., Cheng, C. M. & Lin, Y. W. (2010). Suppression of ERCC1 and Rad51 expression through ERK1/2 inactivation is essential in emodin-mediated cytotoxicity in human non-small cell lung cancer cells. *Biochemical Pharmacology*, *79* (4), 655-664. doi: https://doi.org/10.1016/j.bcp.2009.09.024.

[69] Jang, S. Y., Lee, J. K., Jang, E. H., Jeong, S. Y. & Kim, J. H. (2014). Shikonin blocks migration and invasion of human breast cancer cells through inhibition of matrix metalloproteinase-9 activation. *Oncology Reports*, *31* (6), 2827-2833. doi: 10.3892/or.2014.3159.

[70] Zhang, F. Y., Hu, Y., Que, Z. Y., Wang, P., Liu, Y. H., Wang, Z. H. & Xue, Y. X. (2015). Shikonin Inhibits the Migration and Invasion of Human Glioblastoma Cells by Targeting Phosphorylated β-Catenin and Phosphorylated PI3K/Akt: A Potential Mechanism for the Anti-Glioma Efficacy of a Traditional Chinese Herbal Medicine. *International Journal of Molecular Sciences*, *16* (10), 23823-23848. doi: 10.3390/ijms161023823.

[71] Peng, X., Zhou, Y., Tian, H., Yang, G., Li, C., Geng, Y., Wu, S. & Wu, W. (2015). Sulforaphane inhibits invasion by phosphorylating ERK1/2 to regulate E-cadherin and CD44v6 in human prostate cancer DU145 cells. *Oncology Reports*, *34* (3), 1565-1572. doi: 10.3892/or.2015.4098.

[72] Lan, F., Yang, Y., Han, J., Wu, Q., Yu, H. & Yue, X. (2016). Sulforaphane reverses chemo-resistance to temozolomide in glioblastoma cells by NF-κB-dependent pathway downregulating MGMT expression. *International journal of oncology*, *48* (2), 559-568. doi: 10.3892/ijo.2015.3271.

[73] Mondal, A., Biswas, R., Rhee, Y. H., Kim, J. & Ahn, J. C. (2016). Sulforaphene promotes Bax/Bcl2, MAPK-dependent human gastric cancer AGS cells apoptosis and inhibits migration via EGFR, p-ERK1/2 down-regulation. *General physiology and biophysics*, *35* (1), 25-34. doi: 10.4149/gpb_2015033.

[74] Leung, H. W., Zhao, S. M., Yue, G. G. L., Lee, J. K. M., Fung, K. P., Leung, P. C., Tan, N. H. & Lau, C. B. S. (2015). RA-XII inhibits tumour growth and metastasis in breast tumour-bearing mice via reducing cell adhesion and invasion and promoting matrix degradation. *Scientific Reports*, *5*, 16985. doi: 10.1038/srep16985. https://www.nature.com/articles/srep16985#supplementary-information.

[75] Gutiérrez-Grijalva, E .P., Contreras-Angulo, L. A., Emus-Medina, A., Vázquez-Olivo, G. & Heredia, J. B. (2019). "Anticancer Properties of Medicinal Plants Listed in the Herbal Pharmacopoeia of The United Mexican States." In *Medicinal Plants: Chemistry, Pharmacology, and Therapeutic Applications*, edited by Mallappa Kumara Swamy, Jayanta Kumar Patra and Gudepalya Renukaiah Rudramurthy. CRC Press.

[76] Alonso-Castro, A. J., Villarreal, M. L., Salazar-Olivo, L. A., Gomez-Sanchez, M., Dominguez, F. & Garcia-Carranca, A. (2011). Mexican medicinal plants used for cancer treatment: Pharmacological, phytochemical and ethnobotanical studies. *Journal of Ethnopharmacology*, *133* (3), 945-972. doi: https://doi.org/10.1016/j.jep.2010.11.055.

[77] Jangam, G. B. & Badole, S. L. (2014). "Chapter 46 - Garlic (*Allium sativum*): Role in Metabolic Disorder." In *Polyphenols in Human Health and Disease*, edited by Ronald Ross Watson, Victor R. Preedy and Sherma Zibadi, 611-614. San Diego: Academic Press.

[78] Ozcan, M. M. (2016). Bioactive properties of garlic (Allium sativum L.) - A review. *Zeitschrift Fur Arznei- & Gewurzpflanzen*, *21* (4), 174-182.

[79] Desai, G., Schelske-Santos, M., Nazario, C. M., Rosario-Rosado, R. V., Mansilla-Rivera, I., Ramírez-Marrero, F., Nie, J., Myneni, A.A., Zhang, Z. F., Freudenheim, J. L. & Mu, L. (2019). Onion and Garlic Intake and Breast Cancer, a Case-Control Study in Puerto Rico. *Nutrition and Cancer*, 1-10. doi: 10.1080/01635581.2019.1651349.

[80] Shi, X. M., Lv, Y. B., Mao, C., Yuan, J. Q., Yin, Z. X., Gao, X. & Zhang, Z. F. (2019). Garlic Consumption and All-Cause Mortality among Chinese Oldest-Old Individuals: A Population-Based Cohort Study. *Nutrients*, *11* (7), 14. doi: 10.3390/nu11071504.

[81] Kim, J. Y. & Kwon, O. (2008). Garlic intake and cancer risk: an analysis using the Food and Drug Administration's evidence-based review system for the scientific evaluation of health claims. *The American Journal of Clinical Nutrition*, *89* (1), 257-264. doi: 10.3945/ajcn.2008.26142.

[82] Jin, Z. Y., Wallar, G., Zhou, J. Y., Yang, J., Han, R. Q., Wang, P. H., Liu, A. M., Gu, X. P., Zhang, X. F., Wang, X. S., Su, M., Hu, X., Sun, Z., Li, G., Mu, L. N., Lu, Q. Y., Liu, X., Li, L. M., He, N., Wu, M., Zhao, J. K. & Zhang, Z. F. (2019). Consumption of garlic and its interactions with tobacco smoking and alcohol drinking on esophageal cancer in a Chinese population. *European Journal of Cancer Prevention*, *28* (4), 278-286. doi: 10.1097/ cej. 0000000000000456.

[83] Alkhatib, M. H., Binsiddiq, B. M. & Backer, W. S. (2017). *In vivo* evaluation of the anticancer activity of a water-in-garlic oil nanoemulsion loaded with docetaxel. *International Journal of Pharmaceutical Sciences and Research*, *8* (12), 5373-5379. doi: 10.13040/ijpsr.0975-8232.8(12).5373-79.

[84] Li, S. Y., Yang, G., Zhu, X. S., Cheng, L., Sun, Y. Y. & Zhao, Z. X. (2017). Combination of rapamycin and garlic-derived S-allylmercaptocysteine induces colon cancer cell apoptosis and suppresses tumor growth in xenograft nude mice through autophagy/p62/Nrf2 pathway. *Oncology Reports*, *38* (3), 1637-1644. doi: 10.3892/or.2017.5849.

[85] Jiang, X., Zhu, X., Huang, W., Xu, H., Zhao, Z., Li, S., Li, S., Cai, J. & Cao, J. (2017). Garlic-derived organosulfur compound exerts antitumor efficacy via activation of MAPK pathway and modulation of cytokines in SGC-7901 tumor-bearing mice. *International Immunopharmacology*, *48*, 135-145. doi: https://doi.org/10.1016 / j.intimp.2017.05.004.

[86] Chung, K. S., Shin, S. J., Lee, N. Y., Cheon, S. Y., Park, W., Sun, S. H. & An, H. J. (2016). Anti-Proliferation Effects of Garlic (*Allium sativum* L.) on the Progression of Benign Prostatic Hyperplasia. *Phytotherapy Research*, *30* (7), 1197-1203. doi: 10.1002/ptr.5637.

[87] Ma, J. L., Zhang, L., Brown, L. M., Li, J. Y., Shen, L., Pan, K. F., Liu, W. D., Hu, Y., Han, Z. X., Crystal-Mansour, S., Pee, D., Blot, W. J., Fraumeni, J. F., Jr. You, W. C. & Gail, M. H. (2012). Fifteen-Year Effects of Helicobacter pylori, Garlic, and Vitamin Treatments on Gastric Cancer Incidence and Mortality. *JNCI: Journal of the*

National Cancer Institute, *104* (6), 488-492. doi: 10.1093/jnci/djs003.

[88] Sánchez-Machado, D .I., López-Cervantes, J., Sendón, R. & Sanches-Silva, A. (2017). Aloe vera: Ancient knowledge with new frontiers. *Trends in Food Science & Technology*, *61*, 94-102. doi: https://doi.org/10.1016/j.tifs.2016.12.005.

[89] Rodríguez, E. R., Martín, J. D. & Romero, C. D. (2010). Aloe vera as a Functional Ingredient in Foods. *Critical Reviews in Food Science and Nutrition*, *50* (4), 305-326. doi: 10.1080/10408390802544454.

[90] Kocik, J., Balan, B. J., Zdanowski, R., Jung, L., Skopinska-Rozewska, E. & Skopinski, P. (2014). Feeding mice with Aloe vera gel diminishes L-1 sarcoma-induced early neovascular response and tumor growth. *Central European Journal of Immunology*, *39* (1), 14-18. doi: 10.5114/ceji.2014.42116.

[91] Tomasin, R., de Andrade, R. S. & Gomes-Marcondes, M. C. C. (2015). Oral Administration of *Aloe vera* (L.) Burm. f. (Xanthorrhoeaceae) and Honey Improves the Host Body Composition and Modulates Proteolysis Through Reduction of Tumor Progression and Oxidative Stress in Rats. *Journal of Medicinal Food*, *18* (10), 1128-1135. doi: 10.1089/jmf.2014.0129.

[92] Pandiri, A. R., Sills, R. C., Hoenerhoff, M. J., Peddada, S. D., Ton, T. V. T., Hong, H. H. L., Flake, G. P., Malarkey, D. E., Olson, G. R., Pogribny, I. P., Walker, N. J. & Boudreau, M. D. (2011). Aloe vera Non-Decolorized Whole Leaf Extract-Induced Large Intestinal Tumors in F344 Rats Share Similar Molecular Pathways with Human Sporadic Colorectal Tumors. *Toxicologic Pathology*, *39* (7), 1065-1074. doi: 10.1177/0192623311422081.

[93] Može, Š., Polak, T., Gašperlin, L., Koron, D., Vanzo, A., Poklar Ulrih, N. & Abram, V. (2011). Phenolics in Slovenian Bilberries (Vaccinium myrtillus L.) and Blueberries (Vaccinium corymbosum L.). *Journal of Agricultural and Food Chemistry*, *59* (13), 6998-7004. doi: 10.1021/jf200765n.

[94] Ashour, O. M., Elberry, A. A., Alahdal, A. M., Al Mohamadi, A. M., Nagy, A. A., Abdel-Naim, A. B., Abdel-Sattar, E. A. & Mohamadin, A. M. (2011). Protective effect of bilberry (*Vaccinium myrtillus*) against doxorubicin-induced oxidative cardiotoxicity in rats. *Medical Science Monitor*, *17* (4), BR110-BR115. doi: 10.12659/msm.881711.

[95] Choi, E. H., Park, J. H., Kim, M. K. & Chun, H. S. (2010). Alleviation of doxorubicin-induced toxicities by anthocyanin-rich bilberry (*Vaccinium myrtillus* L.) extract in rats and mice. *BioFactors*, *36* (4), 319-327. doi: 10.1002/biof.106.

[96] Perng, D. S., Tsai, Y. H., Cherng, J., Wang, J. S., Chou, K. S., Shih, C. W. & Cherng, J. M. (2016). Discovery of a novel anticancer agent with both anti-topoisomerase I and II activities in hepatocellular carcinoma SK-Hep-1 cells *in vitro* and *in vivo*: Cinnamomum verum component 2-methoxycinnamaldehyde. *Drug Design Development and Therapy*, *10*, 141-153. doi: 10.2147/dddt.S93599.

[97] Yang, S. M., Tsai, K. D., Wong, H. Y., Liu, Y. H., Chen, T. W., Cherng, J., Hsu, K. C., Ang, Y. U. & Cherng, J. M. (2016). Molecular Mechanism of Cinnamomum verum Component Cuminaldehyde Inhibits Cell Growth and Induces Cell Death in Human Lung Squamous Cell Carcinoma NCI-H520 Cells *In Vitro* and *In Vivo*. *Journal of Cancer*, *7* (3), 251-261. doi: 10.7150/jca.13689.

[98] Hong, J. W., Yang, G. E., Kim, Y. B., Eom, S. H., Lew, J. H. & Kang, H. (2012). Anti-inflammatory activity of cinnamon water extract *in vivo* and *in vitro* LPS-induced models. *Bmc Complementary and Alternative Medicine*, *12*, 8. doi: 10.1186/1472-6882-12-237.

[99] Wong, H. Y., Tsai, K. d., Liu, Y. H., Yang, S. m., Chen, T. W., Cherng, J., Chou, K. S., Chang, C. M., Yao, B. T. & Cherng, J. M. (2016). *Cinnamomum verum* Component 2-Methoxycinnamaldehyde: A Novel Anticancer Agent with Both Anti-Topoisomerase I and II Activities in Human Lung Adenocarcinoma A549 Cells *In*

Vitro and *In Vivo*. *Phytotherapy Research*, *30* (2), 331-340. doi: 10.1002/ptr.5536.

[100] Zhang, K., Han, E. S., Dellinger, T. H., Lu, J., Nam, S., Anderson, R. A., Yim, J. H. & Wen, W. (2017). Cinnamon extract reduces VEGF expression via suppressing HIF-1α gene expression and inhibits tumor growth in mice. *Molecular Carcinogenesis*, *56* (2), 436-446. doi: 10.1002/mc.22506.

[101] Baliga, M. S., Haniadka, R., Pereira, M. M., D'Souza, J. J., Pallaty, P. L., Bhat, H. P. & Popuri, S. (2011). Update on the Chemopreventive Effects of Ginger and its Phytochemicals. *Critical Reviews in Food Science and Nutrition*, *51* (6), 499-523. doi: 10.1080/10408391003698669.

[102] Butt, M. S. & Sultan, M. T. (2011). Ginger and its Health Claims: Molecular Aspects. *Critical Reviews in Food Science and Nutrition*, *51* (5), 383-393. doi: 10.1080/10408391003624848.

[103] Zick, S. M., Turgeon, D. K., Vareed, S. K., Ruffin, M. T., Litzinger, A. J., Wright, B. D., Alrawi, S., Normolle, D. P., Djuric, Z. & Brenner, D. E. (2011). Phase II Study of the Effects of Ginger Root Extract on Eicosanoids in Colon Mucosa in People at Normal Risk for Colorectal Cancer. *Cancer Prevention Research*, *4* (11), 1929-1937. doi: 10.1158/1940-6207.Capr-11-0224.

[104] Elwood, P. C., Gallagher, A. M., Duthie, G. G., Mur, L. A. J. & Morgan, G. (2009). Aspirin, salicylates, and cancer. *The Lancet*, *373* (9671), 1301-1309. doi: 10.1016/S0140-6736(09)60243-9.

[105] Guahk, G. H., Ha, S. K., Jung, H. S., Kang, C., Kim, C. H., Kim, Y. B. & Kim, S. Y. (2010). Zingiber officinale Protects HaCaT cells and C57BL/6 Mice from Ultraviolet B-Induced Inflammation. *Journal of Medicinal Food*, *13* (3), 673-680. doi: 10.1089/jmf.2009.1239.

In: Phytochemicals from Mexican … ISBN: 978-1-53617-397-0
Editors: J. Basilio Heredia et al.

Chapter 7

CARDIOVASCULAR DISEASES

Nayely Leyva-López[1,2], Crisantema Hernández[2], Cynthia E. Lizárraga-Velázquez, Erick P. Gutiérrez-Grijalva[1,3] and J. Basilio Heredia[3,*]

[1]Cátedras CONACYT-CIAD, A. C., Culiacán, México
[2]Laboratory of Nutrition and Food Plant, Centro de Investigación en Alimentación y Desarrollo, A. C., Mazatlán, Sinaloa, México
[3]Laboratory of Functional Foods and Nutraceuticals, Centro de Investigación en Alimentación y Desarrollo, A. C., Culiacán, Sinaloa, México

ABSTRACT

Cardiovascular disease (CVD) is one of the major causes of death worldwide. These illnesses are linked to oxidative stress and inflammation. Several drugs have been developed in order to treat CVDs. Nevertheless, these medicines have serious side effects and are expensive. This situation promotes the search for alternative therapies to prevent and treat CVDs. Some Mexican medicinal plants, such as *Aesculus hippocastanum*, *Allium sativum*, *Cynara scolymus,* and *Ginkgo*

* Corresponding Author's Email: jbheredia@ciad.mx.

biloba, among others, are known for their cardioprotective effects and have been used to prevent and alleviate CVDs. The bioactivity of these plants is related to their content of phytochemicals, such as phenolic compounds, terpenes, and alkaloids. Therefore, in this chapter, current evidence that justifies the use of several Mexican herbal medicines against CVDs and their main active phytochemical constituents will be discussed.

Keywords: cardioprotective effects, herbal treatments, bioactive compounds

INTRODUCTION

Cardiovascular diseases (CVDs) are one of the main causes of death and disability in most regions of the world. CVDs are a group of ailments of the heart and blood vessels, as well as coronary heart disease, cardiomyopathies, cardiac arrhythmias, cerebrovascular disease, peripheral artery disease, congenital and rheumatic heart diseases, deep vein thrombosis and pulmonary embolism [1].

The main risk factors for CVDs are listed in Table 1. The non-modifiable factors include genetics, age, gender, and ethnicity, among others [2]. For example, in a study carried out in Finland, it was found that men and women older than 45 and 55 years, respectively, are more likely to suffer a death from CVDs [3]. Among the modifiable risk factors are unhealthy diets (high in saturated fats and salt), overweight and obesity, lack of exercise, tobacco and alcohol consumption. In addition to lifestyle-related factors, CVDs risk increases in patients with hypertension, high levels of cholesterol and triglycerides, as well as diabetes [4].

Furthermore, atherosclerosis is linked to oxidative stress, which is the top regular reason responsible for CVDs [5], such as ischaemic heart disease, cerebrovascular disease, and diseases of the aorta and arteries, including hypertension and peripheral vascular disease [1]. Atherosclerosis origins as the deterioration in blood vessels, due to the cholesterol deposition in arteries and augmentation of atherosclerotic plaques. These affect vessels and block blood flow and culminates as coronary heart disease or cerebrovascular diseases [6].

Many biomarkers have been proposed in order to predict or identify some cardiovascular conditions. These can be classified as follows [7-9]:

1. Markers of inflammation. These include the C-reactive protein (CRP), myeloperoxidase (MPO), and some cytokines, such as interleukin-6 (IL-6), tumor necrosis factor-α (TNF-α) and monocyte chemoattractant protein-1 (MCP-1), among others.
2. Markers of hemodynamic stress. Components of the natriuretic peptide system mainly represent them.
3. Markers of hemostasis. These include fibrinogen and fibrin degradation products, as well as the von-Willebrand factor.
4. Markers of necrosis and ischemia. These comprise troponin, ischemia-modified albumin, and free fatty acids (FFA).
5. Markers of platelet function. Include the soluble CD40 ligand and P-selectin.

Table 1. Cardiovascular diseases risk factors

Non-modifiable	Modifiable
Genetics	*By changing lifestyle
Gender	Unhealthy diet
Age	Physical inactivity
Ethnicity	Obesity
Social status	Smoking
Low educational status	Alcohol abuse
	Stress
	*By drugs
	Hypertension
	Diabetes
	Dyslipidemia (high levels of cholesterol and hypertriglyceridemia)

Table 2. Drugs usually employed for the treatment of cardiovascular diseases

Drug	Effect
ACE inhibitors	Inhibit secretion of inflammatory cytokines (IL-6, TNF-α) Reduce deployment and maturation of dendritic cells from the spleen to peripheral circulation
Alpha-blockers	Cause vasodilation Impede the union of norepinephrine to the smooth muscle receptors Inhibit cholesterol synthesis
Amiodarone	Inhibits abnormal rhythms of the heart
ARBs	Reduces high blood pressure Improve postischemic cardiac function
Aspirin (acetylsalicylic acid)	Inhibits activation of NF-κB and COXs Blocks platelet aggregation Reduce circulating FFA and triacylglycerols
Beta-blockers	Target the beta receptors Diminish the cardiovascular response to an adrenergic stimulus Reduce ischemia and protect the infarcted myocardium Reduce circulating levels of FFA
Calcium channel blockers	Dilate peripheral and coronary arteries Suppress myocardial carbohydrate use
Central sympathetic inhibitors	Diminish levels of FFA Enhance the secretion of insulin Reduce cholesterol levels
Ivabradine	Inhibit the funny (I_f) current in the sinoatrial node Reduces heart rate
Nitrates	Augment oxidation ration of lipids to carbohydrates
	Dilate large and medium-size coronary arteries and venous capacitance vessels

ACE: Angiotensin-converting enzyme; ARBs: Angiotensin II receptor blockers; NF-κB: nuclear factor kappa-light-chain-enhancer of activated B cells; COXs: Ciclo-oxygenases; FFA: Free fatty acids.

Numerous drugs have been developed in order to treat CVDs (Table 2) [10], including aspirin [11, 12], amiodarone [13, 14], alpha [15] and beta-blockers [16], calcium channel blockers [17], central sympathetic inhibitors [18], ivabradine [19], organic nitrates [20], angiotensin-converting enzyme (ACE) inhibitors [21], and angiotensin II receptor blockers (ARBs) [22], among many others.

Table 3. Major phytochemicals identified in medicinal plants with cardioprotective effects

Medicinal plant	Main constituents	References
Agrimony (*A. eupatoria*)	Agrimoniin, apigenin 7-*O*-glucuronide, catechin, luteolin 7-*O*-glucuronide, procyanidin B3, quercetin 3-*O*-galactoside, quercetin 3-*O*-glucoside	[41]
Aloe (*A. vera*)	Aloe-emodin, aloin, barbaloin,, aloe-barbendol, aeloesaponarin II, aloesin, aloesone, 8-C-glucosyl-noreugenin, aloeresin E, chrysophanol, feroxidin, caffeic acid, chlorogeni acid, isovitexin, isorhamnetin, daidzenin, genistein	[196, 73]
Artichoke (*C. scolymus*)	Cynaroside, scolymoside, luteolin, apigenin, Inositol, choline, trigonelline, aguerin B, cynarin, chlorogenic acid, 1,5-dicaffeoylquinic acid, *O*-caffeoylquinic acid, (E)-cinammaldehyde, *O*-feruloylquinic acid, cynarasaponin E, cynarasaponin C	[103, 104]
Bilberry (*V. myrtillus*)	Cyanidin-3-*O*-glucoside, cyanidin-3-*O*-arabinoside, delphinidin-3-*O*-glucoside, and several delphinidin, cyanidin, petunidin, peonidin and malvidin derivatives, chlorogenic acid, ellagic acid pentoside, quercetin-3-*O*-galactoside and quercetin-3-*O*-glucuronide	[183, 182, 197]
Cardamom (*E. cardamomum*)	Quercetin, kaempferol, rutin, α-terpenyl acetate, 1,8-cineole, linalool, α-terpineol, β-terpineol, myrcene, α -pinene, linalyl acetate, sabinene, 4-terpineol, feranesol, nerolidol	[115, 114, 113]
Cinnamon (*C. verum* or *C. zeylanicum*)	Proanthocyanidins, cinnamic acid, coumarin, cinnamaldehyde, 4-methoxycinnamaldehyde, δ-cadinene, α-copaene, α-muurolene	[198, 199]
Garlic (*A. sativum*)	Allicin, allyl isothiocyanate, gallic acid, chlorogenic acid, naringenin, rutin, quercetin, hyperoside, diallyl mono-, di-, tri- and tetra-sulfides, alliin, GSPC, GSAC	[59-61]
Ginger (*Z. officinale*)	6-Shogaol, 6-gingerol and 8-gingerol	[200-202]
Ginkgo (*G. biloba*)	Ginkgolides (A, B, C), bilobalide, quercetin, kaempferol, isorhamnetin, rutin, quercetin-3-*O*-glucoside, kaempferol-3-*O*-rutinoside, isorhamnetin-3-*O*-rutinoside, quercetin-3-*O*-rhamnoside	[203, 124, 204]
Rossell (*H. sabdariffa*)	Gallic, protocatechuic, chlorogenic and caffeic acids, galloyl ester, quercetin derivative, tilirosid, delphinidin-3-sambubioside, cyanidin-3-sambubioside, chlorogenic acid, tetra-*O*-methyljeediflavanone, quercetin-3-sambubioside, myricetin-3- glucoside	[205-208]

Even though there are several cardiovascular drugs, their use has shown side effects. For instance, amiodarone can induce hyper- and hypothyroidism [23]. ACE inhibitors, such as captopril, can cause bronchospasm and dry cough [24]. Besides, ACE inhibitors, in combination with ARBs, can provoke renal dysfunction and symptomatic hypotension in patients with symptomatic left ventricular dysfunction [25]. Furthermore, beta-blockers have been contraindicated in diabetic patients [26], and calcium channel blockers might cause vomiting, nausea, and gut ischemia [27].

Another drawback of cardiovascular drugs is that these are not always economically accessible for people from low- and middle-income countries. Therefore, the population has turned to more affordable and less harmful alternative therapies to prevent, alleviate, and treat these ailments, such as herbal remedies.

Epidemiological studies have demonstrated that a diet abundant in vegetable products, such as fruits, whole grains, herbs, and spices are linked to a reduced risk of CVDs [28]. Numerous dietary plants and herbs have been reported for their cardioprotective effect [29].

Potential therapeutic plants used in Mexico, include garlic (*Allium sativum*), artichoke (*Cynara scolymus*), horse chestnut seed (*Aesculus hippocastanum*), *Ginkgo biloba*, cardamom (*Elettaria cardamomum*), roselle (*Hibiscus sabdariffa*), *Aloe vera* and many others. The beneficial health effects of these and other plants are associated with their phytochemical composition. The main phytochemical constituents in a diverse group of medicinal plants, used to ameliorate cardiovascular-related ailments, are mentioned in Table 3.

In the next section, evidence that supports the use of Mexican medicinal plants will be addressed, as well as the main bioactive phytochemicals responsible for their cardioprotective activities.

MEXICAN MEDICINAL PLANTS AGAINST CARDIOVASCULAR AILMENTS

Aesculus hippocastanum L.

Aesculus hippocastanum L. is commonly known as horse chestnut. Horse chestnut seed extract (HCSE) has been traditionally used for the treatment of chronic venous insufficiency (CVI) [30]. Obesity is associated with increased cardiovascular risk through the detonation of several risk factors such as insulin resistance, hyperlipidemia, and hypertension [31]. On this matter, Avci et al. [32] evaluated the anti-hyperlipidemic effect of HCSE extract on plasma total cholesterol (TC), triglycerides (TG), high-density lipoprotein (HDL) and low-density lipoprotein (LDL) levels of mice with diet-induced obesity. The results of this study reported an increase in HDL concentration, which has been related to a lower incidence of CVDs, such as atherosclerosis [33]. HDL can protect against atherosclerosis through the inhibition of the expression of adhesion molecules in endothelial cells involved in the recruitment of blood monocytes into the arterial wall [34]. Also, HDL may provide antioxidant protection of LDL *in vivo*. Consequently, it inhibits the generation of proinflammatory oxidized lipids [35].

Phytochemical Constituents of A. hippocastanum

Aesculus hippocastanum has long been used to alleviate circulatory-related ailments. Main phytoconstituents of HCSE are aescin (a mixture of triterpenoid saponins), tannins, glycosides of quercetin, kaempferol, and coumarin derivatives [36]. The therapeutic potential of HCSE is attributed mainly to aescin which exists in two forms, α, and β. The β-aescin is the active molecular form. This molecule is characterized by exhibit anti-edematous, venotonic, anti-inflammatory, and antioxidant properties, which have been widely investigated for the treatment of CVI [37]. On the contrary, the researches on the use of HCSE or its active compound on the prevention of CVDs are scarce.

***Agrimonia eupatoria* L.**

Agrimony (*A. eupatoria* L.) tea has been traditionally used for the Mexican population to treat circulatory problems. In this context, Kuczmannová et al. [38] evaluated the effect of *A. eupatoria* infusions on the aorta relaxation of male Wistar rats. Rats were treated with streptozotocin (STZ) to induce a diabetic state, later the infusions (200 mg/L) were administered to the organisms for five weeks. The results showed a significantly improved aorta relaxation in rats treated with *A. eupatoria* water extracts; therefore, this plant might exert activity as a vasodilator. Furthermore, Ivanova, Vankova, and Nashar [39] investigated the effect of tea from aerial parts of *A. eupatoria* on pro-inflammatory cytokines and metabolic markers of healthy male (5) and female (14) individuals. After one month of tea consumption (200 mL of tea twice a day), plasma levels of high-density lipoprotein (HDL) significantly increased while IL-6 concentration decreased. According to the authors, these results showed the potential of this plant for the prevention of CVDs.

Phytochemical Constituents of A. eupatoria

The phenolic profile and content of *A. eupatoria* have been previously reported. Ivanova et al. [40] reported that infusions of agrimony had 702.29 ± 6.82 μM quercetin equivalents (QE). Granica et al. [41] identified 24 polyphenolic compounds, including phenolic acids (0.6-0.9 mg/g), tannins (6.3-10.9 mg/g), especially agrimoniin, and flavonoids (8.2-10.9 mg/g) in aqueous extracts from aerial parts of agrimony. Furthermore, Kuczmannová et al. [42] reported that tea-like extracts from this plant contained a high content of polyphenols (7.87%), especially quercetin and procyanidin derivatives. It has been demonstrated that aqueous injection of quercetin (5 mg/kg) can reduce 30% of the arterial blood pressure in hypertensive Wistar rats [43]. On the other hand, intragastric administration of procyanidin B2 (30 mg/kg) showed the capacity of reducing myocardial fibrosis by downregulating the action of cardiac fibroblast in mice. Additionally, the organisms treated with this tannin showed reduced serum levels of cholesterol and triglycerides [44].

Therefore, these compounds might be responsible for the cardioprotective effect of *A. eupatoria*.

Allium sativum L.

Garlic (*A. sativum*) is one of the ancient plants used for their medicinal properties; its hypolipidemic, antihypertensive, antiatherosclerotic, and cardioprotective effects have been extensively studied [45-49]. In this context, Lee et al. [50] evaluated the effect of an aqueous garlic extract on HDL levels and regulation of hepatic apolipoprotein A-I (apoA-I) gene expression in male Sprague-Dawley rats fed with a high-fat diet. Supplementation of 20 mg of garlic extract/kg diet for five weeks caused a significant reduction of LDL, TG, atherogenic index and aspartate aminotransferase (AST) levels in plasma of rats from 34.97 ± 2.76 mg/dL to 26.49 ± 3.40 mg/dL, from 50.37 ± 3.51 mg/dL to 40.64 ± 3.94 mg/dL, from 1.75 ± 0.12 to 0.97 ± 0.18 and from 65.45 ± 8.29 IU/L to 45.07 ± 3.98 UI/L, respectively. Hepatic lipids, such as TG and TC, were also reduced. The administration of garlic extract also increased plasma levels of HDL from 25.42 ± 1.21 mg/dL to 40.35 ± 6.30 mg/dL. Besides, gene expression of ATP-binding cassette transporter A1 (ABCA1) and lecithin: cholesterol acyltransferase, involved in HDL synthesis, were too up-regulated. These results show that garlic extract reduces hyperlipidemia in high fat fed-rats and has potential as a preventive agent of CVDs.

Furthermore, garlic aqueous extract supplementation at a dose of 250 mg/kg to Wistar albino rats reduced systolic blood pressure from 125.67 ± 12.04 mmHg to 117.67 ± 12.89 mmHg after three weeks. Plasma levels of lactate dehydrogenase (LDH), a marker of tissue damage, were lessened from 111.67 ± 12.11 U/L to 104.50 ± 7.97 U/L, while creatine kinase-MB (CK-MB) levels, a myocardium-specific marker, were diminished from 205.00 ± 17.32 U/L to 198.17 ± 8.70 U/L. Besides, the heart tissue showed increased levels of total antioxidant capacity (TAC) and Na^+/K^+-ATPase activity and reduced levels of total peroxides (TP) and Ca_2^+. These discoveries suggest that supplementation of garlic extract protects the heart of rats by lowering oxidative stress and regulating cardiac Na^+/K^+-ATPase activity and Ca_2^+ levels [51].

Several studies conducted in humans have established that extracts, oil, or powder of garlic affect numerous biomarkers related to cardiovascular risks, such as CRP [52], LDL [53], HDL, TC and malondialdehyde (MDA) [54, 55], among others. On the other hand, some reports mention that garlic might not have significant effects on lipid profile, blood pressure arterial stiffness, CRP levels, and other markers of endothelial function [56, 57]; so the cardioprotective action of garlic extracts might still be controversial. Recently, Szulińska et al. [58] determined the effect of garlic extract capsules on the arterial stiffness and markers of endothelial function in obese subjects. Administration of two capsules of garlic extract (400 mg) daily during three months significantly reduced arterial stiffness index (↓4%), diminished LDL (↓26.6%) and TG (↓13.2%) levels, CRP concentration (↓55.3%) and increased total antioxidant status (↑11.7%) in the serum blood of subjects. The findings in this research demonstrate that the administration of garlic extract capsules to obese individuals positively alters endothelial biomarkers linked to CVDs risks. Furthermore, garlic extract capsules might be able to inhibit chronic inflammation in obese subjects [58].

Phytochemical Constituents of A. sativum

The main phytochemical constituents of *A. sativum* are sulfuric compounds, phenolic acids, and flavonoids, as well as ascorbic acid. For example, garlic varieties showed a content of ascorbic acid from 0.268 mg/g to 0.658 mg/g, an allicin content ranging from 0.411% to 1.105%. Furthermore, several phenolic acids, flavanones, flavanols, flavones, and flavan-3-ols have also been identified in garlic extracts [59]. The essential oil from garlic bulbs mainly contained diallyl sulfides. Diallyl disulfide content was 29.08 mg/100 mg of essential oil, while diallyl trisulfide content was 21.56 mg/100 mg of essential oil [60]. Alliin (23.63 mg/g of dry weight), γ-L-glutamyl-S-(trans-L-propenyl)-L-cysteine (GSPC) (28.56 mg/g), γ-L-glutamyl-S-allyl-L-cysteine (GSAC) (13.76 mg/g) and allicin (5.58 mg/g) are the main organosulphur compounds reported in aqueous extracts of *A. sativum* [61].

Chan et al. [62] demonstrated that allicin (0.3 to 10 μM) treatment, in H9c2 cardiomyoblast cells, diminish cardiovascular risk by reducing the intracellular reactive oxygen species (ROS) augment, that is to say diminishing free radical-induced myocardial cell apoptosis in ischemic conditions. Furthermore, Chen et al. [63] established that treatment of human endothelial cells (HUVEC) with allicin 30 and 100 μM significantly reduced LDL-induced endothelial damage and cell death by inhibiting NADPH-oxidase signaling pathway. Additionally, it has been mentioned that allicin administration (40 mg/kg per day) during six weeks to male Wistar rats enhances renal and cardiac functions and reduces hypertension. Besides, allicin treatment also diminishes the vascular reactivity to angiotensin II and increases Nrf2 expression, which reduces oxidative stress and increases superoxide dismutase (SOD), catalase (CAT) and glutathione peroxidase (GPx) expression levels [64]. Recently, Liu et al. [65] evaluated the effect of allicin supplementation on carotid artery intima-media thickness (IMT) and plasma homocysteine (Hcy) in coronary heart disease patients. Allicin treatment (40 mg three times a day, for 12 weeks) significantly reduced TC (↓9.,75%), TG (↓24%), and Hcy (↓26.4%) serum levels of subjects when compared with the untreated group. Intima media thickness also was diminished by allicin supplementation (↓8.13%). These findings show that the use of allicin might be beneficial for the prevention of atherosclerosis, partially by modulating serum lipid profile, reducing IMT, and diminishing Hcy levels [65]. Allicin has demonstrated a reductive effect on blood lipids and the aggregation of platelets. Therefore garlic preparations may play a role in preventing CVDs and atherosclerosis [66].

Aloe vera L. Burm

Aloe vera has been described to possess antioxidant, hypolipidemic, antiatherosclerotic, and cardioprotective effects [67, 68]. In this regard,

Dana et al. [69] evaluated the effect of *A. vera* leaf gel on markers of lipid metabolism and inflammation as well as on fatty streak formation in hypercholesterolemic rabbits. The administration of *A. vera* leaf gel in the drinking water of rabbits (3.2% v/v) for 30 days reduced serum levels of TC and CRP when compared with the control group. Cholesterol concentration was reduced from 1059 mg/dL to 576.8 mg/dL and CRP levels were diminished from 2.5 mg/dL to 2.07 mg/dL. Triglyceride concentration in the serum of rabbits was not significantly affected. Besides, dietary *A. vera* significantly decreased fatty streak formation in the aorta of hypercholesterolemic rabbits, which means the potential of this plant to reduce the formation of atherosclerotic lesions [69].

Similarly, Kumar et al. [70] assessed the effect of *A. vera* gel supplementation on lipid profile in hypercholesterolemic male Wistar Rats. Orally administration of 500 mg of *A. vera* gel/kg per day during 45 days significantly reduced serum levels of TC, triacylglycerol, LDL, HDL, and very low-density lipoprotein (VLDL) when compared with the hypercholesterolemic control group. This study showed the hypolipidemic effect of *A. vera* gel in rats with hypercholesterolemia. Thus this plant has the potential to be used as a therapy to reduce cholesterol levels and diminish the risk of CVDs. In a clinical trial, the effect of *A. vera* leaf gel on lipid levels of hyperlipidemic type 2 diabetic patients was evaluated [71]. Patients ingested two aloe gel capsules per day containing 300 mg of aloe gel powder for two months. Glucose levels, as well as TC and LDL levels in serum, were significantly reduced. *Aloe vera* gel could be suggested as a natural treatment to remedy obesity-linked alterations.

Phytochemical Constituents of A. vera

Aloe is a very rich source of phytochemical compounds from different groups, such as anthraquinones (emodin), anthrones (aloins), chromones (aloesins), coumarins, pyrans, pyrones, alkaloids, benzene and naphthalene derivative, flavonoids, carbohydrates (acemannan), amino acids and enzymes and lipoid components [72]. It has been reported that aloe gel contains a total phenolic content of 217 mg GAE/100g and a flavonoids content of 10 mg RE/100 g [73]. Furthermore, Aloe leaf exudates have been reported to contain a total phenolic and flavonoids content of 320 mg GAE/100g and 140 mg RE/100 g, respectively [74].

It has been demonstrated that emodin regulates lipid metabolism in mice. Administration of 80 mg/kg per day for six weeks significantly reduced TC and TG hepatic levels and the size of white and brown adipocytes in obese C57BL/6 mice [75]. According to these findings, emodin might prevent obesity and related metabolic diseases in high-fat diet-fed mice [75]. Furthermore, the effect of the *A. vera* phytochemical aloe-emodin on cardiac injuries induced by myocardial infarction *in vivo* and oxidate insults *in vitro* was evaluated [76]. Aloe-emodin was administrated to Kunming mice at a dose of 100 mg/kg per day for two weeks, while cardiomyocytes isolated from Sprague Dawley rats were treated with aloe-emodin (10, 50 and 100 μM) for 12 h and then died with H_2O_2 for four hours. Aloe-emodin treatment diminished the cell death and ROS production induced by H_2O_2 on the cardiomyocytes.

Furthermore, aloe-emodin administration significantly lowered oxidative stress, infarct size, impaired cardiac function, and therefore reduced cardiac apoptosis on Kunming mice, most likely by up-regulating miR-133, inhibiting ROS production and suppressing caspase-3 apoptotic signaling pathway [76]. It has also been mentioned that emodin improves cardiac fibrosis, and that inhibits the proliferation of vascular smooth muscle cells, which reinforce the potential of *A. vera* as a cardioprotective agent [77, 78]. Emodin might be responsible for the cardioprotective effects of *A. vera*; nevertheless, there are several phytochemicals in this plant that might be exerting synergistic effects.

Cinnamomum verum J. Presl.

Common cinnamon is a medicinal plant that has been traditionally used to reduce arterial blood pressure, plasma glucose, obesity, and to alleviate dyslipidemia [79]. The antihypertensive and vasorelaxant effects of cinnamon bark extracts have been previously demonstrated in rat models [80, 81].

Askari, Rashidkhani, and Hekmatdoost [82] reported that the oral administration of capsules containing 1.5 g of cinnamon powder in non-alcoholic fatty liver disease (NAFLD) patients for 12 weeks significantly diminishes TC, TG, and LDL serum levels, but did not affect HDL concentration. Levels of CRP were also reduced in treated patients (2.1 ± 1.0 mg/mL) when compared to the control group (4.6 ± 2.6 mg/mL) [82]. Recently, it was demonstrated that oral administration of cinnamon powder capsules (3 g per day) for 16 weeks significantly reduced waist circumference, blood pressure, serum levels of cholesterol, and TG in Asian Indians individuals. Furthermore, ingesta of these cinnamon capsules increased HDL and reduced LDL cholesterol in the subjects [83]. Similar results were obtained by Borzoei et al. [84] when 1.5 g of cinnamon powder were administrated to women with polycystic ovary syndrome (PCOS) for eight weeks. Serum levels of TC were higher in the control group (173.52 mg/dl) than in the cinnamon-treated group (169.19 mg/dl); the same trend was shown for the triacylglycerol. High-density lipoprotein levels were higher (44.59 mg/dl) in the treated individuals than in the untreated patients (41.0 mg/dl), while LDL serum levels were lower (105.94 mg/dl) in the cinnamon-treated group than in the control group (108.26 mg/dl). Besides, serum levels of total antioxidant status significantly augmented in the cinnamon-treated group by 9.28%, while MDA levels diminished in 7.87% in the cinnamon-treated group when compared with the control group after the eight weeks trial [84].

These results show that cinnamon exerts a modulatory lipid profile effect as well as anti-inflammatory and antioxidant activities, which might provide to this plant the potential to be used as a cardioprotective agent.

Phytochemical Constituents of C. verum

Several phytochemicals, including polyphenolic compounds and terpenes, have been identified in cinnamon. For instance, cinnamaldehyde (38.78 mg/g), coumarin (4.44 mg/g), cinnamic acid (0.68 mg/g) and cinnamyl alcohol (0.27 mg/g) were the main components in cinnamon bark methanolic extracts. Catechin, *p*-coumaric, and protocatechuic acids were also found [85]. It has been reported that aqueous extracts from cinnamon bark contained polyphenolic compounds, alkaloids, and saponins [80]. Mateos-Martín et al. [86] identified proanthocyanidins, containing epigallocatechin and epicatechin gallate units, as the main constituents on the commercial cinnamon extract. On the other hand, Li et al. [87] identified 22 compounds in extracts of commercial cinnamon. These compounds are: "5-hydroxyethyl salicylate, syringaldehyde, *p*-hydroxybenzoic acid, isovanillic acid, protocatechuic acid, protocatechualdehyde, vanillin, vanillic acid, *trans*-cinnamaldehyde, cinnamyl alcohol, cinnamic acid, sinapaldehyde, litseachromolaevanes A, 4-hydroxy-1,10-seco-muurol-5-ene-1,10-dione, pinoresinol, syringaresinol, coumarin, 4-hydroxymellein, kaempferol, (-)-(2*R*,3*R*)-5,7-dimethoxy-3′,4′-methylenedioxy-flavan-3-ol, decumbic acid, and β-sitosterol" [87].

It has been demonstrated that procyanidins (100 mg/kg) fed to male Wistar rats for four weeks diminished serum levels of TC, TG, and LDL based on a sugary fatty cream diet-model [88]. This effect might be due to the inhibition of intestinal absorption of cholesterol by polyphenolic compounds [89]. On the other hand, cinnamaldehyde (40 μM) increase adipose tissue lipolysis, by enhancing the expression level of lipase and repressing the expression of glycerol-3-phosphate dehydrogenase, and reduce inflammation, by diminishing expression of IL-1β, IL-6, MCP-1, TNF-α and COX in high fat fed-mice [90]. These studies show the potential for phytochemicals from *C. verum* to improve the different risk factors linked to CVDs.

Cynara scolymus L.

Cynara scolymus, also known as artichoke, has been traditionally used as a remedy for several ailments. Among its several beneficial effects, the lowering-lipid property has been vastly studied and reviewed [91-93]. Also, *C. scolymus* extracts have shown *in vitro* moderate (31% at 100 μg/mL) [94] to high (IC_{50} value = 356.62 mg/mL) [95] ACE inhibitory activity, which might provide to this plant the potential to be used as antihypertensive agents.

Studies have demonstrated that oral administration of aqueous extracts of artichoke leaves can reduce serum levels of TC, TG, LDL, VLDL, and MDA as well as increase plasma antioxidant activity in either diabetic or hypercholesterolemic rat models [96, 97]. For instance, Mocelin et al. [98] assessed the hypolipidemic and antiatherogenic effects of an artichoke leaf aqueous extract (150, 300, and 600 mg/kg per day during four weeks) in a hypercholesterolemic male Wistar rat model. Administration of artichoke extracts significantly reduced serum levels of cholesterol and LDL when compared with the untreated hypercholesterolemic control group. Furthermore, the concentration of IL-6, IL-1, TNF, IFN, and CRP was decreased in the artichoke-treated rats. These results suggest that artichoke exerts both hypolipidemic and antiatherogenic effects due to the enhancement of the lipid profile and inhibition of inflammatory cytokines in rats. Recently, Abdulkhaleq et al. [99] evaluated the effect of aqueous artichoke leaves extracts (600 and 1500 mg/kg per day for three weeks) on lipid profile and oxidative stress of high fat diet-fed male albino rats. Administration of 1500 mg/ kg of artichoke extract significantly reduced serum levels of TC, TG, and LDL when compared with the high fat diet-treated group. Furthermore, serum levels of antioxidant enzymes SOD and CAT, as well as total antioxidant capacity and reduced glutathione (GSH), were significantly increased in the Artichoke-treated group when compared with the high fat diet-fed group, while levels of MDA were reduced. These findings suggest that aqueous extracts of artichoke might be useful as a dietary supplement in order to prevent obesity induced by a high ingesta of lipids; besides, artichoke extracts might exert protective effects against oxidative stress-related diseases, such as CVDs [99].

Recent reports stated that artichoke leaf extract (1.8 g/day during 12 weeks) significantly reduced levels of TG and ox-LDL in patients with metabolic syndrome without significant changes in MDA, SOD, and GPx serum levels [100] or cholesteryl ester transfer protein [101]. Also, supplementation of artichoke leaf extract capsules at a dose of 500 mg/day for eight weeks on overweight individuals significantly reduced TC and LDL serum levels, while HDL levels were increased [102]. These findings suggest that artichoke extract could be used as an agent to reduce mild hypercholesterolemia. More clinical trials with a more significant number of individuals are needed in order to establish the hypolipidemic and antioxidant effect of artichoke in humans, and therefore its potential to be used as cardioprotective agents.

Phytochemical Constituents of C. scolymus

Artichoke is a source of diverse phytochemicals compounds, such as alkaloids, sesquiterpene lactones, flavones, and cinnamic acid derivatives [103]. For instance, it has been reported that artichoke leaf aqueous extracts contain a total phenolic content of 359 ± 2.8 mgGAE/g while the total flavonoid content is 48.07 ± 2.8 mgQE/g [98]. Values of 1,5-dicaffeoylquinic acid content in artichoke leaves are 3.6-62 mg/g, for cynaroside is 22.3-55 mg/g, for scolymoside is 17-49 mg/g [104]. Content of flavonoids, such as luteolin and apigenin in artichoke is 3.2-12.7 mg/g and 4.9 mg/g, respectively; values of cinnamaldehyde and caffeoylquinic acids, such as chlorogenic acid, concentration are 0.3-1.1 mg/g and 0.4-15.1 mg/g, respectively [103]. Cinnamic acids and flavonoids might be responsible for the cardioprotective effects of artichoke leaf extract since they are present in a considerable amount in artichoke leaves.

Chlorogenic acid administration (1, 10 mg/kg per day during four weeks) significantly reduced serum levels of cholesterol and LDL in Sprague Dawley rats fed with a high cholesterol diet. Furthermore, lipid depositions in the liver were diminished in the chlorogenic acid-supplemented rats. The hypocholesterolemic effect of chlorogenic acid might lead to other favorable effects, namely atherosclero-protective and cardioprotective action [105]. Besides, it has been demonstrated that chlorogenic acid administration (40 mg/kg per day during 19 days) alleviates myocardial injury induced by isoproterenol in male albino Wistar rats. Luteolin (5, 10 μM), which glycoside is cynaroside, significantly reduced inflammation and oxidative stress-induced in H9c2 cardiomyocytes by inhibiting NF-κB pathway and activating Nrf2 signaling pathway [106]. Furthermore, this flavonoid (20 mg/kg per day during 15 weeks) also diminished cardiac fibrosis, hypertrophy, and dysfunction in a diabetic-induce C57BL/6 mouse model. These findings suggest that luteolin preserve heart tissue integrity in a diabetic mouse model, and therefore it might have the potential to be used as a therapeutic agent.

Elettaria cardamomum L. Maton

Cardamom (*E. cardamomum*) is traditionally used to treat cardiovascular disorders. The crude extract of cardamom reduced blood pressure in rats [107]. Furthermore, 3 g administration of cardamom powder for 12 weeks to individuals with stage 1 hypertension significantly reduced blood pressure and increased fibrinolytic activity. Total antioxidant status was also increased by cardamom supplementation [108]. Nagashree et al. [109] evaluated the antihypercholesterolemic effect of cardamom in Wistar rats fed with a high-fat diet. Cardamom oil administration (3 g/kg during eight weeks) significantly reduced blood levels of cholesterol, LDL, and TG; furthermore, the cholesterol levels on cardiac muscle were also diminished. Besides, the activity of SOD was increased in the hearts of rats treated with cardamom oil. These results suggest the cardioprotective properties of cardamom by reestablishing the changes in lipid profile under hypercholesterolemic conditions [109].

Clinical trial studies have been conducted in order to demonstrate the effect of cardamom supplementation on lipid profile, blood pressure, and prevent inflammation in individuals. For instance, Daneshi-Maskooni et al. [110] reported that the administration of cardamom powder capsules (3 g/day during three months) significantly reduced serum levels of CRP and IL-6. These data suggest the anti-inflammatory action of *E. cardamomum* and its potential use as a therapeutic agent in inflammation-related diseases, such as CVDs. Furthermore, serum levels of TC, TG, and LDL were diminished by the administration of cardamom capsules under the same conditions mentioned last [111]. On the other hand, Fatemeh et al. [112] evaluated the effect of cardamom on serum lipids and blood pressure in pre-diabetic women. Administration of 3 g of cardamom powder/day for two months slightly reduced serum levels of TC, TG, and

LDL. Even though cardamom corrects these blood parameters, its effects are not significant. More studies are needed to clarify and to understand the effects of *E. cardamomum* in pre-diabetic subjects better.

Phytochemical Constituents of E. cardamomum

Gomaa et al. [113] reported that the total phenolic content of extracts from *E. cardamomum* fruit was 26.17 ± 2.88 μg GAE/mg, while flavonoid content was 9.69 ± 0.23 μg RE/mg. Additionally, methanolic extracts of *E. cardamomum* showed the presence of the flavonols, kaempferol, and quercetin (0.51 μg/mL), and of the flavonoid rutin; while essential oils components were 1,8-cineole (45.6%), α-terphenyl acetate (33.7%), sabinene (3.8%), 4-terpineol (2.4%), myrcene (2.2%) and α-pinene (1.5%), among others [114]. Besides, Sereshti et al. [115] reported the presence of linalyl acetate, linalool, and farnesol.

Cardioprotective effects of α terpineol and linalyl acetate, terpenes identified in *E. cardamomum* have been reported. Kim et al. [116] demonstrated that linalyl acetate (300 μM) exerts a relaxative effect on aorta rings of C57BL/6 mice. Furthermore, Linalyl acetate (1, 10, 100 mg/kg) recovers the heart rate of Sprague-Dawley rats exposed to nicotine as well as reduces serum levels of NO. These data suggest that linalyl acetate might be an effective agent that reduces nicotine-induced cardiovascular risk [116]. On the other hand, it has been demonstrated that α-terpineol causes vasorelaxation mediated by the release of NO and activation of the NO–GMPc pathway [117], therefore it might be used as an anti-hypertensive agent.

The cardioprotective effect of quercetin has been reviewed and reported [118]. Quercetin administration (0.01% *ad livitum*, during 26 weeks) to high-fat diet-fed C57BL/6 mice inhibited weight gain and diminished blood pressure. Furthermore, this flavonol reduced ACE activity and ROS levels, while increased total antioxidant capacity, GSH, and CAT activity in mice heart [119]. Quercetin, an important flavonol present in cardamom, decreases obesity and ameliorates oxidative stress in the heart of mice. Thus this plant may have therapeutic potential to alleviate CVDs.

Ginkgo biloba

Ginkgo biloba, also known as ginkgo, is a popular medicinal herb with anti-atherosclerotic properties. The vascular benefits of ginkgo are related to its capacity to stimulate vasodilation, improve blood flow, suppress platelet aggregation, and to function as an antioxidant. In this context, ginkgo leaves extract (3 mg/kg per day during 12 weeks) administration significantly reduced serum levels of TG (59%), TC (18%), and LDL (15%) in a high-fat diet-fed rabbit model. Additionally, the atherosclerotic lesion area was reduced (53.2%) in the group treated with ginkgo extract compared with the untreated group. Therefore, *G. biloba* leaves extract might exert anti-atherosclerotic action by modulating plasma lipid profile and by suppressing atherosclerotic lesion development [120]. Ginkgo administration (200 mg/kg per day during three weeks) as a phytosome formulation significantly increased the activity of the antioxidant enzymes SOD, CAT, and GPx and reduced the serum levels of MDA in Wistar albino rats; furthermore, it prevented myocardial necrosis induced by isoproterenol [121].

Similarly, ginkgo extract administration (200 mg/kg per day during four weeks) significantly increased SOD activity (38.2%) while reduced MDA (44.9%) serum levels in a myocardial ischemia-rat model [122]. Furthermore, ginkgo could effectively restore the metabolism of energy, fatty acids, and amino acids, which are closely related to its antioxidant activity. These results suggest that the cardioprotective effect of ginkgo is achieved through diverse metabolic pathways [122]. Furthermore, Saini, Taliyan, and Sharma [123] demonstrated that ginkgo extract administration (50 mg/kg per day during three weeks) reduced oxidative stress (thiobarbituric acid reactive substances, superoxide anion generation, and myocardium nitrite) and increased activity of the antioxidant enzyme GSH in an STZ-induced diabetic cardiomyopathy rat-model.

Phytochemical Constituents of G. biloba

Ginkgo biloba has a large diversity of phytochemical constituents; nevertheless, the main components belong to the group of terpenoids and flavonoids. Among the terpenoids, ginkgolides (A, B, and C) and bilobalide are the most representative, while quercetin, kaempferol, and isorhamnetin are the primary flavonoids present in *G. biloba* [124]. Ginkgolide A administration (4 mg/kg per day during 15 days) to Wistar rats improved myocardial function recovery after 60 min of reoxygenation [125]. Furthermore, *in vitro* pre- and post-ischemic perfusion of isolated hearts in the occurrence of ginkgolide A and ginkgolide B (0.05 μg/mL) or bilobalide (0.15 μg/mL) significantly improved left ventricular developed pressure, rate pressure product and coronary flow [125]. Additionally, it has been demonstrated that ginkgolide B pretreatment (0.1, 1, 2, 5 and 10 μM) of ventricular cardiomyocytes enhanced the function of ventricles from ischemia-reperfusion injury and diminished infarct size and release of LDH, which might indicate that ginkgolide B can partly prevent the ischemia-reperfusion injury in rat heart [126]. On the other hand, Liu et al. [127] demonstrated that ginkgolide B inhibits expression of CD40 ligand in thrombin-stimulated platelets, which might inhibit platelet function in atherosclerosis.

Hibiscus sabdariffa L.

Rosell (*H. sabdariffa* L.) is a plant with antioxidant, antihyperlipidemic, antihypertensive, and antiatherosclerotic effects [128, 129].

Multiples researches have focused on the evaluation of the anti-CVDs effects of powder and extracts of *Hibiscus sabdariffa* (HS) calyces. Metabolic syndrome (MS) has been a topic of primary interest because it is associated with a high prevalence of CVDs [130]. The MS involves a set of cardiovascular risk factors including central obesity, dyslipidemia (hypercholesterolemia, hypertriglyceridemia, and low levels of HDL), elevated blood pressure, impaired glucose tolerance [130].

Concerning the above mentioned, different studies have carried out, to evaluate the effect of HS on some risk factors of MS. Asgary et al. [131] reported that the daily consumption of 500 mg of powder of HS calyces, decrease systolic blood pressures and serum level of TG in patients with MS. Besides, it has been reported that daily consumption of 2000 mg of powder HS calyces produces a significant decrease in serum TC, LDL and TG levels on dyslipidemia in obese adolescents [132]. Joven et al. [133] reported that oral administration of extract HS calyces (125 mg/kg) reduces blood pressure in patients with MS and hypertensive rats. Pérez-Torres et al. [134] evaluated the antioxidant effect of the infusion of HS calyces on myocardial protection during ischemia/reperfusion in an MS rat model. The authors reported an increase of the antioxidant enzymatic activities (Glutathione S-Transferase, GST; Cu/Zn-SOD, Cu/Zn-SOD; and Mn-SOD, Mn-SOD, and nonenzymatic antioxidant capacity (NO_3^-/NO_2^-ratio and glutatión, GSH). Besides, many studies have shown that the extract of HS calyces exhibits anti-obesogenic potential (Table 3).

On the other hand, some authors have also shown that aqueous extracts of HS calyces exert cardioprotective effects on cardiotoxicity induced in rats and on primary vascular endothelial cells, through their cytoprotective, anti-inflammatory and antioxidant activities [135-137].

Phytochemical Constituents of H. sabdariffa

Roselle calyxes are rich in polyphenols, including anthocyanins (delphinidin-3-sambubioside, delphinidin-3-monoglucoside, cyanidin-3-monoglucoside, and cyanidin-3-sambubioside), flavonoids and proanthocyanidins, which are attributed to their bioactive potential [129].

Currently is known that the mixture of polyphenols reported in HS calyces exerts a hypotensive effect through 1) vasodilation, via the activation of endothelium-derived nitric oxide/cGMP-relaxant pathway and the increased synthesis/release of endothelium-derived nitric oxide [138]; and 2) ACE (angiotensin 1-converting enzyme) inhibition by the stimulation of new vessel formation [139]. The hypocholesterolemic effect of HS calyces is exerted through 1) inhibiting de HMG-CoA reductase involved in cholesterol biosynthesis and 2) inhibition of triacylglycerol synthesis and a consequent decrease of serum LDL [128]. In addition, HS polyphenols may act indirectly by up-regulating antioxidant enzymes expresión or protein activation, such as SOD, catalase (CAT), glutathione peroxidase (GPX) and glutathione reductase [140].

According to the studies mentioned, it is clear that HS induces a favorable endothelial response, so that should be considered for future studies on prevention and treatment of CVDs, which are related to endothelial dysfunction caused by risk factors such as dyslipidemia, arterial hypertension, and obesity [141]

***Humulus lupulus* L.**

Hops (H. *lupulus* L.) represents an essential ingredient in beer production; nevertheless, it has been reported that this plant exerts an effect as a platelet function regulator. For instance, Luzak et al. [142] reported that *H. lupulus* inhibited ADP-induced platelet aggregation and ameliorated anticoagulant activity of human endothelial cells. The antiplatelet action and the anticoagulant activity of vascular endothelium are a strategy for CVDs prevention. Therefore, the results of this study suggest that *H. lupulus* extract plays a potentially beneficial role in the regulation of platelet and endothelial function.

Phytochemical Constituents of H. lupulus

Hops are well known for being a source of resins, essential oils and phenolic compounds such as proanthocyanidins, and prenylflavonoids, including xanthohumol, isoxanthohumol, and 8-prenylnaringenine [143]. Mainly, xanthohumol isolated from *H. lupulus* inhibits cholesteryl ester transfer protein, which catalyzes cholesterol transfer between lipoproteins, leads to an increase in HDL, which is correlated directly with a low risk of atherosclerosis [144].

Hypericum perforatum L.

Hypericum perforatum (HP) is best known for its antipyretic, analgesic, antispasmodic, antiviral, and antimicrobial activities. However, HP extract also exhibits hypolipidemic and anti-atherosclerotic effects [145].

Hyperlipidemia and oxidative stress (oxidative damage to lipids) are considered as a common risk factor of CVDs because they are related to the formation and development of atherosclerotic plaques (Yang et al. 2017). Therefore, some authors have directed their efforts to the evaluation of the hypolipidemic and antioxidant effects of extracts and infusions of HP. Asgary et al. [145] reported that dietary administration of hydroalcoholic HP extract (150 mg/ kg) reduces levels of TG, TC, HDL, oxidized LDL, MDA, and CRP in hypercholesterolemic rabbits. Hernández-Saavedra et al. [146] reported that oral administration of HP infusion reduces body weight and abdominal fat mass and reduced serum TG, TC, LDL, and CRP levels in diet-induced obese rats. On the other hand, Moghaddam, Roghani, and Maleki [147] showed that the intraperitoneal administration of HP extract (300 mg/kg) reduces LDL and MDA levels in hyperlipidemic rats. According to the results, the extracts and infusions of HP could play an essential role in the prevention or treatment of atherosclerosis. Furthermore, an early report mentioned that dietary administration of HP hydroalcoholic extract reduces atherosclerotic lesions in hypercholesterolemic rabbits [148].

Phytochemical Constituents of H. perforatum

The major active constituents identified in HP are pseudohypericin (naphthodianthrones) hyperforin (phloroglucinol), hypericin, quercetin, and rutin (flavonoids). These compounds might be responsible for its hypolipidemic and anti-atherosclerotic activities [145, 149, 150].

Melissa officinalis L.

Melissa officinalis, commonly known as lemon balm, has been reported to exert anti-dyslipidemia effects. For instance, Asadi et al. [151] reported that the extract *M. officinalis* (rich in flavonoids) increased HDL levels in type 2 diabetic patients. Recently, it is well known that the increase in HDL is directly related to a lower risk of CVDs, such as atherosclerosis. Therefore, these results are relevant because the incidence of CVDs is higher in type II diabetic patients. In this regard, Asadi et al. [152] reported that *M. officinalis* extract is safe and effective in the improvement of ApoA-I, Apo B/ApoA-I, and lipids ratios as key factors promoting of CVDs.

Joukar et al. [153] reported that *M. officinalis* extract in low doses (50 mg/kg) is effective against the isoproterenol-induced cardiac injury to some extent by reducing the arterial pressure, the heart rate, and also by helping to balance the redox system. However, the authors also reported that *M. officinalis* extract in high doses (200 mg/kg) may result from increasing in cardiac contractility, increasing of myocardial oxygen demand, and hence more risk of cardiac injury.

Phytochemical Constituents of M. officinalis

Lemon balm, (*M. officinalis* L.), is a rich source of essential oils (geranial, neral, citronellal, and geraniol), triterpenes (ursolic acid and oleanolic acid), and phenolic compounds (caffeic acid derivatives, luteolin, naringin and hesperidin triterpenes, which have been related to the hypolipidemic effects of this plant [154, 151].

Karimi et al. [155] showed that essential oils (EO) of *M. officinalis* decreased serum lipid (TC and TG) and lipoprotein levels (LDL, VLDL, and HDL) in cholesterol-fed rabbits. The lipid-lowering action implies that *M. officinalis* may be useful in the treatment of a hyperlipidaemic state. Although the mechanisms by which EO of *M. officinalis* exerts it is hypolipidemic effects are unknown, it has been proposed that the hypotriglyceridemic effect of EO of *M. officinalis* may result from reduced translocation of SREBP-1c and, subsequently, reduced expression of genes involved in fatty acid synthesis via decreased PCAF histone acetylase activity, which results in the reduction of hepatic fatty acid synthesis [156].

Mentha x piperita L.

Mentha piperita L., commonly known as peppermint, has been used widely as a remedy for the common cold, inflammation of the mouth, pharynx, liver, as well as disorders in the gastrointestinal tract (dyspepsia and nausea) [157].

Currently, the reports on the anti-CVDs of peppermint are restricted and mainly focused on the hypolipidemic effect of peppermint. Johari et al. [158] showed that peppermint extract decreased TC, TG, and LDL levels in Wistar rats. Additionally, peppermint extract decreases the levels TC, TG, LDL, and atherogenic index as well as increased HDL levels in fructose-fed rats [159]. On the other hand, Barbalho et al. [160] and Barbalho et al. [161] reported that the administration of peppermint juice decreased the levels of TC, TG, and LDL; and increased HDL levels in the plasma by Wistar rats and of university students. The results of all studies were related to the antioxidant mechanisms of chemical constituents of peppermint.

Phytochemical Constituents of M. piperita

The medicinal properties of peppermint are related to its main chemical constituents, including terpenes (β-Cuvebene, carvone, and caryophyllene), tocopherols (Vitamin E) essential oils (menthol, eucalyptol, menthofuran, menthyl acetate, and limonene), phenolic compounds (caffeic acid, flavonoids, and tannins) and fatty acids such as α-Linolenic acid [162, 163].

Origanum vulgare L.

Oregano (*O. vulgare* L.) is an aromatic plant commonly used in traditional medicine. Previous studies have reported the anti-inflammatory effect of *O. vulgare* in the treatment for colitis in mice and cancerous cells [164, 165]. However, information on the anti-inflammatory effect of *O. vulgare* in the treatment of CVDs is scarce. In this sense, Ocaña-Fuentes et al. [166] evaluated the anti-inflammatory effect of essential oils (sabinene hydrate, thymol, and carvacrol) from *O. vulgare*, on oxidized-LDLs activated THP-1 macrophages used as a cellular model of atherogenesis. These authors reported a decrease in the synthesis of pro-inflammatory cytokines TNF-a, IL-1b, and IL-6, as well as an increase in the production of anti-inflammatory cytokine IL-10. These results suggest that *O. vulgare* can be used in the treatment of inflammation-related chronic diseases such as atherosclerosis.

Phytochemical Constituents of O. vulgare

Oregano contains a wide array of phenolic compounds such as rosmarinic, caffeic, chicoric and *p*-coumaric acids, eriodictyol, naringenin, luteolin-7-*O*-glucoside, apigenin-7-*O*-glucoside, apigenin, (-)-epicatechin and kaempferol [167]; and essential oil such as sabinene hydrate, thymol and carvacrol [166]. Essential oils have been reported to possess diverse effects, including modulation of lipid profile, reduction in blood pressure, amelioration os oxidative stress and inflammation, as well as the promotion of vasodilation [168]. Essential oils, as well as phenolic compounds found in oregano, may be promising therapeutic agents for CVDs. Nevertheless, studies evaluating their beneficial effects are needed in order to elucidate and understand their mechanisms or modes of action.

Psidium guajava L.

Psidium guajava L., commonly known as guava, has been used in the treatment of hypertension and in the prevention of some risk factors (hypercholesterolemia, endothelial dysfunction, oxidized LDL) that lead to atherosclerosis.

In vivo and *in vitro* studies have tested the anti-CVDs of guava flour and extract. Maryanto [169] showed that the daily consumption of red guava fruit flour (720 mg/kg), decrease the levels of formation of lipid oxidation products (MDA) and increase SOD activity in the blood of hypercholesterolemic rats. The author suggested that the presence of natural antioxidants (vitamin A, vitamin C, lycopene and phenolic compounds) in red guava fruit flour, can inhibit and/or to reduce lipid oxidation (LDL) related to the development of atherosclerosis, through of it is different mechanisms of antioxidant action, including electron-transfer and hydrogen-atom donation to free radicals; and modulation of endogenous antioxidant enzymes (SOD, CAT, and GPX) [170].

Rai, Mehta, and Watal [171] reported that the oral administration of unripe guava fruit peel extract (400 mg/kg), decrease TG, TC, HDL, VLDL and LDL levels in streptozotocin-induced severely diabetic rats. Additionally, guava pulp juice decrease TC and TG levels, as well as increase HDL levels in Wistar rats [172]. The risk of ischaemic heart disease is directly related to the high levels of TC, TG, LDL, and VLDL, and inversely related to the HDL levels. Therefore, the authors suggested that this extract could be used for lowering TC, TG, LDL, and VLDL levels and reducing, thereby the risk of CVDs by increasing HDL level. Additionally, Brito et al. [173] determined the effects of a lycopene-rich extract (LRE) from red guava fruit (25 mg/kg), on the lipid profile and oxidative stress markers in an experimental dyslipidemia model in hamsters. The authors reported a decrease in the levels of TG, MDA, and myeloperoxidase (an enzyme associated with the reactive oxygen species production). The hypotriglyceridemic effect of extract from red guava fruit may have positive repercussions in the treatment of dyslipidemias, while the decrease of MDA and MPO levels indicates the attenuation of oxidative stress, which is associated with the incidence of CVDs.

Phytochemical Constituents of P. guajava

Guava is a source of different phytochemicals, such as phenolic acids, flavonoids, carotenoids, triterpenes, and essential oils [174]. The cardioprotective effect of guava has been associated with the presence of these compounds. For instance, Ademiluyi et al. [175] reported that the mixture of major phenolic compounds identified (rosmarinic and caffeic acids, eugenol, carvacrol, and catechin) in extracts of leaves and pulp of four varieties guava (giant white, small white, stripped and pink), inhibited ACE activity and sodium nitroprusside (SNP) induced lipid peroxidation reactions in the heart of mice, so that extracts evaluated could be used in the management and prevention of hypertension. Another study, reported that the mixture of flavonoids (quercetin, kaempferol, catechin, quercitrin, rutin, luteolin, epicatechin) and phenolic acids (caffeic, chlorogenic and gallic acids) of guava leaves extract could be used to manage hypertension and attenuation of oxidative stress (cause and consequence of hypertension), because they effectively inhibit ACE, and Fe (2+) induced lipid peroxidation in hypertensive rats [176]. Díaz-de-Cerio et al. [177] reported that the administration of guava leaves extract (5 mg/kg) improves endothelial dysfunction in mice with diet-induced obesity. This result was associated with the vasodilator effects that can exert the phenolic compounds identified (including gallocatechin, procyanidins, hyperin, quercetin, and reynoutrin) in guava leaves extract.

Vaccinium myrtillus **L.**

Vaccinium myrtillus, commonly named bilberries, have shown cardioprotective effects by reducing risk factors of CVDs, such as improving blood pressure and HDL cholesterol levels [178]. For example, administration of 2 g/day of bilberries powder (containing 180 mg of anthocyanins) for four weeks affects the lipid profile of blood serum in an alloxan diabetes-induced Wistar rat model. Serum levels of TC, LDL, VLDL and TG were significantly reduced from 136.20 ± 23.49 mg/dL to 81.33 ± 21.34 mg/dL, from 25.20 ± 1.64 mg/dL to 17.22 ± 2.72 mg/dL, from 29.92 ± 3.82 mg/dL to 9.82 ± 4.27 mg/dL and from 149.60 ± 19.13 mg/dL to 49.12 ± 21.39 mg/dL respectively, in the bilberry-treated group when compared with the diabetic group. Even though HDL and CRP serum concentrations were not significantly affected, these parameters showed an increase from 34.80 ± 4.49 mg/dL to 44.85 ± 9.58 mg/dL and a reduction from 54.40 ± 4.77 mg/dL to 43.37 ± 7.34 mg/dL, respectively in the bilberry-treated group in contrast with the diabetic group [179]. The results of the study conducted by Asgary et al. [179] indicate the anti-hyperlipidemic effect of *V. myrtillus*.

Furthermore, Habanova et al. [180] reported that a daily ingesta of 150 g frozen bilberries three times per week during six weeks significantly reduced TC, TG, and LDL, while HDL serum levels increased in the experimental volunteers. These authors conclude that the intake of bilberries might reduce CVDs risk factors by enhancing the blood plasma lipid profile [180]. In the same context, Furlan et al. [181] investigated the cardiometabolic protective effect of *V. myrtillus* incorporation on individuals fed with a high-fat breakfast meal. Consumption of 100 g of bilberry significantly reduced the post-prandial serum levels of TC, LDL and triacylglycerides from 44.7 mmol/L·min to 42.0 mmol/L·min, from 39.4 mmol/L·min to 36.4 mmol/L·min and from 144.1 mmol/L·min to 122.4 mmol/L·min, respectively. Nevertheless, the HDL level was not significantly affected. These results show the potential CVDs risk-modulating effect of bilberries, mainly by improving the lipid profile in humans.

Phytochemical Constituents of V. myrtillus

Bilberries, *V. myrtillus*, have a high content of phenolic compounds (20752.50 mg/kg), including anthocyanins (3038.25 mg/kg), which are responsible for their blue-red coloring [180]. It has been reported that total phenolic and flavonoid content of *V. myrtillus* is 13.7 ± 0.2 mg/100 g, containing chlorogenic acid (6.1 ± 1.8 mg/100g), quercetin-3-*O*-galactoside (4.4 ± 0.8 mg/100 g) and quercetin-3-*O*-glucuronide (3.2 ± 1.1 mg/100 g) [182]. Also 12 anthocyanins (568.8 ± 8.8 mg/100 g) were identified in V. myrtillus, including delphinidin-3-*O*-galactoside (92.1 ±

4.7 mg/100 g), delphinidin-3-*O*-glucoside (86.6 ± 10.5 mg/100 g), delphinidin-3-*O*-arabinoside (59.1 ± 4.8 mg/100 g), petunidin-3-*O*-glucoside (53.4 ± 5.9 mg/100 g) and malvidin-3-*O*-glucoside (50.8 ± 3.3 mg/100 g), among others. On the other hand, Primetta et al. [183] reported that the total anthocyanin content of bilberries varied from 17080 to 36470 mg/kg, containing anthocyanidin glycosides, such as delphinidin, cyanidin, petunidin, peonidin, and malvidin. The anthocyanin content and profile of bilberries may vary from the region and maturity stage at which these berries are collected [184].

Since anthocyanins are the main constituents of *V. myrtillus*, these phytochemicals might be responsible for the bilberries cardioprotective effect. For instance, a high intake of anthocyanins has been related to a diminished risk of myocardial infarction in young and middle-aged women [185]. Several anthocyanins (0.1 μM), such as cyanidin-3-arabinoside, delphinidin-3-glucoside, and peonidin-3-glucoside, reduced P-selectin positive platelets by 10.2%, 10.8%, and 10.8%, respectively; whereas cyanidin-3-galactoside and cyanidin-3-arabinoside diminished platelet-neutrophil aggregation from blood samples of seven healthy individuals [186]. These results show that in addition to their lipid profile improvement, anthocyanins might exert their cardioprotective effect by modulating markers of platelet activation, and propose a potential use of these compounds as anti-atherosclerotic agents. Furthermore, it has been demonstrated that feeding anthocyanidins (40 mg/kg body weight) to C57BL/6J mice for 14 weeks alleviate the high fat diet-induced obesity, reduced the concentration of TG and the expression of TNF-α and MCP-1 in the liver. These results suggest that anthocyanin ingesta inhibits liver lipid deposition and inflammation [187].

Zingiber officinale **Roscoe**

Ginger (*Zingiber officinale* Roscoe) has been commonly consumed as herbal medicine. In recent years, have demonstrated that ginger possesses the potential to prevent and manage CVDs [188].

Several studies have reported that ginger can decrease the levels of blood lipids, blood pressure, and attenuate oxidative stress, contributing to the protection of CVDs. Ginger extract (two varieties, red and white) decrease the levels of TC, TG, VLDL, LDL, and MDA content in the liver and heart tissues of rats fed a high cholesterol diet [189]. The mechanisms were related to ACE inhibitory activity. Additionally, TC, TG, and LDL concentrations and arterial blood pressure were decreased by ginger extract in Wistar rats and rabbits [190]. These results were attributed to the inhibiting cellular cholesterol biosynthesis and a cardio-depressant activity on the rate and force of spontaneous contractions. Khosravani et al. [191] reported that the combination of ginger extract and aerobic exercise decreased the concentration of TC, TG, and LDL, and increased the levels de HDL in rats fed a high-fat. The ginger extract also reduces the levels of TC, TG, and VLDL in rats fed a high-fat. The hypolipidemic effect is related to the reduced expressions of peroxisome proliferator-activated receptors (PPARα and PPARγ) in the liver [192].

The cardioprotective effects of ginger extracts have also been evaluated. Amran et al. [193] reported that ginger extract exerts cardioprotective potential, through the decreased cardiac enzyme activities (cardiac troponin, creatine kinase MB isoenzyme, lactate dehydrogenase, alanine transaminase, and aspartate transaminase) and the reduction of damage to the myocardium in isoproterenol-induced myocardial infarction in a rat model. These effects may be due to the antioxidant properties exhibited by ginger. Moreover, the ginger extract increased vasoprotection via suppression of NO synthase and cyclooxygenase on porcine coronary arteries [194].

Phytochemical Constituents of Z. officinale

Phenolic compounds such as gingerols, shogaols, and paradols have been identified in ginger. However, gingerols (6-gingerol and 8-gingerol) and shogaols (8-shogaol) are considered as the compounds responsible for specific biological activities, including cardiotonic properties of the ginger [195].

Conclusion

Several commonly used medicinal plants in Mexico are valued for their medicinal properties. Cardioprotective effects of the plants reviewed in this chapter are related to their capacity to successfully modulate the blood lipids profile *in vivo* and due to their antioxidant and anti-inflammatory activities. However, there are scarce pharmacological studies that can shed light on understanding the therapeutic mechanisms and the safety of medicinal plants consumption.

Even though there is clinical evidence for the use of several medicinal plants, studies to better understand the mechanisms underlying their effects are needed in order to use them as controlled therapies.

References

[1] Mendis, S., Puska, P., Norrving, B., and Organization, W. H. 2011. *Global atlas on cardiovascular disease prevention and control*: Geneva: World Health Organization.

[2] Lennon, R. P., Claussen, K. A., and Kuersteiner, K. A. (2018). State of the Heart: An Overview of the Disease Burden of Cardiovascular Disease from an Epidemiologic Perspective. *Primary Care: Clinics in Office Practice* 45 (1):1-15. doi: https://doi.org/10.1016/j.pop.2017.11.001.

[3] Jousilahti, P., Vartiainen, E., Tuomilehto, J., and Puska, P. (1999). Sex, Age, Cardiovascular Risk Factors, and Coronary Heart Disease. *Circulation* 99 (9):1165-1172. doi: doi:10.1161/01.CIR.99.9.1165.

[4] Perumareddi, P. (2019). Prevention of Hypertension Related to Cardiovascular Disease. *Primary Care: Clinics in Office Practice* 46 (1):27-39. doi: https://doi.org/10.1016/j.pop.2018.10.005.

[5] Kaneto, H., Katakami, N., Matsuhisa, M., and Matsuoka, T.-a. (2010). Role of Reactive Oxygen Species in the Progression of Type

2 Diabetes and Atherosclerosis. *Mediators of Inflammation* 2010:11. doi: 10.1155/2010/453892.

[6] Shah, P. K. (2019). Inflammation, infection and atherosclerosis. *Trends in Cardiovascular Medicine*. doi: https://doi.org/10.1016/j.tcm.2019.01.004.

[7] Eapen, D. J., Manocha, P., Patel, R. S., Hammadah, M., Veledar, E., Wassel, C., Nanjundappa, R. A., Sikora, S., Malayter, D., Wilson, P. W. F., Sperling, L., Quyyumi, A. A., and Epstein, S. E. (2013). Aggregate Risk Score Based on Markers of Inflammation, Cell Stress, and Coagulation Is an Independent Predictor of Adverse Cardiovascular Outcomes. *Journal of the American College of Cardiology* 62 (4):329-337. doi: https://doi.org/10.1016/j.jacc.2013.03.072.

[8] Ruwanpathirana, T., Owen, A., and Reid, C. M. (2015). Review on Cardiovascular Risk Prediction. *Cardiovascular Therapeutics* 33 (2):62-70. doi: 10.1111/1755-5922.12110.

[9] Cohn, J. N. (2018). Cardiovascular Disease Progression: A Target for Therapy? *The American Journal of Medicine* 131 (10):1170-1173. doi: https://doi.org/10.1016/j.amjmed.2018.03.032.

[10] Fragasso, G., Margonato, A., Spoladore, R., and Lopaschuk, G. D. (2019). Metabolic effects of cardiovascular drugs. *Trends in Cardiovascular Medicine* 29 (3):176-187. doi: https://doi.org/10.1016/j.tcm.2018.08.001.

[11] Wooles, W. R., Borzelleca, J. F., and Branham, G. W. (1967). Effect of acute and prolonged salicylate administration on liver and plasma triglyceride levels and diet-induced hypercholesterolemia. *Toxicology and Applied Pharmacology* 10 (1):1-7. doi: https://doi.org/10.1016/0041-008X(67)90122-6.

[12] Kopp, E., and Ghosh, S. (1994). Inhibition of NF-κB by sodium salicylate and aspirin. *Science* 265 (5174):956-959. doi: 10.1126/science.8052854.

[13] Singh, B. N., Singh, S. N., Reda, D. J., Tang, X. C., Lopez, B., Harris, C. L., Fletcher, R. D., Sharma, S. C., Atwood, J. E., Jacobson, A. K., Lewis, H. D., Raisch, D. W., and Ezekowitz, M. D.

(2005). Amiodarone versus Sotalol for Atrial Fibrillation. *The New England Journal of Medicine* 352 (18):1861-1872. doi: 10.1056/NEJMoa041705.

[14] Vamos, M., and Hohnloser, S. H. (2016). Amiodarone and dronedarone: An update. *Trends in Cardiovascular Medicine* 26 (7):597-602. doi: https://doi.org/10.1016/j.tcm.2016.03.014.

[15] Jackevicius, C. A., Ghaznavi, Z., Lu, L., and Warner, A. L. (2018). Safety of Alpha-Adrenergic Receptor Antagonists in Heart Failure. *JACC: Heart Failure* 6 (11):917-925. doi: https://doi.org/10.1016/j.jchf.2018.06.015.

[16] Day, J. L. (1975). The metabolic consequences of adrenergic blockade: A review. *Metabolism* 24 (8):987-996. doi: https://doi.org/10.1016/0026-0495(75)90090-6.

[17] Vider, E., and Zada, I. (2019). The Use of Calcium Channel Blockers in Heart Failure. *The Journal for Nurse Practitioners* 15 (1):149-150.e1. doi: https://doi.org/10.1016/j.nurpra.2018.07.009.

[18] Kario, K. 2018. "26 - Central Sympathetic Agents and Direct Vasodilators." In *Hypertension: A Companion to Braunwald's Heart Disease (Third Edition)*, edited by George L. Bakris and Matthew J. Sorrentino, 254-260. Elsevier.

[19] Skalidis, E. I., Hamilos, M. I., Chlouverakis, G., Zacharis, E. A., and Vardas, P. E. (2011). Ivabradine improves coronary flow reserve in patients with stable coronary artery disease. *Atherosclerosis* 215 (1):160-165. doi: https://doi.org/10.1016/j.atherosclerosis.2010.11.035.

[20] Münzel, T., and Gori, T. (2013). Nitrate therapy and nitrate tolerance in patients with coronary artery disease. *Current Opinion in Pharmacology* 13 (2):251-259. doi: https://doi.org/10.1016/j.coph.2012.12.008.

[21] Ma, Y., Yuan, J., Hu, J., Gao, W., Zou, Y., and Ge, J. (2019). ACE inhibitor suppresses cardiac remodeling after myocardial infarction by regulating dendritic cells and AT2 receptor-mediated mechanism in mice. *Biomedicine & Pharmacotherapy* 114:108660. doi: https://doi.org/10.1016/j.biopha.2019.108660.

[22] Yoshiyama, M., Kim, S., Yamagishi, H., Omura, T., Tani, T., Yanagi, S., Toda, I., Teragaki, M., Akioka, K., Takeuchi, K., and Takeda, T. (1994). Cardioprotective effect of the angiotensin II type 1 receptor antagonist TCV-116 on ischemia-reperfusion injury. *American Heart Journal* 128 (1):1-6. doi: https://doi.org/10.1016/0002-8703(94)90002-7.

[23] Trohman, R. G., Sharma, P. S., McAninch, E. A., and Bianco, A. C. (2019). Amiodarone and thyroid physiology, pathophysiology, diagnosis and management. *Trends in Cardiovascular Medicine* 29 (5):285-295. doi: https://doi.org/10.1016/j.tcm.2018.09.005.

[24] Agrawal, N., Akella, A., and Deshpande, S.B. (2016). Captopril augments acetylcholine-induced bronchial smooth muscle contractions *in vitro* via kinin-dependent mechanisms. *Indian Journal of Experimental Biology* 54 (6):365-369.

[25] Phillips, C. O., Kashani, A., Ko, D. K., Francis, G., and Krumholz, H. M. (2007). Adverse Effects of Combination Angiotensin II Receptor Blockers Plus Angiotensin-Converting Enzyme Inhibitors for Left Ventricular Dysfunction: A Quantitative Review of Data From Randomized Clinical Trials. *JAMA Internal Medicine* 167 (18):1930-1936. doi: 10.1001/archinte.167.18.1930.

[26] Tsujimoto, T., Sugiyama, T., Shapiro, M. F., Noda, M., and Kajio, H. (2017). Risk of Cardiovascular Events in Patients with Diabetes Mellitus on β-Blockers. *Hypertension (Dallas, Tex. 1979)* 70 (1):103-110. doi: 10.1161/HYPERTENSIONAHA.117.09259.

[27] Perbet, S., Constantin, J. M., Guérin, R., Faure, M., Brugère, C., Da Inès, D., Joubert, J., and Bazin, J. E. (2009). Non-occlusive colonic ischemia induced by verapamil ER overdose. *Intensive Care Medicine* 35 (5):956. doi: 10.1007/s00134-008-1375-2.

[28] Casas, R., Castro-Barquero, S., Estruch, R., and Sacanella, E. (2018). Nutrition and Cardiovascular Health. *International journal of molecular sciences* 19 (12):3988. doi: 10.3390/ijms19123988.

[29] Akinmoladun, A. C., Olaleye, M. T., and Farombi, E. O. 2014. "13 - Cardiotoxicity and Cardioprotective Effects of African Medicinal

Plants." In *Toxicological Survey of African Medicinal Plants*, edited by Victor Kuete, 395-421. Elsevier.

[30] Pittler, M. H., and Ernst, E. (2012). Horse chestnut seed extract for chronic venous insufficiency. *Cochrane Database of Systematic Reviews* (11). doi: 10.1002/14651858.CD003230.pub4.

[31] Kachur, S., Lavie, C. J., Milani, R., and Ventura, H. (2017). Obesity and cardiovascular diseases. *Minerva Medica* 108 (3):212-228.

[32] Avci, G., Küçükkurt, I., Küpeli Akkol, E., and Yeşilada, E. (2010). Effects of escin mixture from the seeds of Aesculus hippocastanum on obesity in mice fed a high fat diet. *Pharmaceutical Biology* 48 (3):247-252. doi: 10.3109/13880200903085466.

[33] Kosmas, C. E., Martinez, I., Sourlas, A., Bouza, K. V., Campos, F. N., Torres, V., Montan, P. D., and Guzman, E. (2018). High-density lipoprotein (HDL) functionality and its relevance to atherosclerotic cardiovascular disease. *Drugs in Context* 7:212525-212525. doi: 10.7573/dic.212525.

[34] Barter, P. J., Nicholls, S., Rye, K.-A., Anantharamaiah, G. M., Navab, M., and Fogelman, A. M. (2004). Antiinflammatory Properties of HDL. *Circulation Research* 95 (8):764-772. doi: doi:10.1161/01.RES.0000146094.59640.13.

[35] Kontush, A., and Chapman, M.J. (2010). Antiatherogenic function of HDL particle subpopulations: focus on antioxidative activities. *Current Opinion in Lipidology* 21 (4):312-318.

[36] Küçükkurt, I., Ince, S., Keleş, H., Akkol, E. K., Avcı, G., Yeşilada, E., and Bacak, E. (2010). Beneficial effects of Aesculus hippocastanum L. seed extract on the body's own antioxidant defense system on subacute administration. *Journal of Ethnopharmacology* 129 (1):18-22.

[37] Dutra, L. S., Leite, M. N., Brandão, M. A., de Almeida, P. A., Vaz, F. A., and de Oliveira, M. A. (2013). A Rapid Method for Total B-Escin Analysis in Dry, Hydroalcoholic and Hydroglycolic Extracts of Aesculus hippocastanum L. by Capillary Zone Electrophoresis. *Phytochemical Analysis* 24 (6):513-519.

[38] Kuczmannová, A., Balažová, A., Račanská, E., Kameníková, M., Fialová, S., Majerník, J., Nagy, M., Gál, P., and Mučaji, P. (2016). Agrimonia eupatoria L. and Cynara cardunculus L. Water Infusions: Comparison of Anti-Diabetic Activities. *Molecules* 21 (5):564. doi: https://doi.org/10.3390/molecules21050564.

[39] Ivanova, D., Vankova, D., and Nashar, M. (2013). Agrimonia eupatoria tea consumption in relation to markers of inflammation, oxidative status and lipid metabolism in healthy subjects. *Archives of Physiology and Biochemistry* 119 (1):32-37. doi: 10.3109/13813455.
2012.729844.

[40] Ivanova, D., Gerova, D., Chervenkov, T., and Yankova, T. (2005). Polyphenols and antioxidant capacity of Bulgarian medicinal plants. *Journal of Ethnopharmacology* 96 (1):145-150. doi: https://doi.org/10.1016/j.jep.2004.08.033.

[41] Granica, S., Krupa, K., Kłębowska, A., and Kiss, A. K. (2013). Development and validation of HPLC-DAD-CAD–MS3 method for qualitative and quantitative standardization of polyphenols in *Agrimoniae eupatoriae* herba (Ph. Eur). *Journal of Pharmaceutical and Biomedical Analysis* 86:112-122. doi: https://doi.org/10.1016/j.jpba.2013.08.006.

[42] Kuczmannová, A., Gál, P., Varinská, L., Treml, J., Kováč, I., Novotný, M., Vasilenko, T., Dall'Acqua, S., Nagy, M., and Mučaji, P. (2015). *Agrimonia eupatoria* L. and *Cynara cardunculus* L. Water Infusions: Phenolic Profile and Comparison of Antioxidant Activities. *Molecules* 20 (11):20538-20550.

[43] Porcu, E. P., Cossu, M., Rassu, G., Giunchedi, P., Cerri, G., Pourová, J., Najmanová, I., Migkos, T., Pilařová, V., Nováková, L., Mladěnka, P., and Gavini, E. (2018). Aqueous injection of quercetin: An approach for confirmation of its direct in vivo cardiovascular effects. *International Journal of Pharmaceutics* 541 (1):224-233. doi: https://doi.org/10.1016/j.ijpharm.2018.02.036.

[44] Luan, S. S., Yu, F., Li, B. Y., Qin, R. J., Li, X. L., Cai, Q., Yin, W. B., Cheng, M., and Gao, H. Q. (2014). Quantitative proteomics

study of protective effects of grape seed procyanidin B2 on diabetic cardiomyopathy in db/db mice. *Bioscience, Biotechnology, and Biochemistry* 78 (9):1577-1583. doi: 10.1080/09168451.2014.930320.

[45] Asdaq, S. M., and Inamdar, M. N. (2010). Potential of garlic and its active constituent, S-allyl cysteine, as antihypertensive and cardioprotective in presence of captopril. *Phytomedicine* 17 (13):1016-1026. doi: https://doi.org/10.1016/j.phymed.2010.07.012.

[46] El-Sayyad, H. I., Abou-El-Naga, A. M., Gadallah, A. A., and Bakr, I. H. (2010). Protective effects of *Allium sativum* against defects of hypercholesterolemia on pregnant rats and their offspring. *International Journal of Clinical and Experimental Medicine* 3 (2):152-163.

[47] Xiong, X. J., Wang, P. Q., Li, S. J., Li, X. K., Zhang, Y. Q., and Wang, J. (2015). Garlic for hypertension: A systematic review and meta-analysis of randomized controlled trials. *Phytomedicine* 22 (3):352-361. doi: https://doi.org/10.1016/j.phymed.2014.12.013.

[48] Karagodin, V. P., Sobenin, I. A., and Orekhov, A. N. (2016). Antiatherosclerotic and Cardioprotective Effects of Time-Released Garlic Powder Pills. *Current Pharmaceutical Design* 22 (2):196-213. doi: http://dx.doi.org/10.2174/1381612822666151112153351.

[49] Sahebkar, A., Serban, C., Ursoniu, S., and Banach, M. (2016). Effect of garlic on plasma lipoprotein(a) concentrations: A systematic review and meta-analysis of randomized controlled clinical trials. *Nutrition* 32 (1):33-40. doi: https://doi.org/10.1016/j.nut.2015.06.009.

[50] Lee, S., Joo, H., Kim, C. T., Kim, I. H., and Kim, Y. (2012). High hydrostatic pressure extract of garlic increases the HDL cholesterol level via up-regulation of apolipoprotein A-I gene expression in rats fed a high-fat diet. *Lipids in Health and Disease* 11 (1):77. doi: 10.1186/1476-511X-11-77.

[51] Gomaa, A. M. S., Abdelhafez, A. T., and Aamer, H. A. (2018). Garlic (*Allium sativum*) exhibits a cardioprotective effect in experimental chronic renal failure rat model by reducing oxidative

stress and controlling cardiac Na+/K+-ATPase activity and Ca2+ levels. *Cell Stress and Chaperones* 23 (5):913-920. doi: 10.1007/s12192-018-0898-x.

[52] Kumar, R., Chhatwal, S., Arora, S., Sharma, S., Singh, J., Singh, N., Bhandari, V., and Khurana, A. (2013). Antihyperglycemic, antihyperlipidemic, anti-inflammatory and adenosine deaminase-lowering effects of garlic in patients with type 2 diabetes mellitus with obesity. *Diabetes, metabolic syndrome and obesity: targets and therapy* 6:49-56. doi: 10.2147/DMSO.S38888.

[53] Dhawan, V., and Jain, S. (2004). Effect of garlic supplementation on oxidized low density lipoproteins and lipid peroxidation in patients of essential hypertension. *Molecular and Cellular Biochemistry* 266 (1):109-115. doi: 10.1023/b:Mcbi.0000049146.89059.53.

[54] Durak, İ., Aytaç, B., Atmaca, Y., Devrim, E., Avcı, A., Erol, Ç., and Oral, D. (2004). Effects of garlic extract consumption on plasma and erythrocyte antioxidant parameters in atherosclerotic patients. *Life Sciences* 75 (16):1959-1966. doi: https://doi.org/10.1016/j.lfs.2004.04.015.

[55] Durak, İ., Kavutcu, M., Aytaç, B., Avcı, A., Devrim, E., Özbek, H., and Öztürk, H. S. (2004). Effects of garlic extract consumption on blood lipid and oxidant/antioxidant parameters in humans with high blood cholesterol. *The Journal of Nutritional Biochemistry* 15 (6):373-377. doi: https://doi.org/10.1016/j.jnutbio.2004.01.005.

[56] Turner, B., Mølgaard, C., and Marckmann, P. (2004). Effect of garlic (*Allium sativum*) powder tablets on serum lipids, blood pressure and arterial stiffness in normo-lipidaemic volunteers: a randomised, double-blind, placebo-controlled trial. *British Journal of Nutrition* 92 (4):701-706. doi: 10.1079/BJN20041255.

[57] Atkin, M., Laight, D., and Cummings, M. H. (2016). The effects of garlic extract upon endothelial function, vascular inflammation, oxidative stress and insulin resistance in adults with type 2 diabetes at high cardiovascular risk. A pilot double blind randomized placebo controlled trial. *Journal of Diabetes and its Complications* 30 (4):723-727. doi: https://doi.org/10.1016/j.jdiacomp.2016.01.003.

[58] Szulińska, M., Kręgielska-Narożna, M., Świątek, J., Styś, P., Kuźnar-Kamińska, B., Jakubowski, H., Walkowiak, J., and Bogdański, P. (2018). Garlic extract favorably modifies markers of endothelial function in obese patients –randomized double blind placebo-controlled nutritional intervention. *Biomedicine & Pharmacotherapy* 102:792-797. doi: https://doi.org/10.1016/j.biopha.2018.03.131.

[59] Fratianni, F., Ombra, M. N., Cozzolino, A., Riccardi, R., Spigno, P., Tremonte, P., Coppola, R., and Nazzaro, F. (2016). Phenolic constituents, antioxidant, antimicrobial and anti-proliferative activities of different endemic Italian varieties of garlic (*Allium sativum* L.). *Journal of Functional Foods* 21:240-248. doi: https://doi.org/10.1016/j.jff.2015.12.019.

[60] Casella, S., Leonardi, M., Melai, B., Fratini, F., and Pistelli, L. (2013). The Role of Diallyl Sulfides and Dipropyl Sulfides in the In Vitro Antimicrobial Activity of the Essential Oil of Garlic, *Allium sativum* L., and Leek, *Allium porrum* L. *Phytotherapy Research* 27 (3):380-383. doi: 10.1002/ptr.4725.

[61] Kim, S., Kim, D. B., Jin, W., Park, J., Yoon, W., Lee, Y., Kim, S., Lee, S., Kim, S., Lee, O. H., Shin, D., and Yoo, M. (2018). Comparative studies of bioactive organosulphur compounds and antioxidant activities in garlic (*Allium sativum* L.), elephant garlic (*Allium ampeloprasum* L.) and onion (*Allium cepa* L.). *Natural Product Research* 32 (10):1193-1197. doi: 10.1080/14786419.2017.1323211.

[62] Chan, J. Y. Y., Tsui, H. T., Chung, I. Y. M., Chan, R. Y. K., Kwan, Y. W., and Chan, S. W. (2014). Allicin protects rat cardiomyoblasts (H9c2 cells) from hydrogen peroxide-induced oxidative injury through inhibiting the generation of intracellular reactive oxygen species. *International Journal of Food Sciences and Nutrition* 65 (7):868-873. doi: 10.3109/09637486.2014.925428.

[63] Chen, X., Pang, S., Lin, J., Xia, J., and Wang, Y. (2016). Allicin prevents oxidized low-density lipoprotein-induced endothelial cell injury by inhibiting apoptosis and oxidative stress pathway. *BMC*

Complementary and Alternative Medicine 16:133-133. doi: 10.1186/s12906-016-1126-9.

[64] Garcia-Trejo, E. M., Arellano-Buendia, A. S., Arguello-Garcia, R., Loredo-Mendoza, M. L., Garcia-Arroyo, F. E., Arellano-Mendoza, M. G., Castillo-Hernandez, M. C., Guevara-Balcazar, G., Tapia, E., Sanchez-Lozada, L. G., and Osorio-Alonso, H. (2016). Effects of Allicin on Hypertension and Cardiac Function in Chronic Kidney Disease. *Oxidative Medicine and Cellular Longevity* 2016:3850402. doi: 10.1155/2016/3850402.

[65] Liu, D. S., Wang, S. L., Li, J. M., Liang, E. S., Yan, M. Z., and Gao, W. (2017). Allicin improves carotid artery intima-media thickness in coronary artery disease patients with hyperhomocysteinemia. *Experimental and Therapeutic Medicine* 14 (2):1722-1726. doi: 10.3892/etm.2017.4698.

[66] Mousa, S. A. 2010. "Antithrombotic Effects of Naturally Derived Products on Coagulation and Platelet Function." In *Anticoagulants, Antiplatelets, and Thrombolytics: Second Edition*, edited by Shaker A. Mousa, 229-240. Totowa, NJ: Humana Press.

[67] Radha, M. H., and Laxmipriya, N. P. (2014). Evaluation of biological properties and clinical effectiveness of *Aloe vera*: A systematic review. *Journal of Traditional and Complementary Medicine* 5 (1):21-26. doi: 10.1016/j.jtcme.2014.10.006.

[68] Rahoui, W., Merzouk, H., El Haci, I. A., Bettioui, R., Azzi, R., and Benali, M. (2018). Beneficial effects of Aloe vera gel on lipid profile, lipase activities and oxidant/antioxidant status in obese rats. *Journal of Functional Foods* 48:525-532. doi: https://doi.org/10.1016/j.jff.2018.07.050.

[69] Dana, N., Javanmard, S. H., Asgary, S., Asnaashari, H., and Abdian, N. (2012). The effect of *Aloe vera* leaf gel on fatty streak formation in hypercholesterolemic rabbits. *Journal of Research in Medical Sciences: The Official Journal of Isfahan University of Medical Sciences* 17 (5):439-442.

[70] Kumar, M., Rakesh, S., Nagpal, R., Hemalatha, R., Ramakrishna, A., Sudarshan, V., Ramagoni, R., Shujauddin, M., Verma, V.,

Kumar, A., Tiwari, A., Singh, B., and Kumar, R. (2013). Probiotic *Lactobacillus rhamnosus* GG and *Aloe vera* gel improve lipid profiles in hypercholesterolemic rats. *Nutrition* 29 (3):574-579. doi: https://doi.org/10.1016/j.nut.2012.09.006.

[71] Huseini, H. F., Kianbakht, S., Hajiaghaee, R., and Dabaghian, F. H. (2012). Anti-hyperglycemic and anti-hypercholesterolemic effects of *Aloe vera* leaf gel in hyperlipidemic type 2 diabetic patients: a randomized double-blind placebo-controlled clinical trial. *Planta Medica* 78 (4):311-6. doi: 10.1055/s-0031-1280474.

[72] Kumar, R., Singh, A. K., Gupta, A., Bishayee, A., and Pandey, A. K. (2019). Therapeutic potential of *Aloe vera*—A miracle gift of nature. *Phytomedicine*: 152996. doi: https://doi.org/10.1016/j.phymed.2019.152996.

[73] Lucini, L., Pellizzoni, M., Pellegrino, R., Molinari, G. P., and Colla, G. (2015). Phytochemical constituents and in vitro radical scavenging activity of different Aloe species. *Food Chemistry* 170:501-507. doi: https://doi.org/10.1016/j.foodchem.2014.08.034.

[74] Cardarelli, M., Rouphael, Y., Pellizzoni, M., Colla, G., and Lucini, L. (2017). Profile of bioactive secondary metabolites and antioxidant capacity of leaf exudates from eighteen Aloe species. *Industrial Crops and Products* 108:44-51. doi: https://doi.org/10.1016/j.indcrop.2017.06.017.

[75] Li, J., Ding, L., Song, B., Xiao, X., Qi, M., Yang, Q., Yang, Q., Tang, X., Wang, Z., and Yang, L. (2016). Emodin improves lipid and glucose metabolism in high fat diet-induced obese mice through regulating SREBP pathway. *European Journal of Pharmacology* 770:99-109. doi: https://doi.org/10.1016/j.ejphar.2015.11.045.

[76] Yu, Y., Liu, H., Yang, D., He, F., Yuan, Y., Guo, J., Hu, J., Yu, J., Yan, X., Wang, S., and Du, Z. (2019). Aloe-emodin attenuates myocardial infarction and apoptosis via up-regulating miR-133 expression. *Pharmacological Research* 146:104315. doi: https://doi.org/10.1016/j.phrs.2019.104315.

[77] Xu, K., Al-ani, M. K., Wang, C., Qiu, X., Chi, Q., Zhu, P., and Dong, N. (2018). Emodin as a selective proliferative inhibitor of

vascular smooth muscle cells versus endothelial cells suppress arterial intima formation. *Life Sciences* 207:9-14. doi: https://doi.org/10.1016/j.lfs.2018.05.042.

[78] Xiao, D., Zhang, Y., Wang, R., Fu, Y., Zhou, T., Diao, H., Wang, Z., Lin, Y., Li, Z., Wen, L., Kang, X., Kopylov, P., Shchekochikhin, D., Zhang, Y., and Yang, B. (2019). Emodin alleviates cardiac fibrosis by suppressing activation of cardiac fibroblasts via upregulating metastasis associated protein 3. *Acta Pharmaceutica Sinica B* 9 (4):724-733. doi: https://doi.org/10.1016/j.apsb.b 2019.04.003.

[79] Mollazadeh, H., and Hosseinzadeh, H. (2016). Cinnamon effects on metabolic syndrome: a review based on its mechanisms. *Iranian Journal of Basic Medical Sciences* 19 (12):1258-1270. doi: 10.22038/ijbms.2016.7906.

[80] Nyadjeu, P., Dongmo, A., Nguelefack Télesphore, B., and Kamanyi, A. (2011). Antihypertensive and Vasorelaxant Effects of *Cinnamomum zeylanicum* Stem Bark Aqueous Extract in Rats. *Journal of Complementary and Integrative Medicine* 8 (1). doi: 10.2202/1553-3840.1490.

[81] Nyadjeu, P., Nguelefack-Mbuyo, E. P., Atsamo, A. D., Nguelefack, T. B., Dongmo, A. B., and Kamanyi, A. (2013). Acute and chronic antihypertensive effects of *Cinnamomum zeylanicum* stem bark methanol extract in L-NAME-induced hypertensive rats. *BMC Complementary and Alternative Medicine* 13 (1):27. doi: 10.1186/1472-6882-13-27.

[82] Askari, F., Rashidkhani, B., and Hekmatdoost, A. (2014). Cinnamon may have therapeutic benefits on lipid profile, liver enzymes, insulin resistance, and high-sensitivity C-reactive protein in nonalcoholic fatty liver disease patients. *Nutrition Research* 34 (2):143-148. doi: https://doi.org/10.1016/j.nutres.2013.11.005.

[83] Gupta, S., Puri, S., Misra, A., Gulati, S., and Mani, K. (2017). Effect of oral cinnamon intervention on metabolic profile and body composition of Asian Indians with metabolic syndrome: a

randomized double -blind control trial. *Lipids in Health and Disease* 16 (1):113. doi: 10.1186/s12944-017-0504-8.

[84] Borzoei, A., Rafraf, M., Niromanesh, S., Farzadi, L., Narimani, F., and Doostan, F. (2018). Effects of cinnamon supplementation on antioxidant status and serum lipids in women with polycystic ovary syndrome. *Journal of Traditional and Complementary Medicine* 8 (1):128-133. doi: https://doi.org/10.1016/j.jtcme.2017.04.008.

[85] Khuwijitjaru, P., Sayputikasikorn, N., Samuhasaneetoo, S., Penroj, P., Siriwongwilaichat, P., and Adachi, S. (2012). Subcritical water extraction of flavoring and phenolic compounds from cinnamon bark *Cinnamomum zeylanicum*. *Journal of Oleo Science* 61 (6):349-355. doi: 10.5650/jos.61.349.

[86] Mateos-Martín, M. L., Fuguet, E., Quero, C., Pérez-Jiménez, J., Torres, J. L. J. A., and Chemistry, B. (2012). New identification of proanthocyanidins in cinnamon (*Cinnamomum zeylanicum* L.) using MALDI-TOF/TOF mass spectrometry. *Molecules* 402 (3):1327-1336. doi: 10.1007/s00216-011-5557-3.

[87] Li, A. L., Li, G. H., Li, Y. R., Wu, X. Y., Ren, D. M., Lou, H. X., Wang, X.-N., and Shen, T. (2019). Lignan and flavonoid support the prevention of cinnamon against oxidative stress related diseases. *Phytomedicine* 53:143-153. doi: https://doi.org/10.1016/j.phymed.2018.09.022.

[88] Vega-Vargas, F., Guisado-Requena, I. M., Canadas-De la Fuente, G., Castillo, R., and Fernandez-Sanchez, E. (2017). Effects of Combinational Procyanidins in Grape Seed Extract and Exercise on the Levels of Glucose and Blood Lipid Profile. *Journal of Diabetes and Metabolism* 8 (754):2.

[89] Shalaby, M. A., and Saifan, H. Y. (2014). Some pharmacological effects of cinnamon and ginger herbs in obese diabetic rats. *Journal of Intercultural Ethnopharmacology* 3 (4):144-149. doi: 10.5455/jice.20140818050741.

[90] Khare, P., Jagtap, S., Jain, Y., Baboota, R. K., Mangal, P., Boparai, R. K., Bhutani, K. K., Sharma, S. S., Premkumar, L. S., Kondepudi, K. K., Chopra, K., and Bishnoi, M. (2016). Cinnamaldehyde

supplementation prevents fasting-induced hyperphagia, lipid accumulation, and inflammation in high-fat diet-fed mice. *Biofactors* 42 (2):201-11. doi: 10.1002/biof.1265.

[91] Bundy, R., Walker, A. F., Middleton, R. W., Wallis, C., and Simpson, H. C. R. (2008). Artichoke leaf extract (*Cynara scolymus*) reduces plasma cholesterol in otherwise healthy hypercholesterolemic adults: A randomized, double blind placebo controlled trial. *Phytomedicine* 15 (9):668-675. doi: https://doi.org/10.1016/j.phymed.2008.03.001.

[92] Santos, H. O., Bueno, A. A., and Mota, J. F. (2018). The effect of artichoke on lipid profile: A review of possible mechanisms of action. *Pharmacological Research* 137:170-178. doi: https://doi.org/10.1016/j.phrs.2018.10.007.

[93] Mahboubi, M. (2018). *Cynara scolymus* (artichoke) and its efficacy in management of obesity. *Bulletin of Faculty of Pharmacy, Cairo University* 56 (2):115-120. doi: https://doi.org/10.1016/j.bfopcu.2018.10.003.

[94] Villiger, A., Sala, F., Suter, A., and Butterweck, V. (2015). In vitro inhibitory potential of *Cynara scolymus, Silybum marianum, Taraxacum officinale,* and *Peumus boldus* on key enzymes relevant to metabolic syndrome. *Phytomedicine* 22 (1):138-144. doi: https://doi.org/10.1016/j.phymed.2014.11.015.

[95] Khan, M.Y., and Kumar, V. (2019). Mechanism & inhibition kinetics of bioassay-guided fractions of Indian medicinal plants and foods as ACE inhibitors. *Journal of Traditional and Complementary Medicine* 9 (1):73-84. doi: https://doi.org/10.1016/j.jtcme.2018.02.001.

[96] Heidarian, E., and Soofiniya, Y. J. J. o. m. p. r. (2011). Hypolipidemic and hypoglycemic effects of aerial part of *Cynara scolymus* in streptozotocin-induced diabetic rats. *Journal of Medicinal Plants Research* 5 (13):2717-2723.

[97] Küskü-Kiraz, Z., Mehmetçik, G., Doğru-Abbasoğlu, S., and Uysal, M. (2010). Artichoke leaf extract reduces oxidative stress and

lipoprotein dyshomeostasis in rats fed on high cholesterol diet. *Phytotherapy Research* 24 (4):565-570. doi: 10.1002/ptr.2985.

[98] Mocelin, R., Marcon, M., Santo, G. D., Zanatta, L., Sachett, A., Schönell, A. P., Bevilaqua, F., Giachini, M., Chitolina, R., Wildner, S. M., Duarte, M. M. M. F., Conterato, G. M. M., Piato, A. L., Gomes, D. B., and Roman Junior, W. A. (2016). Hypolipidemic and antiatherogenic effects of *Cynara scolymus* in cholesterol-fed rats. *Revista Brasileira de Farmacognosia* 26:233-239.

[99] Abdulkhaleq, M. M., Hussein, R. H., Alnahdi, H. S. J. I. J. o. P. R., and Sciences, A. (2019). Artichoke Leaves Water Extract Attenuate Oxidative Stress and Regulates Lipid Profile in Rats Fed High Fat Diet. *International Journal of Pharmaceutical Research and Allied Sciences* 8 (1):138-145.

[100] Rezazadeh, K., Aliashrafi, S., Asghari-Jafarabadi, M., and Ebrahimi-Mameghani, M. (2018). Antioxidant response to artichoke leaf extract supplementation in metabolic syndrome: A double-blind placebo-controlled randomized clinical trial. *Clinical Nutrition* 37 (3):790-796. doi: https://doi.org/10.1016/j.clnu.2017.03.017.

[101] Rezazadeh, K., Rezazadeh, F., and Ebrahimi-Mameghani, M. (2018). The effect of artichoke leaf extract supplementation on lipid and CETP response in metabolic syndrome with respect to Taq 1B CETP polymorphism: A randomized placebo-controlled clinical trial. *European Journal of Integrative Medicine* 17:112-118. doi: https://doi.org/10.1016/j.eujim.2017.12.008.

[102] Rondanelli, M., Giacosa, A., Opizzi, A., Faliva, M. A., Sala, P., Perna, S., Riva, A., Morazzoni, P., and Bombardelli, E. (2013). Beneficial effects of artichoke leaf extract supplementation on increasing HDL-cholesterol in subjects with primary mild hypercholesterolaemia: a double-blind, randomized, placebo-controlled trial. *International Journal of Food Sciences and Nutrition* 64 (1):7-15. doi: 10.3109/09637486.2012.700920.

[103] Farag, M. A., Elsebai, M. F., and Khattab, A. R. (2018). Metabolome based classification of artichoke leaf: A prospect for phyto-equivalency of its different leaf origins and commercial

preparations. *Journal of Pharmaceutical and Biomedical Analysis* 158:151-159. doi: https://doi.org/10.1016/j.jpba.2018.05.046.

[104] El Senousy, A. S., Farag, M. A., Al-Mahdy, D. A., and Wessjohann, L. A. (2014). Developmental changes in leaf phenolics composition from three artichoke cvs. (*Cynara scolymus*) as determined via UHPLC–MS and chemometrics. *Phytochemistry* 108:67-76. doi: https://doi.org/10.1016/j.phytochem.2014.09.004.

[105] Wan, C. W., Wong, C. N. Y., Pin, W. K., Wong, M. H. Y., Kwok, C. Y., Chan, R. Y. K., Yu, P. H. F., and Chan, S. W. (2013). Chlorogenic Acid Exhibits Cholesterol Lowering and Fatty Liver Attenuating Properties by Up-regulating the Gene Expression of PPAR-α in Hypercholesterolemic Rats Induced with a High-Cholesterol Diet. *Phytotherapy Research* 27 (4):545-551. doi: 10.1002/ptr.4751.

[106] Li, L., Luo, W., Qian, Y., Zhu, W., Qian, J., Li, J., Jin, Y., Xu, X., and Liang, G. (2019). Luteolin protects against diabetic cardiomyopathy by inhibiting NF-κB-mediated inflammation and activating the Nrf2-mediated antioxidant responses. *Phytomedicine* 59:152774. doi: https://doi.org/10.1016/j.phymed.2018.11.034.

[107] Gilani, A. H., Jabeen, Q., Khan, A.-u., and Shah, A. J. (2008). Gut modulatory, blood pressure lowering, diuretic and sedative activities of cardamom. *Journal of Ethnopharmacology* 115 (3):463-472. doi: https://doi.org/10.1016/j.jep.2007.10.015.

[108] Verma, S. K., Jain, V., and Katewa, S. S. (2009). Blood pressure lowering, fibrinolysis enhancing and antioxidant activities of cardamom (*Elettaria cardamomum*). *Indian Journal of Biochemistry and Biophysics* 46 (6):503-506.

[109] Nagashree, S., Archana, K. K., Srinivas, P., Srinivasan, K., and Sowbhagya, H. B. (2017). Anti-hypercholesterolemic influence of the spice cardamom (*Elettaria cardamomum*) in experimental rats. *Journal of the Science Food and Agriculture* 97 (10):3204-3210. doi: 10.1002/jsfa.8165.

[110] Daneshi-Maskooni, M., Keshavarz, S. A., Qorbani, M., Mansouri, S., Alavian, S. M., Badri-Fariman, M., Jazayeri-Tehrani, S. A., and

Sotoudeh, G. (2018). Green cardamom increases Sirtuin-1 and reduces inflammation in overweight or obese patients with non-alcoholic fatty liver disease: a double-blind randomized placebo-controlled clinical trial. *Nutrition & Metabolism* 15:63-63. doi: 10.1186/s12986-018-0297-4.

[111] Daneshi-Maskooni, M., Keshavarz, S. A., Qorbani, M., Mansouri, S., Alavian, S. M., Badri-Fariman, M., Jazayeri-Tehrani, S. A., and Sotoudeh, G. (2019). Green cardamom supplementation improves serum irisin, glucose indices, and lipid profiles in overweight or obese non-alcoholic fatty liver disease patients: a double-blind randomized placebo-controlled clinical trial. *BMC Complementary and Alternative Medicine* 19 (1):59-59. doi: 10.1186/s12906-019-2465-0.

[112] Fatemeh, Y., Siassi, F., Rahimi, A., Koohdani, F., Doostan, F., Qorbani, M., and Sotoudeh, G. (2017). The effect of cardamom supplementation on serum lipids, glycemic indices and blood pressure in overweight and obese pre-diabetic women: a randomized controlled trial. *Journal of Diabetes and Metabolic Disorders* 16:40-40. doi: 10.1186/s40200-017-0320-8.

[113] Gomaa, A. A., Makboul, R. M., El-Mokhtar, M. A., Abdel-Rahman, E. A., Ahmed, I. A., and Nicola, M. A. (2019). Terpenoid-rich *Elettaria cardamomum* extract prevents Alzheimer-like alterations induced in diabetic rats via inhibition of GSK3β activity, oxidative stress and pro-inflammatory cytokines. *Cytokine* 113:405-416. doi: https://doi.org/10.1016/j.cyto.2018.10.017.

[114] Masoumi-Ardakani, Y., Mandegary, A., Esmaeilpour, K., Najafipour, H., Sharififar, F., Pakravanan, M., and Ghazvini, H. (2016). Chemical Composition, Anticonvulsant Activity, and Toxicity of Essential Oil and Methanolic Extract of *Elettaria cardamomum*. *Planta Medica* 82 (17):1482-1486. doi: 10.1055/s-0042-106971.

[115] Sereshti, H., Rohanifar, A., Bakhtiari, S., and Samadi, S. (2012). Bifunctional ultrasound assisted extraction and determination of *Elettaria cardamomum* Maton essential oil. *Journal of*

Chromatography A 1238:46-53. doi: https://doi.org/10.1016/j.chroma.2012.03.061.

[116] Kim, J. R., Kang, P., Lee, H. S., Kim, K. Y., and Seol, G. H. (2017). Cardiovascular effects of linalyl acetate in acute nicotine exposure. *Environmental Health and Preventive Medicine* 22 (1):42-42. doi: 10.1186/s12199-017-0651-6.

[117] Ribeiro, T. P., Porto, D. L., Menezes, C. P., Antunes, A. A., Silva, D. F., De Sousa, D. P., Nakao, L. S., Braga, V. A., and Medeiros, I. A. (2010). Unravelling the cardiovascular effects induced by alpha-terpineol: a role for the nitric oxide-cGMP pathway. *Clinical and Experimental Pharmacology and Physiology* 37 (8):811-816. doi: 10.1111/j.1440-1681.2010.05383.x.

[118] Patel, R. V., Mistry, B. M., Shinde, S. K., Syed, R., Singh, V., and Shin, H.-S. (2018). Therapeutic potential of quercetin as a cardiovascular agent. *European Journal of Medicinal Chemistry* 155:889-904. doi: https://doi.org/10.1016/j.ejmech.2018.06.053.

[119] Cheserek, M. J., Wu, G., Li, L., Li, L., Karangwa, E., Shi, Y., and Le, G. (2016). Cardioprotective effects of lipoic acid, quercetin and resveratrol on oxidative stress related to thyroid hormone alterations in long-term obesity. *The Journal of Nutritional Biochemistry* 33:36-44. doi: https://doi.org/10.1016/j.jnutbio.2016.02.008.

[120] Wei, J. M., Wang, X., Gong, H., Shi, Y.J., and Zou, Y. (2013). Ginkgo suppresses atherosclerosis through downregulating the expression of connexin 43 in rabbits. *Archives of Medical Science : AMS* 9 (2):340-346. doi: 10.5114/aoms.2013.34416.

[121] Panda, V. S., and Naik, S. R. (2008). Cardioprotective activity of Ginkgo biloba Phytosomes in isoproterenol-induced myocardial necrosis in rats: A biochemical and histoarchitectural evaluation. *Experimental and Toxicologic Pathology* 60 (4):397-404. doi: https://doi.org/10.1016/j.etp.2008.03.010.

[122] Wang, Z., Zhang, J., Ren, T., and Dong, Z. (2016). Targeted metabolomic profiling of cardioprotective effect of *Ginkgo biloba* L. extract on myocardial ischemia in rats. *Phytomedicine* 23 (6):621-631. doi: https://doi.org/10.1016/j.phymed.2016.03.005.

[123] Saini, A. S., Taliyan, R., and Sharma, P. L. (2014). Protective effect and mechanism of *Ginkgo biloba* extract-EGb 761 on STZ-induced diabetic cardiomyopathy in rats. *Pharmacognosy Magazine* 10 (38):172-178. doi: 10.4103/0973-1296.131031.

[124] Wang, M., Zhao, J., Avula, B., Wang, Y.-H., Avonto, C., Chittiboyina, A. G., Wylie, P. L., Parcher, J. F., and Khan, I. A. (2014). High-Resolution Gas Chromatography/Mass Spectrometry Method for Characterization and Quantitative Analysis of Ginkgolic Acids in *Ginkgo biloba* Plants, Extracts, and Dietary Supplements. *Journal of Agricultural and Food Chemistry* 62 (50):12103-12111. doi: 10.1021/jf503980f.

[125] Pietri, S., Maurelli, E., Drieu, K., and Culcasi, M. (1997). Cardioprotective and Anti-oxidant Effects of the Terpenoid Constituents of *Ginkgo biloba* Extract (EGb 761). *Journal of Molecular and Cellular Cardiology* 29 (2):733-742. doi: https://doi.org/10.1006/jmcc.1996.0316.

[126] Hao, Y., Sun, Y., Xu, C., Jiang, X., Sun, H., Wu, Q., Yan, C., and Gu, S. (2009). Improvement of contractile function in isolated cardiomyocytes from ischemia-reperfusion rats by ginkgolide B pretreatment. *Journal of Cardiovascular Pharmacology* 54 (1):3-9. doi: 10.1097/FJC.0b013e3181a91410.

[127] Liu, X., Yan, Y., Bao, L., Chen, B., Zhao, Y., and Qi, R. (2014). Ginkgolide B inhibits platelet release by blocking Syk and p38 MAPK phosphorylation in thrombin-stimulated platelets. *Thrombosis Research* 134 (5):1066-1073. doi: https://doi.org/10.1016/j.thromres.2014.08.025.

[128] Hopkins, A. L., Lamm, M. G., Funk, J. L., and Ritenbaugh, C. (2013). *Hibiscus sabdariffa* L. in the treatment of hypertension and hyperlipidemia: A comprehensive review of animal and human studies. *Fitoterapia* 85:84-94. doi: https://doi.org/10.1016/j.fitote.2013.01.003.

[129] Ojulari, O. V., Lee, S. G., and Nam, J.-O. (2019). Beneficial Effects of Natural Bioactive Compounds from *Hibiscus sabdariffa* L. on

Obesity. *Molecules (Basel, Switzerland)* 24 (1):210. doi: 10.3390/molecules24010210.

[130] Alshehri, A. M. (2010). Metabolic syndrome and cardiovascular risk. *Journal of Family & Community Medicine* 17 (2):73-78. doi: 10.4103/1319-1683.71987.

[131] Asgary, S., Soltani, R., Zolghadr, M., Keshvari, M., and Sarrafzadegan, N. 2016. Evaluation of the effects of roselle (*Hibiscus sabdariffa* L.) on oxidative stress and serum levels of lipids, insulin and hs-CRP in adult patients with metabolic syndrome: a double-blind placebo-controlled clinical trial. In *Journal of Complementary and Integrative Medicine*.

[132] Sabzghabaee, A. M., Ataei, E., Kelishadi, R., Ghannadi, A., Soltani, R., Badri, S., and Shirani, S. (2013). Effect of *Hibiscus sabdariffa* calices on dyslipidemia in obese adolescents: a triple-masked randomized controlled trial. *Materia Socio-Medica* 25 (2):76.

[133] Joven, J., March, I., Espinel, E., Fernández-Arroyo, S., Rodríguez-Gallego, E., Aragonès, G., Beltrán-Debón, R., Alonso-Villaverde, C., Rios, L., and Martin-Paredero, V. (2014). *Hibiscus sabdariffa* extract lowers blood pressure and improves endothelial function. *Molecular Nutrition & Food Research* 58 (6):1374-1378.

[134] Pérez-Torres, I., Torres-Narváez, J. C., Guarner-Lans, V., Díaz-Díaz, E., Perezpeña-Diazconti, M., Palacios, A. R., and Manzano-Pech, L. (2019). Myocardial Protection from Ischemia-Reperfusion Damage by the Antioxidant Effect of *Hibiscus sabdariffa* Linnaeus on Metabolic Syndrome Rats. *Oxidative Medicine and Cellular Longevity* 2019.

[135] Micucci, M., Malaguti, M., Gallina Toschi, T., Di Lecce, G., Aldini, R., Angeletti, A., Chiarini, A., Budriesi, R., and Hrelia, S. (2015). Cardiac and vascular synergic protective effect of *Olea europea* L. leaves and *Hibiscus sabdariffa* L. flower extracts. *Oxidative Medicine and Cellular Longevity* 2015.

[136] Obouayeba, A. P., Meite, S., Boyvin, L., Yeo, D., Kouakou, T. H., and N'Guessan, J. D. (2015). Cardioprotective and anti-

inflammatory activities of a polyphenols enriched extract of *Hibiscus sabdariffa* petal extracts in wistar rats. *Journal of Pharmacognosy and Phytochemistry* 4 (1).

[137] Si, L. Y.-N., Ali, S. A. M., Latip, J., Fauzi, N. M., Budin, S. B., and Zainalabidin, S. (2017). Roselle is cardioprotective in diet-induced obesity rat model with myocardial infarction. *Life Sciences* 191:157-165.

[138] Ajay, M., Chai, H. J., Mustafa, A. M., Gilani, A. H., and Mustafa, M. R. (2007). Mechanisms of the anti-hypertensive effect of *Hibiscus sabdariffa* L. calyces. *Journal of Ethnopharmacology* 109 (3):388-393. doi: https://doi.org/10.1016/j.jep.2006.08.005.

[139] Herrera-Arellano, A., Miranda-Sánchez, J., Ávila-Castro, P., Herrera-Álvarez, S., Jiménez-Ferrer, J. E., Zamilpa, A., Román-Ramos, R., Ponce-Monter, H., and Tortoriello, J. (2007). Clinical effects produced by a standardized herbal medicinal product of *Hibiscus sabdariffa* on patients with hypertension. A randomized, double-blind, lisinopril-controlled clinical trial. *Planta Medica* 73 (01):6-12.

[140] Herranz-López, M., Olivares-Vicente, M., Encinar, J. A., Barrajón-Catalán, E., Segura-Carretero, A., Joven, J., and Micol, V. (2017). Multi-Targeted Molecular Effects of *Hibiscus sabdariffa* Polyphenols: An Opportunity for a Global Approach to Obesity. *Nutrients* 9 (8):907. doi: 10.3390/nu9080907.

[141] Sitia, S., Tomasoni, L., Atzeni, F., Ambrosio, G., Cordiano, C., Catapano, A., Tramontana, S., Perticone, F., Naccarato, P., Camici, P., Picano, E., Cortigiani, L., Bevilacqua, M., Milazzo, L., Cusi, D., Barlassina, C., Sarzi-Puttini, P., and Turiel, M. (2010). From endothelial dysfunction to atherosclerosis. *Autoimmunity Reviews* 9 (12):830-834. doi: https://doi.org/10.1016/j.autrev.2010.07.016.

[142] Luzak, B., Golanski, J., Przygodzki, T., Boncler, M., Sosnowska, D., Oszmianski, J., Watala, C., and Rozalski, M. (2016). Extract from spent hop (*Humulus lupulus* L.) reduces blood platelet aggregation and improves anticoagulant activity of human endothelial cells in

vitro. *Journal of Functional Foods* 22:257-269. doi: https://doi.org/10.1016/j.jff.2016.01.029.

[143] Olsovska, J., Bostikova, V., Dusek, M., Jandovska, V., Bogdanova, K., Cermak, P., Bostik, P., Mikyska, A., and Kolar, M. (2016). *Humulus lupulus* L.(hops)-A valuable source of compounds with bioactive effects for future therapies. *Military Medical Science Letters (Voj Zdrav Listy)* 85 (1):19-30.

[144] Hirata, H., Takazumi, K., Segawa, S., Okada, Y., Kobayashi, N., Shigyo, T., and Chiba, H. (2012). Xanthohumol, a prenylated chalcone from *Humulus lupulus* L., inhibits cholesteryl ester transfer protein. *Food Chemistry* 134 (3):1432-1437. doi: https://doi.org/10.1016/j.foodchem.2012.03.043.

[145] Asgary, S., Solhpour, A., Parkhideh, S., Madani, H., Mahzouni, P., and Kabiri, N. (2012). Effect of hydroalcoholic extract of *Hypericum perforatum* on selected traditional and novel biochemical factors of cardiovascular diseases and atherosclerotic lesions in hypercholesterolemic rabbits: A comparison between the extract and lovastatin. *Journal of Pharmacy and Bioallied Sciences* 4 (3):212-218. doi: 10.4103/0975-7406.99044.

[146] Hernández-Saavedra, D., Pérez-Ramírez, I. F., Ramos-Gómez, M., Mendoza-Díaz, S., Loarca-Pina, G., and Reynoso-Camacho, R. (2016). Phytochemical characterization and effect of *Calendula officinalis, Hypericum perforatum,* and *Salvia officinalis* infusions on obesity-associated cardiovascular risk. *Medicinal Chemistry Research* 25 (1):163-172.

[147] Moghaddam, M. H. G., Roghani, M., and Maleki, M. (2016). Effect of *Hypericum perforatum* aqueous extracts on serum lipids, aminotransferases, and lipid peroxidation in hyperlipidemic rats. *Research in Cardiovascular Medicine* 5 (2).

[148] Kabiri, N., Asgary, S., and Setorki, M. (2011). The effects of concurrent hydroalcoholic extract of Amaranthus caudatus L. and *Hypericum perforatum* L. of fatty streak formation in hypercholesterolemic animals. *African Journal of Pharmacy and Pharmacology* 5 (16):1911-1919.

[149] Tolonen, A., Hohtola, A., and Jalonen, J. (2003). Fast high-performance liquid chromatographic analysis of naphthodianthrones and phloroglucinols from *Hypericum perforatum* extracts. *Phytochemical Analysis* 14 (5):306-309. doi: 10.1002/pca.720.

[150] Schulz, H.-U., Schürer, M., Bässler, D., and Weiser, D. (2005). Investigation of the bioavailability of hypericin, pseudohypericin, hyperforin and the flavonoids quercetin and isorhamnetin following single and multiple oral dosing of a hypericum extract containing tablet. *Arzneimittelforschung* 55 (01):15-22.

[151] Asadi, A., Shidfar, F., Safari, M., Hosseini, A. F., Fallah Huseini, H., Heidari, I., and Rajab, A. (2019). Efficacy of *Melissa officinalis* L. (lemon balm) extract on glycemic control and cardiovascular risk factors in individuals with type 2 diabetes: A randomized, double-blind, clinical trial. *Phytotherapy Research* 33 (3):651-659. doi: 10.1002/ptr.6254.

[152] Asadi, A., shidfar, F., Safari, M., Malek, M., Hosseini, A. F., Rezazadeh, S., Rajab, A., Shidfar, S., and Hosseini, S. (2018). Safety and efficacy of *Melissa officinalis* (lemon balm) on ApoA-I, Apo B, lipid ratio and ICAM-1 in type 2 diabetes patients: A randomized, double-blinded clinical trial. *Complementary Therapies in Medicine* 40:83-88. doi: https://doi.org/10.1016/j.ctim.2018.07.015.

[153] Joukar, S., Asadipour, H., Sheibani, M., Najafipour, H., and Dabiri, S. (2016). The effects of *Melissa officinalis* (lemon balm) pretreatment on the resistance of the heart to myocardial injury. *Pharmaceutical Biology* 54 (6):1005-1013. doi: 10.3109/13880209.2015.1091845.

[154] Shakeri, A., Sahebkar, A., and Javadi, B. (2016). *Melissa officinalis* L. – A review of its traditional uses, phytochemistry and pharmacology. *Journal of Ethnopharmacology* 188:204-228. doi: https://doi.org/10.1016/j.jep.2016.05.010.

[155] Karimi, I., Hayatgheybi, H., RAZMJO, M., YOUSEFI, M., DADYAN, A., and HADIPOUR, M. (2010). Anti-hyperlipidaemic effects of an essential oil of *Melissa officinalis*. L in cholesterol-fed

rabbits. *Journal of Applied Biological Sciences (JABS) E-ISSN: 2146-0108* 4 (1):17-22.

[156] Jun, H. J., Lee, J. H., Jia, Y., Hoang, M. H., Byun, H., Kim, K. H., and Lee, S. J. (2012). *Melissa officinalis* Essential Oil Reduces Plasma Triglycerides in Human Apolipoprotein E2 Transgenic Mice by Inhibiting Sterol Regulatory Element-Binding Protein-1c–Dependent Fatty Acid Synthesis. *The Journal of Nutrition* 142 (3):432-440. doi: 10.3945/jn.111.152538.

[157] Trevisan, S. C. C., Menezes, A. P. P., Barbalho, S. M., and Guiguer, É. L. (2017). Properties of *Mentha piperita*: a brief review. *World Journal of Pharmaceutical and Medical Research* 3 (1):309-313.

[158] Johari, N. Z., Ismail, I. S., Sulaiman, M. R., Abas, F., and Shaari, K. (2008). Original Research Acute toxicity and metabolomics analysis of hypocholesterolemic effect of *Mentha piperita* aqueous extract in Wistar rats. *International Journal of Applied Research in Natural Products* 8 (1):1-11.

[159] Mani Badal, R., Badal, D., Badal, P., Khare, A., Shrivastava, J., and Kumar, V. (2011). Pharmacological Action of *Mentha piperita* on Lipid Profile in Fructose-Fed Rats. *Iranian Journal of Pharmaceutical Research* 10 (4):843-848.

[160] Barbalho, S. M., Spada, A. P. M., Oliveira, E. P. d., Paiva-Filho, M. E., Martuchi, K. A., Leite, N. C., Deus, R. M., Sasaki, V., Braganti, L. S., and Oshiiwa, M. (2009). *Mentha piperita* effects on wistar rats plasma lipids. *Brazilian Archives of Biology and Technology* 52 (5):1137-1143.

[161] Barbalho, S. M., Machado, F. M. V. F., Oshiiwa, M., Abreu, M., Guiger, E. L., Tomazela, P., and Goulart, R. A. (2011). Investigation of the effects of peppermint (*Mentha piperita*) on the biochemical and anthropometric profile of university students. *Food Science and Technology* 31 (3):584-588.

[162] Gavahian, M., Farahnaky, A., Farhoosh, R., Javidnia, K., and Shahidi, F. (2015). Extraction of essential oils from *Mentha piperita* using advanced techniques: Microwave versus ohmic assisted

hydrodistillation. *Food and Bioproducts Processing* 94:50-58. doi: https://doi.org/10.1016/j.fbp.2015.01.003.

[163] Hossain, M. A., Al-Hdhrami, S. S., Weli, A. M., Al-Riyami, Q., and Al-Sabahi, J. N. (2014). Isolation, fractionation and identification of chemical constituents from the leaves crude extracts of *Mentha piperita* L grown in Sultanate of Oman. *Asian Pacific Journal of Tropical Biomedicine* 4:S368-S372. doi: https://doi.org/10.12980/APJTB.4.2014C1051.

[164] Bukovská, A., Cikoš, Š., Juhás, Š., Il'ková, G., Rehák, P., and Koppel, J. (2007). Effects of a combination of thyme and oregano essential oils on TNBS-induced colitis in mice. *Mediators of Inflammation* 2007.

[165] Kogiannou, D. A., Kalogeropoulos, N., Kefalas, P., Polissiou, M. G., and Kaliora, A. C. (2013). Herbal infusions; their phenolic profile, antioxidant and anti-inflammatory effects in HT29 and PC3 cells. *Food and Chemical Toxicology* 61:152-159.

[166] Ocaña-Fuentes, A., Arranz-Gutiérrez, E., Señorans, F. J., and Reglero, G. (2010). Supercritical fluid extraction of oregano (*Origanum vulgare*) essentials oils: Anti-inflammatory properties based on cytokine response on THP-1 macrophages. *Food and Chemical Toxicology* 48 (6):1568-1575. doi: https://doi.org/10.1016/j.fct.2010.03.026.

[167] Gutiérrez-Grijalva, E. P., Picos-Salas, M. A., Leyva-López, N., Criollo-Mendoza, M. S., Vazquez-Olivo, G., and Heredia, J. B. (2018). Flavonoids and Phenolic Acids from Oregano: Occurrence, Biological Activity and Health Benefits. *Plants* 7 (1):2.

[168] Seol, G. H., and Shin, Y. K. (2018). Essential Oils and Factors Related to Cardiovascular Diseases. *Journal of Potential of Essential Oils*: 129.

[169] Maryanto, S. (2013). The effects of red guava (*Psidium guajava* L) fruits on lipid peroxidation in hypercholesterolemic rats. *Basic Research Journals of Medicine and Clinical Science* 2 (11):116-121.

[170] Soto, M. E., Zuñiga-Muñoz, A., Guarner Lans, V., Duran-Hernández, E. J., and Pérez-Torres, I. (2016). Infusion of *Hibiscus*

sabdariffa L. Modulates Oxidative Stress in Patients with Marfan Syndrome. *Mediators of Inflammation* 2016:8625203-8625203. doi: 10.1155/2016/8625203.

[171] Rai, P. K., Mehta, S., and Watal, G. (2010). Hypolipidaemic & hepatoprotective effects of *Psidium guajava* raw fruit peel in experimental diabetes. *The Indian Journal of Medical Research* 131 (820):820-4.

[172] Farinazzi-Machado, F. M., Elen, L., de Souza, M. d. S. S., dos Santos Bueno, P. C., Mendes, C. G., Araujo, A. C., de Lara Lima, L. M., Marim, N. S., Brunnati, A. C. S., and da Silva, P. M. R. (2012). Effects of *Psidium guajava* on the metabolic profile of Wister rats. *Journal of Medicinal Plants Research* 6 (18):3450-3454.

[173] Brito, A. K. d. S., Lima, G. d. M., Farias, L. M. d., Rodrigues, L. A. R. L., Carvalho, V. B. L. d., Pereira, C. F. d. C., Frota, K. d. M. G., Conde-Júnior, A. M., Silva, A. M. O., Rizzo, M. d. S., Fonseca, C. M. B., Moura, R. C. D., Santos, R. C. d., Leite, J. R. d. S. d. A., Santos, M. A. P. d., Nunes, P. H. M., Arcanjo, D. D. R., and Martins, M. d. C. d. C. e. (2019). Lycopene-Rich Extract from Red Guava (*Psidium guajava* L.) Decreases Plasma Triglycerides and Improves Oxidative Stress Biomarkers on Experimentally-Induced Dyslipidemia in Hamsters. *Nutrients* 11 (2):393.

[174] de Souza, T. d. S., Ferreira, M. F. d. S., Menini, L., Souza, J. R. C. d. L., Bernardes, C. d. O., and Ferreira, A. (2018). Chemotype diversity of *Psidium guajava* L. *Phytochemistry* 153:129-137. doi: https://doi.org/10.1016/j.phytochem.2018.06.006.

[175] Ademiluyi, A. O., Oboh, G., Ogunsuyi, O. B., and Oloruntoba, F. M. (2016). A comparative study on antihypertensive and antioxidant properties of phenolic extracts from fruit and leaf of some guava (*Psidium guajava* L.) varieties. *Comparative Clinical Pathology* 25 (2):363-374.

[176] Irondi, E. A., Agboola, S. O., Oboh, G., Boligon, A. A., Athayde, M. L., and Shode, F. O. (2016). Guava leaves polyphenolics-rich extract inhibits vital enzymes implicated in gout and hypertension in vitro. *Journal of Intercultural Ethnopharmacology* 5 (2):122.

[177] Díaz-de-Cerio, E., Rodríguez-Nogales, A., Algieri, F., Romero, M., Verardo, V., Segura-Carretero, A., Duarte, J., and Galvez, J. (2017). The hypoglycemic effects of guava leaf (*Psidium guajava* L.) extract are associated with improving endothelial dysfunction in mice with diet-induced obesity. *Food Research International* 96:64-71.

Erlund, I., Koli, R., Alfthan, G., Marniemi, J., Puukka, P., Mustonen, P., Mattila, P., and Jula, A. (2008). Favorable effects of berry consumption on platelet function, blood pressure, and HDL cholesterol. *The American Journal of Clinical Nutrition* 87 (2):323-331. doi: 10.1093/ajcn/87.2.323.

[178] Asgary, S., RafieianKopaei, M., Sahebkar, A., Shamsi, F., and Golimalekabadi, N. (2016). Anti-hyperglycemic and anti-hyperlipidemic effects of *Vaccinium myrtillus* fruit in experimentally induced diabetes (antidiabetic effect of Vaccinium myrtillus fruit). *Journal of the Science Food and Agriculture* 96 (3):764-768. doi: 10.1002/jsfa.7144.

[179] Habanova, M., Saraiva, J. A., Haban, M., Schwarzova, M., Chlebo, P., Predna, L., Gažo, J., and Wyka, J. (2016). Intake of bilberries (*Vaccinium myrtillus* L.) reduced risk factors for cardiovascular disease by inducing favorable changes in lipoprotein profiles. *Nutrition Research* 36 (12):1415-1422. doi: https://doi.org/10.1016/j.nutres.2016.11.010.

[180] Furlan, C. P. B., Valle, S. C., Maróstica, M. R., Östman, E., Björck, I., and Tovar, J. (2019). Effect of bilberries, lingonberries and cinnamon on cardiometabolic risk-associated markers following a hypercaloric-hyperlipidic breakfast. *Journal of Functional Foods* 60:103443. doi: https://doi.org/10.1016/j.jff.2019.103443.

[181] Prencipe, F. P., Bruni, R., Guerrini, A., Rossi, D., Benvenuti, S., and Pellati, F. (2014). Metabolite profiling of polyphenols in Vaccinium berries and determination of their chemopreventive properties. *Journal of Pharmaceutical and Biomedical Analysis* 89:257-267. doi: https://doi.org/10.1016/j.jpba.2013.11.016.

[182] Primetta, A. K., Jaakola, L., Ayaz, F. A., Inceer, H., and Riihinen, K. R. (2013). Anthocyanin fingerprinting for authenticity studies of

bilberry (*Vaccinium myrtillus* L.). *Food Control* 30 (2):662-667. doi: https://doi.org/10.1016/j.foodcont.2012.09.009.

[183] Åkerström, A., Jaakola, L., Bång, U., and Jäderlund, A. (2010). Effects of Latitude-Related Factors and Geographical Origin on Anthocyanidin Concentrations in Fruits of *Vaccinium myrtillus* L. (Bilberries). *Journal of Agricultural and Food Chemistry* 58 (22):11939-11945. doi: 10.1021/jf102407n.

[184] Cassidy, A., Mukamal, K. J., Liu, L., Franz, M., Eliassen, A. H., and Rimm, E. B. (2013). High Anthocyanin Intake Is Associated With a Reduced Risk of Myocardial Infarction in Young and Middle-Aged Women. *Circulation* 127 (2):188-196. doi: doi:10.1161/CIRCULATIONAHA.112.122408.

[185] Krga, I., Vidovic, N., Milenkovic, D., Konic-Ristic, A., Stojanovic, F., Morand, C., and Glibetic, M. (2018). Effects of anthocyanins and their gut metabolites on adenosine diphosphate-induced platelet activation and their aggregation with monocytes and neutrophils. *Archives of Biochemistry and Biophysics* 645:34-41. doi: https://doi.org/10.1016/j.abb.2018.03.016.

[186] Daveri, E., Cremonini, E., Mastaloudis, A., Hester, S. N., Wood, S. M., Waterhouse, A. L., Anderson, M., Fraga, C. G., and Oteiza, P. I. (2018). Cyanidin and delphinidin modulate inflammation and altered redox signaling improving insulin resistance in high fat-fed mice. *Redox Biology* 18:16-24. doi: https://doi.org/10.1016/j.redox.2018.05.012.

[187] Mao, Q. Q., Xu, X. Y., Cao, S. Y., Gan, R. Y., Corke, H., Beta, T., and Li, H.-B. (2019). Bioactive Compounds and Bioactivities of Ginger (*Zingiber officinale* Roscoe). *Foods (Basel, Switzerland)* 8 (6):185. doi: 10.3390/foods8060185.

[188] Akinyemi, A. J., Ademiluyi, A. O., and Oboh, G. (2014). Inhibition of angiotensin-1-converting enzyme activity by two varieties of ginger (*Zingiber officinale)* in rats fed a high cholesterol diet. *Journal of Medicinal Food* 17 (3):317-323.

[189] Tende, J. A., Ayo, J. O., Mohammed, A., and Zezi, A. U. (2015). Blood pressure lowering and cardio-protective effects of garlic

(*Allium sativum*) and ginger (*Zingiber officinale*) extracts in some laboratory animals. *International Journal of Medicine and Medical Sciences* 7 (1):8-13.

[190] Khosravani, M., Azarbayjani, M., Abolmaesoomi, M., Yusof, A., Abidin, N. Z., Rahimi, E., Feizolahi, F., Akbari, M., Seyedjalali, S., and Dehghan, F. (2016). Ginger extract and aerobic training reduces lipid profile in high-fat fed diet rats. *European Review for Medical and Pharmacological Sciences* 20 (8):1617-1622.

[191] de las Heras, N., Valero-Muñoz, M., Martín-Fernández, B., Ballesteros, S., López-Farré, A., Ruiz-Roso, B., and Lahera, V. (2016). Molecular factors involved in the hypolipidemic- and insulin-sensitizing effects of a ginger (*Zingiber officinale* Roscoe) extract in rats fed a high-fat diet. *Applied Physiology, Nutrition, and Metabolism* 42 (2):209-215. doi: 10.1139/apnm-2016-0374.

[192] Amran, A. Z., Jantan, I., Dianita, R., and Buang, F. (2015). Protective effects of the standardized extract of *Zingiber officinale* on myocardium against isoproterenol-induced biochemical and histopathological alterations in rats. *Pharmaceutical Biology* 53 (12):1795-1802. doi: 10.3109/13880209.2015.1008147.

[193] Wu, H. C., Horng, C. T., Tsai, S. C., Lee, Y. L., Hsu, S. C., Tsai, Y. J., Tsai, F. J., Chiang, J. H., Kuo, D. H., and Yang, J. S. (2018). Relaxant and vasoprotective effects of ginger extracts on porcine coronary arteries. *International Journal of Molecular Medicine* 41 (4):2420-2428. doi: 10.3892/ijmm.2018.3380.

[194] Semwal, R. B., Semwal, D. K., Combrinck, S., and Viljoen, A. M. (2015). Gingerols and shogaols: Important nutraceutical principles from ginger. *Phytochemistry* 117:554-568. doi: https://doi.org/10.1016/j.phytochem.2015.07.012.

[195] Tan, Z., Li, F., and Xing, J. (2011). Separation and Purification of Aloe Anthraquinones Using PEG/Salt Aqueous Two-Phase System. *Separation Science and Technology* 46 (9):1503-1510. doi: 10.1080/01496395.2011.563766.

[196] Babova, O., Occhipinti, A., Capuzzo, A., and Maffei, M. E. (2016). Extraction of bilberry (*Vaccinium myrtillus*) antioxidants using

supercritical/subcritical CO2 and ethanol as co-solvent. *The Journal of Supercritical Fluids* 107:358-363. doi: https://doi.org/10.1016/j.supflu.2015.09.029.

[197] Jiao, L., Zhang, X., Huang, L., Gong, H., Cheng, B., Sun, Y., Li, Y., Liu, Q., Zheng, L., and Huang, K. (2013). Proanthocyanidins are the major anti-diabetic components of cinnamon water extract. *Food and Chemical Toxicology* 56:398-405. doi: https://doi.org/10.1016/j.fct.2013.02.049.

[198] Hajimonfarednejad, M., Nimrouzi, M., Heydari, M., Zarshenas, M. M., Raee, M. J., and Jahromi, B. N. (2018). Insulin resistance improvement by cinnamon powder in polycystic ovary syndrome: A randomized double-blind placebo controlled clinical trial. *Phytotherapy Research* 32 (2):276-283. doi: 10.1002/ptr.5970.

[199] Hong, Y. J., Huang, X. M., Liu, X. B., Zhang, C. Y., Zhang, L., and Xu, X. L. (2013). 6-Shogaol protects against oxidized LDL-induced endothelial injruries by inhibiting oxidized LDL-evoked LOX-1 signaling. *Evidence-Based Complementary and Alternative Medicine* 2013.

[200] Imani, H., Tabibi, H., Najafi, I., Atabak, S., Hedayati, M., and Rahmani, L. (2015). Effects of ginger on serum glucose, advanced glycation end products, and inflammation in peritoneal dialysis patients. *Nutrition* 31 (5):703-707. doi: https://doi.org/10.1016/j.nut.2014.11.020.

[201] Azimi, P., Ghiasvand, R., Feizi, A., Hosseinzadeh, J., Bahreynian, M., Hariri, M., and Khosravi-Boroujeni, H. (2016). Effect of cinnamon, cardamom, saffron and ginger consumption on blood pressure and a marker of endothelial function in patients with type 2 diabetes mellitus: A randomized controlled clinical trial. *Blood pressure* 25 (3):133-140.

[202] Ma, Y.-C., Mani, A., Cai, Y., Thomson, J., Ma, J., Peudru, F., Chen, S., Luo, M., Zhang, J., Chapman, R. G., and Shi, Z. T. (2016). An effective identification and quantification method for *Ginkgo biloba* flavonol glycosides with targeted evaluation of adulterated products.

Phytomedicine 23 (4):377-387. doi: https://doi.org/10.1016/j.phymed.2016.02.003.

[203] Zhang, Q., Chen, L. J., Ye, H. Y., Gao, L., Hou, W., Tang, M., Yang, G., Zhong, Z., Yuan, Y., and Peng, A. (2007). Isolation and purification of ginkgo flavonol glycosides from *Ginkgo biloba* leaves by high-speed counter-current chromatography. *Journal of Separation Science* 30 (13):2153-2159. doi: 10.1002/jssc.200600544.

[204] Peng, C. H., Chyau, C. C., Chan, K. C., Chan, T. H., Wang, C. J., and Huang, C. N. (2011). *Hibiscus sabdariffa* polyphenolic extract inhibits hyperglycemia, hyperlipidemia, and glycation-oxidative stress while improving insulin resistance. *Journal of Agricultural and Food Chemistry* 59 (18):9901-9909.

[205] Morales-Luna, E., Pérez-Ramírez, I. F., Salgado, L. M., Castaño-Tostado, E., Gómez-Aldapa, C. A., and Reynoso-Camacho, R. (2019). The main beneficial effect of roselle (*Hibiscus sabdariffa*) on obesity is not only related to its anthocyanin content. *Journal of the Science of Food and Agriculture* 99 (2):596-605. doi: 10.1002/jsfa.9220.

[206] Herranz-López, M., Fernández-Arroyo, S., Pérez-Sanchez, A., Barrajón-Catalán, E., Beltrán-Debón, R., Menéndez, J. A., Alonso-Villaverde, C., Segura-Carretero, A., Joven, J., and Micol, V. (2012). Synergism of plant-derived polyphenols in adipogenesis: Perspectives and implications. *Phytomedicine* 19 (3):253-261. doi: https://doi.org/10.1016/j.phymed.2011.12.001.

[207] Villalpando-Arteaga, E. V., Mendieta-Condado, E., Esquivel-Solís, H., Canales-Aguirre, A. A., Gálvez-Gastélum, F. J., Mateos-Díaz, J. C., Rodríguez-González, J. A., and Márquez-Aguirre, A. L. (2013). *Hibiscus sabdariffa* L. aqueous extract attenuates hepatic steatosis through down-regulation of PPAR-γ and SREBP-1c in diet-induced obese mice. *Food & Function* 4 (4):618-626.

In: Phytochemicals from Mexican … ISBN: 978-1-53617-397-0
Editors: J. Basilio Heredia et al.

Chapter 8

BIOACTIVE COMPOUNDS FROM MEDICINAL PLANTS: GREEN EXTRACTION TECHNOLOGIES

Alexis Emus-Medina[1], Laura A. Contreras-Angulo[1], Salma A. Enriquez-Valencia[2] and Leticia X. Lopez-Martínez[2,*]

[1]Centro de Investigación en Alimentación y Desarrollo, AC, Culiacán, Sinaloa, México

[2]CONACYT-Centro de Investigación en Alimentación y Desarrollo A.C., Hermosillo, Sonora, Mexico

ABSTRACT

At present, effective extraction methods of bioactive compounds are required to use medicinal plant extracts as natural sources with possible applications in human health and the food industry. Many factors play a vital role in the extraction efficiency, such as concentration and type of extraction solvent, extraction temperature, extraction time, and extraction

* Corresponding Author's Email: leticia.lopez@ciad.mx.

pH. The selection of solvent extraction is based on the chemical nature and polarity of the bioactive compounds to be extracted; however, it is necessary that extraction methods are faster and more efficient than the traditional ones (liquid-liquid or solid-liquid extraction), which most of the times produce low extraction yields, consuming high amount of solvents and time inefficiency and inferior extract quality due to traces of organic solvents present in them. Those technologies that are referred to as non-conventional clean or green technologies include microwave-assisted extraction (MAE), ultrasound-assisted extraction (UAE), supercritical carbon dioxide extraction (scCO2), pressurized liquid extraction (PLE), enzyme-assisted extraction (EAE) and Pulsed electric field (PEF).

Keywords: bioactive compounds, green technologies, medicinal plants

1. Introduction

Medicinal plants have been studied since ancient times, when people looked for drugs in nature in search of cures for their diseases. Extensive investigations have revealed that medicinal plants in different shapes, either in crude form or pure molecules isolated from them, represent the most ancient mode of medication. The compounds that have beneficial effects on health can be derived from roots, seeds, barks, leaves, flowers, and fruits [1]. Compounds of medicinal plants that have beneficial effects are phenolic compounds, essential oils, and carotenoids, among others [2-5] that exhibit a wide range of biological effects, such as anti-aging, antibacterial, antiviral, anti-inflammatory, and anticancer [3, 6, 7].

Taking into account the valuable health effects of these components, selection of proper extraction method needs strict evaluation, the commonly used methods are the conventional liquid-liquid or solid-liquid extraction, that require large amounts of organic solvents, high energy expenditure, and are time consuming, which has generated interest in new technologies that are referred to as non-conventional clean or green technologies, which can be defined in relation to green extraction of natural products, as: *Green Extraction is based on the discovery and design of extraction processes which will reduce energy consumption, allows use of alternative solvents and renewable natural products, and ensure a safe and high quality extract/product* [8].

These non-conventional methods have been developed in order to be more environmentally friendly due to decreased use of synthetic and organic chemicals, reduced operational time, and better yield and quality of extract, have been developed during the last 50 years [9]. To enhance the overall yield and selectivity of bioactive components from medicinal plant materials, ultrasound pulsed electric field, enzyme digestion microwave heating supercritical fluids have been studied as green or non-conventional methods (Figure 1).

2. Extraction Methods of Bioactive Compounds from Medicinal Plants

To develop an efficient approach to green extraction for recovery bioactive compounds from medicinal plants is necessary to know the methodologies that have been developed for this purpose and thus decide the most convenient extraction method in order to obtain the best result.

2.1. Ultrasound-Assisted Extraction (UAE) or Sonication Extraction

Ultrasound-assisted extraction implies the use of ultrasound ranging from 20 kHz to 2000 kHz [9]. The mechanic effect of acoustic cavitation (waves passes through a medium by creating compression and expansion) involves the formation, growth, and collapse of bubbles [10].

As a consequence of this collapse, a lot of energy is produced, which occurs uncontrollably in each direction in the media (ultrasonic bath tank). This massive energy release causes extreme changes in temperature (up to 4700°C) and pressure (100 MPa) within bubbles during extraction, causing liquid solutes to leach at speeds of 280 m/s [11, 12]. Upon sonication, ultrasonic waves disrupt the solid particles and remove inert material layers resulting in an increased surface area for mass transfer of solutes during extraction (Figure 1).

The procedure is simple, relatively easy to use; it is versatile and flexible. It requires relatively low-cost technology compared with other extraction techniques, besides can be used in both small and larger scale of phytochemical extraction.

Physical and chemical properties of the materials subjected to ultrasound are altered and disrupt the plant cell wall, facilitating the release of compounds and enhanced the mass transport of solvents to plant cells [13].

In turn, different mechanisms are involved in the UAE that may or may not related to each other, such as fragmentation, erosion, sonocapillary, disintegration and sonoporation (Table 1). The action of these mechanisms is linked to the increase in the yield of the different extractions that use this method [8].

Table 1. Mechanisms related to the extraction of bioactive compounds by UAE

Mechanism	Characteristics	Reference
Fragmentation	Reduction of particle size due to the collision between particles created by the cavitation of bubbles in the liquid	[14]
Erosion	Damage to external cell structures due to cavitation	[14]
Sonocapillarity	Increase in the depth and speed of liquid penetration through channels and pores. Increase in desorption processes and diffusion of solutes outside plant material.	[15]
Sonoporation	Pore formation in cell membranes, causing solutes to enter the environment.	[16]
Shear force	Generation of shear forces within the liquid and near the solid material resulting from the oscillation and collapse of the bubbles within the fluid	[10]

The benefits of UAE are mainly due to a reduction in extraction time and solvent consumption. However, the use of ultrasound energy more than 20 kHz may affect the bioactive characteristics of phytochemicals through the formation of free radicals [10]. Table 2 shows some studies that describe the use of ultrasound to obtain bioactive compounds from medicinal plants.

Table 2. Ultrasound-assisted extraction of bioactive compounds from medicinal plants

Scientific Name	Bioactives	Reference
Allium sativum L.	Allicin	[17]
Cynara scolymus L.	Clorogenic acid	[18]
Vaccinium myrtillus L.	Total anthocyanins, Cyanidin-3-*O*-glucoside, cyanidin-3-*O*-acetyl-hexoside	[19, 20]
Calendula officinalis L.	Rutin	[21]
Turnera diffusa Wild	Phenolic compounds	[22]
Echinacea angustifolia and Echinacea pallida	Caftaric acid; chlorogenic acid, caffeic aci. Echinacoside, ferulic acid, cichoric acid, verbascoside	[23]
Psidium guajava L.	Total phenolic Total flavonoids	[2]
Hypericum perforatum L.	Total phenolic compounds	[24]
Hibiscus sabdariffa L.	Total anthoycanins	[25]
Zingiber officinale Roscoe	Gingerol	[26]
Humulus lupulus L.	Rutin, quercetin, taxifolin, isoquercitin, cloroghenic acid, ferulic acid, caffeic acid	[27]
Melissa officinalis L.	Rosmarinic acid	[28]
Mentha x piperita L.	Total phenolic compounds, total flavonoids	[29]
Melissa officinalis L.	Rosmarinic acid	[28]
Mentha x piperita L.	Total phenolic compounds, total flavonoids	[29]
Citrus aurantium L.	Total phenolic compounds	[30]
Origanum vulgare L.	Total phenolic compounds	[31]

2.2. Microwave-Assisted Extraction (MAE)

The use of microwave for the extraction of bioactive compounds from medicinal plants has demonstrated great research potential (Table 3). MAE utilizes microwave energy to facilitate the extraction of phytochemicals from the sample matrix into the solvent [32]. Fast time of extraction with lees solvent, less energy consumption, in addition, the protection offered to thermolabile phytochemicals could allow that the purity of the extracts to increase, are some of the attractive advantages of MAE [33]. Microwaves are a source of heat without contact with the sample, which can provide more effective heating, accelerating energy transfer, and reducing the thermal gradient. Compounds such as essential oils, flavonoids, carotenes, and other organic compounds can be separated efficiently using this technology [34].

Table 3. Ultrasound-assisted extraction of bioactive compounds from medicinal plants

Scientific name	Bioactives	Reference
Allium sativum L.	Allicin	[39]
Vaccinium myrtillus L.	Caffeic acid, ferulic acid, ellagic acid, chlorogenic acid	[40]
Calendula officinalis L.	Protocatechuic acid, 4- hydroxybenzoic acid, chlorogenic acid, vanillic acid, caffeic acid, syringic acid, ferulic acid, rutin, isoquercetin, hesperidin, rosmarinic acid, quercetin	[41]
Cinnamomum verum J Presi	Trans-cinnamaldehyde	[42]
Elettaria cardamomum L. *Maton*	Total carotenoids	[5]
Echinacea angustifolia and Echinacea pallida	Caftaric acid; chlorogenic acid; caffeic acid, echinacoside; cichoric acid, verbascoside, quercitin, kaempferol,	[23]
Ginkgo biloba	Total phenolic compounds	[43]
Psidium guajava L.	Quercetin	[44]
Hibiscus sabdariffa L.	Total anthocyanins Chlorogenic acid Myrecitin, Kaemferol, quercetin 3-rutinoside, methyl digallate, uercetin-3-sambubioside, Quercetin-3-glucoside, Prodelphinidin B3	[45] [46]
Zingiber officinale Roscoe	Gingerlo	[26]
Melissa officinalis L.	Rosmarinic acid	[28]
Mentha x piperita L.	Total phenolic, total flavonoids	[47]
Citrus aurantium L.	Essential oil	[4]
Origanum vulgare L.	Total phenolic compounds	[31] [36]

The MAE starts with the generation of electromagnetic waves from a cavity magnetron, the tissues and cell walls of the plants interact with the emitted radiation waves. This interaction results in the heating of the moisture trapped inside the plant's matrix due to the absorption of the characteristic photonic energy of the electromagnetic waves that cause the evaporation of the moisture from the plant matrix. This causes considerable pressure to be applied on the walls at the cellular and subcellular level, which results in the swelling of plant cells, this eventually causes structural changes in the plant matrix, promoting higher mass transfer of solutes due to the cell breakdown This, in turn, facilitates the leaching of compounds from the cell-matrix of the plant to the extractor during the MAE [10, 35].

MAE is affected by the number of factors such as the duration of microwave radiation, the power of microwave, properties of the solvent used, the particle size of the sample, the temperature, and solid to sample ratio [36].

Microwave is non-ionizing electromagnetic waves of frequency between 300 MHz to 300 GHz, and the radiation interacts with dipoles of polar and polarizable solvents and samples, causes heating near the surface of the materials and heat is transferred by conduction. Dipole rotation of the molecules induced by microwave electromagnetic disrupts hydrogen bonding, producing a better migration of dissolved ions and helping the penetration of solvents in the matrix [37]. When nonpolar solvents without polarizable groups are used, poor heating occurs [38]. The solvent properties can be modified by combining different solvents, which also leads to varying selectivity of the solvent for different compounds. In general, the higher the dielectric constant and dielectric loss, the higher is the capacity of the solvent to absorb microwave energy, which can lead to a faster rate of heating of the solvent with respect to the plant material.

MAE can be considered as specific methods that favor polar molecules and solvents with high dielectric constant such as water, ethanol, methanol, toluene, acetone, hexane and dichloromethane. This method is limited to small bioactive molecules stables under microwave heating conditions up to 100°C for 20 min [36-38]. For larger molecules such as tannins, anthocyanins, and flavanones, temperature conditions and MAE cycles could result in oxidation and a drastic decrease in its yield. However, solvent combination with relatively lower dielectric properties can be used to ensure that the solvent temperature will remain lower to cool-off the solutes once they are liberated into the solvent. In this case, the microwave energy interacts preferentially with the plant matrix [37], leading to the efficient release of the plant compounds in the cooler solvent. Stalikas et al. [38] described that also pH of the solvent can significantly affect the different efficiency of the extraction step.

2.3. Supercritical Fluid Extraction (SFE)

SFE extraction uses carbon dioxide above its critical point and exhibits liquid-like properties (dissolvent power, negligible surface tension) as well as gas-like properties (matrix penetration and transport) [48]. SFE is extensively used for the extraction of bioactive compounds in medicinal plants (Table 4) since it is a colorless, odorless, non-toxic, non-flammable, safe, highly pure, and cost-effective solvent.

Table 4. Supercritic extraction of bioactive compounds from medicinal plants

Scientific name	Bioactives	Reference
Allium sativum L.	3-Vinyl-4H-1,2-dithiin Allicin	[55] [56]
Vaccinium myrtillus L.	Cafeic acid, ferulic acid, elagic acid, clorogenic acid	[40]
Calendula officinalis L.	Faradiol	[57]
Cinnamomum verum J Presi	Tyrosinase Melanin	[58]
Ginkgo biloba	Flavonoids	[59, 60]
Psidium guajava L.	β-Caryophyllene	[61]
Hypericum perforatum L	Phloroglucinol, benzophenone derivatives	[62]
Hibiscus sabdariffa L.	Chlorogenic acid,myrecitin, kaemferol, quercetin 3-rutinoside, methyl digallate, uercetin-3-sambubioside, quercetin-3-glucoside, Prodelphinidin B3	[46]
Zingiber officinale Roscoe	Gingerol	[60]
Melissa officinalis L.	Gallic acid, protocatechuic acid, p-hydroxybenzoic acid, vanillic acid, syringic acid	[63]
Mentha x piperita L.	Flavonoids	[64]
Citrus aurantium L.	Flavonoids	[65]
Origanum vulgare L.	Total phenolics	[36]

During the process, the raw material (preferably in powder form) is placed in an extraction vessel equipped with temperature and pressure controllers to maintain the required conditions. After this, the extraction vessel is pressurized with the fluid employing a pump. Once the fluid and dissolved compounds are transported to the separators, the products are collected through a tap located at the bottom of the separators. Finally, the fluid is regenerated and recycled or released into the environment. The selection of supercritical fluids is essential for the proper functioning of this process [49].

Generally, in SF, carbon dioxide (CO_2) is utilized for the extraction of the phytochemicals from medicinal plants. CO_2 becomes SF above 31.1°C and 7380 kPa (73.8 bar), is an excellent solvent for non-polar compounds is available in high purity, non-flammable and with low toxicity. Although SC-CO_2 has poor solubility for obtaining polar compounds, for example, glycosylated phenolic compounds, modifications such as adding a small amount of water, ethanol, or methanol and its mixtures allow polar compounds to be obtained. SC-CO_2 also makes it possible to obtain compounds in concentrated form when CO_2 vaporizes at room temperature. The strength of SC solvents can be easily altered by changing the temperature, pressure, or adding modifiers that reduce extraction time [49, 50]. The polarity of CO_2 it is suitable for the extraction of lipophilic compounds such as plant essential oils [51, 52]. Nonetheless, when there are polarity limitations, organic solvents are added to CO_2, gases such as argon can also be used because they are inexpensive and more inert [9].

Because SFE has several properties, this allows it to have certain advantages such as SFs have relatively lower viscosity and higher diffusivity, a fresh fluid is continuously forced to flow through the samples; therefore, it can provide quantitative or complete extraction. In SFE, the solvation power of the fluid can be controlled by changing temperature and/or pressure; it may allow achieving a high selectivity, this regulatable solvation power of SFs is especially useful for the extraction of complex samples such as plant materials, solutes dissolved in supercritical CO_2 can be easily separated by depressurization, SFE is performed at low temperatures, so it may be an ideal technique to study thermally labile compounds [53, 54].

2.4. Pressurized Liquid Extraction (PLE)

PLE is the application of high pressure to remain solvent liquid above their typical atmospheric boiling point. High pressure facilitates the extraction process; PLE is known with different names: high-pressure solvent extraction (HSPE), accelerated fluid extraction (ASE), pressurized fluid extraction (PFE), and enhanced solvent extraction (ESE) [66]. In case when water is used as the extraction solvent, the technique is referred to as Pressurized Hot Water Extraction (PHWE), sub-critical water extraction, or superheated water extraction [67].

In PLE, it is possible to manipulate the processing parameters that will allow selectivity for a particular group of compounds. In this technology, water, ethanol or ethanol-water mixtures can be used as solvents successfully to extract temperature-sensitive phytochemicals from different plant sources [54, 67, 68].

The efficiency of extraction depends on the nature of the sample, the compounds to be extracted, and their location within the matrix of the sample. When the extraction of heterogeneous samples is performed, it is assumed that the matrix is porous and is surrounded by an organic layer, then for extraction and recovery of the compound, it is desorbed from the sample; then diffused through the organic part of the matrix in order to reach the matrix-fluid interface after that, the compound is distributed and diffused through the extraction phase present within the pore and then, reaches the extraction phase. The final stage of the extraction process is the collection of the extracted compound [69, 70].

A critical step in SLE extraction process is the place where the compound is located within the sample matrix; the compounds can be considered in five different points:

1) Adsorbed to the surface of the matrix.
2) Dissolved in a solvent pore and/or adsorbed on the surface.
3) Dissolved/adsorbed in a micro-nano-pore matrix.
4) Chemically bonded to the matrix
5) Dissolved in a solution [69, 70].

PLE allows fast extraction in a closed and inert medium under high pressure and temperature. Once the extraction is performed at high pressures, the solvent can remain at liquid state (so-called subcritical), even when taken to temperatures much over their boiling points. Thus, PLE can operate at high temperatures, where the solubility of the target compounds and desorption kinetics from the matrix are enhanced [67, 69].

Hu et al. [71] examined the potential for pressurized liquid extraction (PLE) of ginger (*Zingiber officinale* Roscoe). Among the solvents tested for PLE (bioethanol, ethanol, and water), the use of bioethanol leads the highest efficiency to recover the components of gingerol-related compounds (6-gingerol, 8-gingerol, and 10-gingerol). The optimized extraction conditions of PLE in order to obtain a yield of 106.8%, 109.3% and 108.0% of 6-gingerol, 8-gingerol and 10 -gingerol, respectively were 70% bioethanol, 1500 psi and 100°C for 20 min and 5 min of static extraction time are recommended as optimized extraction conditions in relation to the corresponding constituent obtained by 8 h of Soxhlet extraction (ethanol as extraction solvent).

In a study to maximize the accelerated solvent extraction (ASE) of the phenolic compounds content from two spices of Lamiaceae family; rosemary and oregano, conditions about extraction temperature (66–129°C) and solvent concentration (32–88% methanol) were identified using response surface methodology (RSM). For the two spices, results showed that 129°C was the optimum temperature to obtain extracts with a high content of phenolics (34.22 mg/g). Optimal methanol concentrations with respect to the content of phenolics (35.015 mg/g) of rosemary extracts were 56%. Oregano exhibited a different response to the effect of methanol concentration and was optimally extracted at 33%. The content of total phenolic compounds yields of the optimal ASE extracts was higher than solid/liquid extracts (9,2 and 41% for rosemary and oregano, respectively) [72].

The subcritical water extraction at different temperatures ranging from 25-200°C has been studied to extract phenolic compounds from rosemary leaves. The results indicate that this method is highly selective extracting compounds such as carnosol, rosmanol, carnosic acid, methyl carnosate and some flavonoids souch as cirsimaritin and genkwanin. When the temperature was increased from 25 to 200°C, a decrease in the extraction ability of subcritical water toward the more polar compounds was observed. Carnosic acid, carnosol, and genkwanin were first extracted at 100°C, decreasing their content in the extract while increasing temperatures for all compounds except carnosic acid, which is favored extracted at 200°C [73].

A study of SWE of phenolic compounds from *E. purpurea* flowers was developed and optimized. A 5- level response surface methodology with a central composite design was useful in estimating the effects of the extraction parameters on the investigated responses.

It demonstrated that there was a high fit degree between the values observed in the experiment and the predicted values confirming the validity of the performed models. The optimized extraction conditions were extraction temperature of 147.56°C and extraction time of 8.43 min, while maximum TP content, TF content were 4.89 g GAE/100 g DW, 3.58 g CE/100 g DW respectively [74].

2.5. Pulsed-Electric Field Extraction (PEF)

PEF technique is the application of short-duration pulses of a high-intensity electric field (0.1–50 kV/cm) at room temperature [75]. The electric fields of few to several hundred microseconds are capable of inducing the formation of pores in the cell membrane (electroporation), which allows the subsequent extraction of bioactive molecules.

The principle of PEF is to destroy the cell membrane structure for increasing extraction. During the suspension of a living cell in the electric field, an electric potential passes through the membrane of that cell. Based on the dipole nature of membrane molecules, electric potential separates molecules according to their charge in the cell membrane. After exceeding a critical value of approximately 1V of transmembrane potential, repulsion occurs between the charge carrying molecules that form pores in weak areas of the membrane and causes a drastic increase of permeability [76].

The advantages of PEF include its non-thermal nature, selective extraction, shorter extraction time, and clean process, which are useful for thermally sensitive compounds such as polyphenols. Operating parameters such as electric field strength, number, and duration of pulses, the size of pores on the membrane can be manipulated to obtain the desired components and keeping other components inside the cell. Under optimum conditions, the recovery factor is significantly increased with the minimum extraction time [77].

It has been shown that the application of the PEF technique improves the extraction of compounds of different matrices with water, increases the rate of the extracted compounds, reduces the temperature and the amount of solvents used [78]. It facilitates the selective recovery of compounds without deteriorating the matrix, favoring their separation, and purification for later stages [79].

Frozen/thawed European blueberry (*Vaccinium myrtillus* L.) fruits were treated with pulsed electric field in order to enhance the cell membrane permeability and the quality of blueberry juice during the subsequent pressing process. Results show that, in comparison with the pressing of frozen/thawed sample, the application of PEF pretreatment may promote polyphenols extraction, which increased from 7.4% after PEF treatment at 1 Kv/cm and to up to 8.0% after PEF treatment at 3 Kv/cm. Therefore, only pressing of the freeze-thawed sample was not sufficient to ensure the total release of polyphenols in the juice, and the application of PEF can be of value since they may induce an enhancement in the yield of extraction of bioactive compounds, probably being able to permeabilize also inner cell membranes. Although anthocyanin content slightly increased after PEF treatment was no significant difference for the control [80].

Luengo et al. [81] studied the influence of PEF treatment on the extraction by pressing of total phenolics and flavonoids (naringin and hesperin) from orange peel. A treatment time of 20 pulses of 3 μs each achieved the highest cell disintegration at the different electric field strengths tested. After 30 min of pressurization at 5 bars, the total phenolic extraction increased 20%, 129%, 153%, and 159% for orange peel PEF treated at 1, 3, 5, and 7 kV / cm, respectively. A PEF treatment of 5 kV/cm to the orange peels increased the quantity of naringin and hesperidin in the extract of 100 g of orange peels from 1 to 3.1 and from 1.3 to 4.6 mg/100 g orange peel respectively.

2.6. Enzyme Assisted Extraction

The premise of enzyme-assisted extraction is that the release of phytochemical nature compounds from plant cells can be achieved using enzymes that are capable of causing cell disruption.

The cell wall of the plants contains polysaccharides such as pectins, cellulose, and hemicellulose that act as structural support but also as a barrier to the release of intracellular substances. Enzymes such as pectinases, cellulases, and hemicellulase are widely required to disrupt the structural integrity of the plant cell wall and membranes; these enzymes hydrolyze cell wall components, depolymerizing its polysaccharides thereby increasing cell wall permeability, which results in higher extraction of bioactive compounds from plants. Therefore, these enzymes have been proposed as tools to optimize the extraction of compounds from the plant matrix [82].

The primary step of the extraction method is enzymatic hydrolysis, which is a mixture of the enzymes, sample, and solvent (water mainly), this mixture is incubated around 35–50°C, with adjusted pH. The hydrolysis is stopped by deactivating enzymes at a temperature of 80–90°C. Thus factors such as type of enzyme, enzyme composition, and concentration, the particle size of plant materials, solid to water ratio, hydrolysis time, pH, and temperature are recognized as key factors for extraction [33, 83].

The main advantage of EAE is that is an environmentally friendly process. Water is widely used thus reduces solvent usage besides decreases energy consumption.

However; this technology has some technical limitations:

a) The cost of enzymes is relatively expensive for processing large volumes of raw plants
b) The limitations of the available enzyme preparations since they are not able to completely hydrolyze the cell walls of the plants, which limits the extraction yields of phytochemical compounds
c) Enzyme-assisted extraction can be difficult to be applied at an industrial scale. However, if the above limitations can be overcome, then enzyme-based extraction could provide an opportunity not only to increase extraction yields but also to enhance product quality by enabling the use of milder processing conditions such as lower extraction temperatures.

Despite these limitations, enzyme-based extraction could provide an opportunity not only to increase extraction yields but also to improve the quality of extracts obtained from medicinal plants [84].

Li et al., [2, 85] studied fresh orange peels treated with the following conditions: enzymes: (0-5% w / w) Celluzyme CL;Celluzyme MX 1.5%; and Kleepase (pectinase) AFP 106L; and aqueous extraction (100%), Extraction time: 1-51 h, Temperature: 19, 37, 50, 65 and 80°C. The total phenolic content of orange peels ranged from 76 to 93 mg of GAE /100 g of sample. The variation in the concentration of the total phenolic compounds depended on the extraction conditions. The highest extraction was obtained at a temperature of 50°C, 3 h as extraction time using the enzyme Celluzyme MX. The enzyme concentration showed some variations, but it was revealed that a concentration of 1.5% (w/w) was sufficient to obtain the best extraction.

Wang et al. [85] compared to untreated guava leaves, single xylanase-assisted extraction did not change the composition and yield of soluble phenolics. In contrast, single cellulase or β-glucosidase assisted extraction significantly improved the soluble phenolics content, flavonoids content, by 103.2%, 81.6%, respectively. Notably, the contents of quercetin and kaempferol with higher bioactivity were enhanced by 3.5- and 2.2-fold, respectively.

Chen et al. [86] reported an enzyme involved approach to improving the extraction of phenolic compounds from Ginkgo biloba, in which the enzyme is employed for increasing the solubility of phenolic compounds besides of cell wall degradation but in the ethanol-water extractant. *Penicillium decumbens* cellulase, *Trichoderma reesei* cellulase, and *Aspergillus niger* pectinase were analyzed. *P. decumbens* cellulase transform flavonol aglycones into more polar glucosides (higher solubility) of which improved extraction. The influence of glycosyl donor, solvent, pH, and temperature on the enzymatic transglycosylation was investigated. For three predominant flavonoids in *Ginko biloba* (quercetin, kaempferol, and isorhamnetin), the extraction yield turned to be 28.3 mg/g of dw, 31% higher than that under the pre-optimized conditions, and 102% higher than that under the conditions without enzymes.

The previous studies showed that green extraction technology could be useful for extracting bioactive components from the plant matrix and has good potential for use in the food and pharmaceutical industries.

CONCLUSION

Popularity and demand for medicinal plants as a source of bioactive compounds that benefit health are increasing. The extraction of bioactive compounds involves complex mechanisms, and various techniques can accomplish it. Seeking to improve the extraction yields, reduce processing time and the environmental damage caused by toxic solvents, forces us to develop new extraction methods for recovery of bioactive compounds are emerging with the responsibility to generate clean and green technologies which employ less solvent and can be used either water or ethanol as solvents. The scientific advancements are guiding the development of green technologies by providing methods such as ultrasound-assisted extraction, microwave-assisted extraction, ultrasound-assisted extraction, pulsed electric field-assisted extraction, and enzyme assisted extraction, are promising methods to extract phytochemicals from medicinal plant parts such as roots, seeds, barks, leaves, flowers, and fruits. The impacts of factors like temperature, pressure, solvent, sample particle size, pH, extraction time, frequency (UAE), electric field strength and pulse duration (PEF) are the main factors controlling the mass transfer and solubility of bioactive compounds from medicinal plants.

References

[1] Khan, H. 2014. Medicinal plants in light of history: recognized therapeutic modality. *Journal of Evidence-based Complementary & Alternative Medicine*. 19(3): p. 216-219.

[2] Li, J., C. Wu, F. Li, R. Yu, X. Wu, L. Shen, Y. Liu, and W. Zeng. 2019. Optimization of ultrasound-assisted water extraction of flavonoids from Psidium guajava leaves by response surface analysis. *Preparative Biochemistry and Biotechnology*. 49(1): p. 21-29.

[3] Han, X. and T. L. Parker. 2017. Anti-inflammatory, tissue remodeling, immunomodulatory, and anticancer activities of oregano (*Origanum vulgare*) essential oil in a human skin disease model. *Biochimie Open*. 4: p. 73-77.

[4] Franco-Vega, A., N. Ramírez-Corona, E. Palou, and A. López-Malo. 2016. Estimation of mass transfer coefficients of the extraction process of essential oil from orange peel using microwave assisted extraction. *Journal of Food Engineering*. 170: p. 136-143.

[5] Hamdan, S., H. G. Daood, M. Toth-Markus, and V. Illés. 2008. Extraction of cardamom oil by supercritical carbon dioxide and sub-critical propane. *The Journal of Supercritical Fluids*. 44(1): p. 25-30.

[6] Asadi-Samani, M., M. T. Moradi, M. Bahmani, and M. Shahrani. 2016. Antiviral medicinal plants of Iran: A Review of Ethnobotanical evidence. *International Journal of Pharm Tech Research*. 9(5): p. 427-434.

[7] Moussaoui, F. and T. Alaoui. 2016. Evaluation of antibacterial activity and synergistic effect between antibiotic and the essential oils of some medicinal plants. *Asian Pacific Journal of Tropical Biomedicine*. 6(1): p. 32-37.

[8] Chemat, F., M. A. Vian, and G. Cravotto. 2012. Green extraction of natural products: concept and principles. *International Journal of Molecular Sciences*. 13(7): p. 8615-8627.

[9] Handa, S. S., S. P. S. Khanuja, G. Longo, and D. D. Rakesh. 2008. Extraction technologies for medicinal and aromatic plants (United Nations Industrial Development Organisation and the International Centre for Science and High Technology). *International Centre for Science and High Technology-United Nations Industrial Development Organization, area Science Park Padriciano*. 99: p. 34012.

[10] Azmir, J., I. Zaidul, M. Rahman, K. Sharif, A. Mohamed, F. Sahena, M. Jahurul, K. Ghafoor, N. Norulaini, and A. Omar. 2013. Techniques for extraction of bioactive compounds from plant materials: A review. *Journal of Food Engineering*. 117(4): p. 426-436.

[11] Vilkhu, K., R. Mawson, L. Simons, and D. Bates. 2008. Applications and opportunities for ultrasound assisted extraction in the food industry—A review. *Innovative Food Science & Emerging Technologies*. 9(2): p. 161-169.

[12] Vázquez, M. B., L. R. Comini, R. E. Martini, S. N. Montoya, S. Bottini, and J. L. Cabrera. 2014. Comparisons between conventional, ultrasound-assisted and microwave-assisted methods for extraction of anthraquinones from *Heterophyllaea pustulata* Hook f.(Rubiaceae). *Ultrasonics Sonochemistry*. 21(2): p. 478-484.

[13] Dhanani, T., S. Shah, N. Gajbhiye, and S. Kumar. 2017. Effect of extraction methods on yield, phytochemical constituents and

antioxidant activity of *Withania somnifera*. *Arabian Journal of Chemistry*. 10: p. S1193-S1199.

[14] Xu, D. P., J. Zheng, Y. Zhou, Y. Li, S. Li, and H. B. Li. 2017. Ultrasound-assisted extraction of natural antioxidants from the flower of *Limonium sinuatum*: Optimization and comparison with conventional methods. *Food Chemistry*. 217: p. 552-559.

[15] Mason, T. J., L. Paniwnyk, and J. Lorimer. 1996. The uses of ultrasound in food technology. *Ultrasonics Sonochemistry*. 3(3): p. S253-S260.

[16] Meullemiestre, A., C. Breil, M. Abert-Vian, and F. Chemat. 2016. Microwave, ultrasound, thermal treatments, and bead milling as intensification techniques for extraction of lipids from oleaginous *Yarrowia lipolytica* yeast for a biojetfuel application. *Bioresource Technology*. 211: p. 190-199.

[17] Mathialagan, R., N. Mansor, M. R. Shamsuddin, Y. Uemura, and Z. Majeed. 2017. Optimisation of Ultrasonic-Assisted Extraction (UAE) of Allicin from Garlic (*Allium sativum* L.). *Chemical Engineering Transactions*. 56: p. 1747-1752.

[18] Saleh, I. A., M. Vinatoru, T. J. Mason, N. Abdel-Azim, E. Aboutabl, and F. Hammouda. 2016. A possible general mechanism for ultrasound-assisted extraction (UAE) suggested from the results of UAE of chlorogenic acid from *Cynara scolymus* L. (artichoke) leaves. *Ultrasonics Sonochemistry*. 31: p. 330-336.

[19] Rocha, J. D. C. G., F. R. Procopio, A. C. Mendonca, L. M. Vieira, I. T. Perrone, F. A. R. D. Barros, and P. C. Stringheta. 2018. Optimization of ultrasound-assisted extraction of phenolic compounds from jussara (*Euterpe edulis* M.) and blueberry (*Vaccinium myrtillus*) fruits. *Food Science and Technology*. 38(1): p. 45-53.

[20] Machado, A. P. D. F., A. L. D. Pereira, G. F. Barbero, and J. Martínez. 2017. Recovery of anthocyanins from residues of *Rubus fruticosus*, *Vaccinium myrtillus* and *Eugenia brasiliensis* by ultrasound assisted extraction, pressurized liquid extraction and their combination. *Food Chemistry*. 231: p. 1-10.

[21] Martins, F. S., E. C. da Conceição, E. S. Bandeira, and J. O. C. Silva. 2014. Th effects of extraction method on recovery rutin from *Calendula officinalis* L. (Asteraceae). *Pharmacognosy Magazine*. 10(Suppl 3): p. S569.

[22] Paz, J. E. W., D. B. M. Márquez, G. C. M. Ávila, R. E. B. Cerda, and C. N. Aguilar. 2015. Ultrasound-assisted extraction of polyphenols from native plants in the Mexican desert. *Ultrasonics Sonochemistry*. 22: p. 474-481.

[23] Russo, D., I. Faraone, F. Labanca, C. Sinisgalli, M. Bartolo, P. B. Andrade, P. Valentao, and L. Milella. 2019. Comparison of different green-extraction techniques and determination of the phytochemical profile and antioxidant activity of Echinacea angustifolia L. extracts. *Phytochemical Analysis*.

[24] Nicolai, M., P. Pereira, R. F. Vitor, C. P. Reis, A. Roberto, and P. Rijo. 2016. Antioxidant activity and rosmarinic acid content of ultrasound-assisted ethanolic extracts of medicinal plants. *Measurement*. 89: p. 328-332.

[25] Pinela, J., M. Prieto, E. Pereira, I. Jabeur, M. F. Barreiro, L. Barros, and I. C. Ferreira. 2019. Optimization of heat-and ultrasound-assisted extraction of anthocyanins from *Hibiscus sabdariffa* calyces for natural food colorants. *Food Chemistry*. 275: p. 309-321.

[26] Kou, X., Y. Ke, X. Wang, M. R. T. Rahman, Y. Xie, S. Chen, and H. Wang. 2018. Simultaneous extraction of hydrophobic and hydrophilic bioactive compounds from ginger (*Zingiber officinale* Roscoe). *Food Chemistry*. 257: p. 223-229.

[27] Almeida, A. D. R., M. V. D. O. B. Maciel, M. H. Machado, G. C. Bazzo, R. D. de Armas, V. B. Vitorino, L. Vitali, J. M. Block, and P. L. M. Barreto. 2019. Bioactive compounds and antioxidant activities of Brazilian hop (*Humulus lupulus* L.) extracts. *International Journal of Food Science & Technology*.

[28] Caleja, C., L. Barros, M. Prieto, M. F. Barreiro, M. B. P. Oliveira, and I. C. Ferreira. 2017. Extraction of rosmarinic acid from *Melissa officinalis* L. by heat-, microwave-and ultrasound-assisted extraction

techniques: A comparative study through response surface analysis. *Separation and Purification Technology*. 186: p. 297-308.

[29] Žlabur, J. Š., S. Voća, N. Dobričević, S. Pliestić, A. Galić, A. Boričević, and N. Borić. 2016. Ultrasound-assisted extraction of bioactive compounds from lemon balm and peppermint leaves. *International Agrophysics*. 30(1): p. 95-104.

[30] Hao, K., W. Hu, M. Hou, D. Cao, Y. Wang, Q. Guan, X. Zhang, A. Wang, J. Yu, and B. Guo. 2019. Optimization of Ultrasonic-Assisted Extraction of Total Phenolics from *Citrus aurantium* L. Blossoms and Evaluation of Free Radical Scavenging, Anti-HMG-CoA Reductase Activities. *Molecules*. 24(13): p. 2368.

[31] Nile, S. H., A. S. Nile, and Y. S. Keum. 2017. Total phenolics, antioxidant, antitumor, and enzyme inhibitory activity of Indian medicinal and aromatic plants extracted with different extraction methods. *3 Biotech*. 7(1): p. 76.

[32] Trusheva, B., D. Trunkova, and V. Bankova. 2007. Different extraction methods of biologically active components from propolis: a preliminary study. *Chemistry Central Journal*. 1(1): p. 13.

[33] Manousi, N., I. Sarakatsianos, and V. Samanidou. 2019. *Extraction techniques of phenolic compounds and other bioactive compounds from medicinal and aromatic plants*, in *Engineering Tools in the Beverage Industry*. Elsevier. p. 283-314.

[34] Li, J., H. Lee, and K. J. Vahala. 2013. Microwave synthesizer using an on-chip Brillouin oscillator. *Nature Communications*. 4: p. 2097.

[35] Delazar, A., L. Nahar, S. Hamedeyazdan, and S. D. Sarker. 2012. *Microwave-assisted extraction in natural products isolation*, in *Natural products isolation*. Springer. p. 89-115.

[36] Hithamani, G. and K. Ramalakshmi. 2013. Microwave assisted extraction of phenolics from *Origanum vulgare*. International Journal of Agricultural and Soil Science 1: p. 7-12.

[37] Kaufmann, B. and P. Christen. 2002. Recent extraction techniques for natural products: microwave-assisted extraction and pressurised solvent extraction. *Phytochemical Analysis: An International Journal of Plant Chemical and Biochemical Techniques*. 13(2): p. 105-113.

[38] Stalikas, C. D. 2007. Extraction, separation, and detection methods for phenolic acids and flavonoids. *Journal of Separation Science*. 30(18): p. 3268-3295.

[39] Bose, S., B. Laha, and S. Banerjee. 2014. Quantification of allicin by high performance liquid chromatography-ultraviolet analysis with effect of post-ultrasonic sound and microwave radiation on fresh garlic cloves. *Pharmacognosy Magazine*. 10(Suppl 2): p. S288.

[40] Babova, O., A. Occhipinti, A. Capuzzo, and M. E. Maffei. 2016. Extraction of bilberry (*Vaccinium myrtillus*) antioxidants using supercritical/subcritical CO2 and ethanol as co-solvent. *The Journal of Supercritical Fluids*. 107: p. 358-363.

[41] Bekdeşer, B. 2019. Modeling and optimizing microwave-assisted extraction of antioxidant compounds from marigold *Calendula officinalis* L. using response surface methodology. *Turkish Journal of Chemistry*. 43(5): p. 1457-1471.

[42] Asghari, J., B. Ondruschka, and M. Mazaheritehrani. 2011. Extraction of bioactive chemical compounds from the medicinal Asian plants by microwave irradiation. *Journal of Medicinal Plants Research*. 5(4): p. 495-506.

[43] Kaur, P., A. Chaudhary, B. Singh, and Gopichand. 2012. An Efficient Microwave Assisted Extraction of Phenolic Compounds and Antioxidant Potential of *Ginkgo biloba. Natural product communications*. 7(2): p. 1934578X1200700222.

[44] Huang, J. and Z. Zhang. 2004. Microwave-assisted extraction of quercetin and acid degradation of its glycosides in *Psidium guajava* leaves. *Analytical Sciences*. 20(2): p. 395-397.

[45] Yusoff, N. and C. Leo. 2017. Microwave assisted extraction of defatted roselle (*Hibiscus sabdariffa* L.) seed at subcritical conditions with statistical analysis. *Journal of Food Quality*. 2017.

[46] Pimentel-Moral, S., I. Borrás-Linares, J. Lozano-Sánchez, D. Arráez-Román, A. Martínez-Férez, and A. Segura-Carretero. 2018. Microwave-assisted extraction for *Hibiscus sabdariffa* bioactive compounds. *Journal of Pharmaceutical and Biomedical Analysis*. 156: p. 313-322.

[47] Petkova, N., L. Ivanova, G. Filova, I. Ivanov, and P. Denev. 2017. Antioxidants and carbohydrate content in infusions and microwave extracts from eight medicinal plants. *Journal of Applied Pharmaceutical Science*. 7(10): p. 055-061.

[48] Capuzzo, A., M. E. Maffei, and A. Occhipinti. 2013. Supercritical fluid extraction of plant flavors and fragrances. *Molecules*. 18(6): p. 7194-7238.

[49] Sihvonen, M., E. Järvenpää, V. Hietaniemi, and R. Huopalahti. 1999. Advances in supercritical carbon dioxide technologies. *Trends in food Science & Technology*. 10(6-7): p. 217-222.

[50] Azwanida, N. 2015. A review on the extraction methods use in medicinal plants, principle, strength and limitation. *Medicinal and Aromatic Plants*. 4(196): p. 2167-0412.1000196.

[51] Reverchon, E. and C. Marrone. 1997. Supercritical extraction of clove bud essential oil: isolation and mathematical modeling. *Chemical Engineering Science*. 52(20): p. 3421-3428.

[52] Zeković, Z., O. Bera, S. Đurović, and B. Pavlić. 2017. Supercritical fluid extraction of coriander seeds: Kinetics modelling and ANN optimization. *The Journal of Supercritical Fluids*. 125: p. 88-95.

[53] King, J. W. 2014. Modern supercritical fluid technology for food applications. *Annual Review of Food Science and Technology*. 5: p. 215-238.

[54] Paes, J., R. Dotta, G. F. Barbero, and J. Martínez. 2014. Extraction of phenolic compounds and anthocyanins from blueberry (*Vaccinium myrtillus* L.) residues using supercritical CO_2 and pressurized liquids. *The Journal of Supercritical Fluids*. 95: p. 8-16.

[55] Del Valle, J., C. Mena, and M. Budinich. 2008. Extraction of garlic with supercritical CO_2 and conventional organic solvents. *Brazilian Journal of Chemical Engineering*. 25(3): p. 532-542.

[56] del Valle, J. M., V. Glatzel, and J. L. Martínez. 2012. Supercritical CO_2 extraction of allicin from garlic flakes: Screening and kinetic studies. *Food Research International*. 45(1): p. 216-224.

[57] Hamburger, M., S. Adler, D. Baumann, A. Förg, and B. Weinreich. 2003. Preparative purification of the major anti-inflammatory

triterpenoid esters from Marigold (*Calendula officinalis*). *Fitoterapia*. 74(4): p. 328-338.

[58] Marongiu, B., A. Piras, S. Porcedda, E. Tuveri, E. Sanjust, M. Meli, F. Sollai, P. Zucca, and A. Rescigno. 2007. Supercritical CO_2 extract of *Cinnamomum zeylanicum*: chemical characterization and antityrosinase activity. *Journal of Agricultural and Food Chemistry*. 55(24): p. 10022-10027.

[59] Chiu, K. L., Y. C. Cheng, J. H. Chen, C. J. Chang, and P. W. Yang. 2002. Supercritical fluids extraction of Ginkgo ginkgolides and flavonoids. *The Journal of Supercritical Fluids*. 24(1): p. 77-87.

[60] Balachandran, S., S. Kentish, R. Mawson, and M. Ashokkumar. 2006. Ultrasonic enhancement of the supercritical extraction from ginger. *Ultrasonics Sonochemistry*. 13(6): p. 471-479.

[61] Moura, P., G. Prado, M. Meireles, and C. Pereira. 2012. Supercritical fluid extraction from guava (*Psidium guajava*) leaves: global yield, composition and kinetic data. *The Journal of Supercritical Fluids*. 62: p. 116-122.

[62] Mannila, M., Q. Lang, C. M. Wai, Y. Cui, and C. Y. Ang. 2003. *Supercritical fluid extraction of bioactive components from St. John's Wort (Hypericum perforatum* L.*) and Ginkgo biloba*. ACS Publications.

[63] Karasová, G., J. Lehotay, E. Kłodzinska, R. Gadzała-Kopciuch, and B. Buszewski. 2006. Comparison of several extraction methods for the isolation of benzoic acid derivatives from *Melissa officinalis*. *Journal of Liquid Chromatography & Related Technologies*. 29(11): p. 1633-1644.

[64] Bimakr, M., R. A. Rahman, A. Ganjloo, F. S. Taip, L. M. Salleh, and M. Z. I. Sarker. 2012. Optimization of supercritical carbon dioxide extraction of bioactive flavonoid compounds from spearmint (*Mentha spicata* L.) leaves by using response surface methodology. *Food and Bioprocess Technology*. 5(3): p. 912-920.

[65] He, J. Z., P. Shao, J. H. Liu, and Q. M. Ru. 2012. Supercritical carbon dioxide extraction of flavonoids from Pomelo (*Citrus grandis*

(L.) Osbeck) peel and their antioxidant activity. *International Journal of Molecular Sciences*. 13(10): p. 13065-13078.

[66] Nieto, A., F. Borrull, E. Pocurull, and R. M. Marcé. 2010. Pressurized liquid extraction: a useful technique to extract pharmaceuticals and personal-care products from sewage sludge. *TrAC Trends in Analytical Chemistry*. 29(7): p. 752-764.

[67] Mustafa, A. and C. Turner. 2011. Pressurized liquid extraction as a green approach in food and herbal plants extraction: A review. *Analytica Chimica Acta*. 703(1): p. 8-18.

[68] Wijngaard, H., M. B. Hossain, D. K. Rai, and N. Brunton. 2012. Techniques to extract bioactive compounds from food by-products of plant origin. *Food Research International*. 46(2): p. 505-513.

[69] Camel, V. 2001. Recent extraction techniques for solid matrices—supercritical fluid extraction, pressurized fluid extraction and microwave-assisted extraction: their potential and pitfalls. *Analyst*. 126(7): p. 1182-1193.

[70] Pawliszyn, J. 2003. Sample preparation: quo vadis? *Analytical Chemistry*. 75(11): p. 2543-2558.

[71] Hu, J., Z. Guo, M. Glasius, K. Kristensen, L. Xiao, and X. Xu. 2011. Pressurized liquid extraction of ginger (*Zingiber officinale* Roscoe) with bioethanol: An efficient and sustainable approach. *Journal of Chromatography a*. 1218(34): p. 5765-5773.

[72] Hossain, M., C. Barry-Ryan, A. B. Martin-Diana, and N. Brunton. 2011. Optimisation of accelerated solvent extraction of antioxidant compounds from rosemary (*Rosmarinus officinalis* L.), marjoram (*Origanum majorana* L.) and oregano (*Origanum vulgare* L.) using response surface methodology. *Food Chemistry*. 126(1): p. 339-346.

[73] Ibañez, E., M. Herrero, P. Martín, F. J. Señoráns, G. Reglero, and A. Cifuentes. 2006. Accelerated Solvent Extraction: A New Procedure To Obtain Functional Ingredients from Natural Sources.

[74] Vidović, S., N. Nastić, A. Gavarić, M. Cindrić, and J. Vladić. 2019. Development of green extraction process to produce antioxidant-rich extracts from purple coneflower. *Separation Science and Technology*. 54(7): p. 1174-1181.

[75] Lopez, N., E. Puertolas, S. Condon, J. Raso, and I. Alvarez. 2009. Enhancement of the solid-liquid extraction of sucrose from sugar beet (*Beta vulgaris*) by pulsed electric fields. *LWT-Food Science and Technology*. 42(10): p. 1674-1680.

[76] Bryant, G. and J. Wolfe. 1987. Electromechanical stresses produced in the plasma membranes of suspended cells by applied electric fields. *The Journal of Membrane Biology*. 96(2): p. 129-139.

[77] Teh, S. S., B. E. Niven, A. E. D. A. Bekhit, A. Carne, and E. J. Birch. 2015. Microwave and pulsed electric field assisted extractions of polyphenols from defatted canola seed cake. *International Journal of Food Science & Technology*. 50(5): p. 1109-1115.

[78] Boussetta, N., E. Vorobiev, V. Deloison, F. Pochez, A. Falcimaigne-Cordin, and J. L. Lanoisellé. 2011. Valorisation of grape pomace by the extraction of phenolic antioxidants: application of high voltage electrical discharges. *Food Chemistry*. 128(2): p. 364-370.

[79] Barba, F. J., N. Boussetta, and E. Vorobiev. 2015. Emerging technologies for the recovery of isothiocyanates, protein and phenolic compounds from rapeseed and rapeseed press-cake: effect of high voltage electrical discharges. *Innovative Food Science & Emerging Technologies*. 31: p. 67-72.

[80] Lamanauskas, N., R. Bobinaitė, S. Šatkauskas, P. Viškelis, G. Pataro, and G. Ferrari. 2015. Pulsed electric field-assisted juice extraction of frozen/thawed blueberries. *Zemdirbyste-Agriculture*. 102(1).

[81] Luengo, E., S. Condón-Abanto, S. Condón, I. Álvarez, and J. Raso. 2014. Improving the extraction of carotenoids from tomato waste by application of ultrasound under pressure. *Separation and Purification Technology*. 136: p. 130-136.

[82] Wilkinson, B. and J. Micklefield. 2007. Mining and engineering natural-product biosynthetic pathways. *Nature Chemical Biology*. 3(7): p. 379.

[83] Niranjan, K. and P. Hanmoungjai. 2004. Enzyme-aided aquous extraction. *Nutritionally Enhanced Edible Oil Processing. AOCS Publishing*.

[84] Li, B., B. Smith, and M. M. Hossain. 2006. Extraction of phenolics from citrus peels: II. Enzyme-assisted extraction method. *Separation and Purification Technology*. 48(2): p. 189-196.

[85] Wang, L., Y. Wu, Y. Liu, and Z. Wu. 2017. Complex enzyme-assisted extraction releases antioxidative phenolic compositions from guava leaves. *Molecules*. 22(10): p. 1648.

[86] Chen, S., X. H. Xing, J. J. Huang, and M. S. Xu. 2011. Enzyme-assisted extraction of flavonoids from *Ginkgo biloba* leaves: Improvement effect of flavonol transglycosylation catalyzed by Penicillium decumbens cellulase. *Enzyme and Microbial Technology*. 48(1): p. 100-105.

About the Editors

J. Basilio Heredia, PhD
Research Professor
Centro de Investigación en Alimentación y Desarrollo, A. C.,
Culiacán, Sinaloa, México
Email: jbheredia@ciad.mx

Dr. Heredia is a Research Professor at Centro de Investigación en Alimentación y Desarrollo, A. C. (CIAD) Culiacan Unit. He has a BS degree in Biochemical Engineering (TNM-Tecnológico de Culiacán), an MS in Nutrition and Foods (CIAD-Conacyt), and a PhD (Texas A&M University) with specialization in the study of phytochemicals, bioactive compounds, nutraceuticals, and functional foods. With over 20 years of experience in these fields, he is the head of the Nutraceutical and Functional Foods Research Group at CIAD Culiacan.

Publications within the past 3 years:

1. Gutiérrez-Grijalva, E.P., Antunes-Ricardo, M., Acosta-Estrada, B.A., Gutiérrez-Uribe, J.A., and Heredia, J.B. (2019). Cellular antioxidant activity and in vitro inhibition of α-glucosidase, α-amylase and pancreatic lipase of oregano polyphenols under simulated

gastrointestinal digestion. *Food Research International* 116:676-686. doi: 10.1016/j.foodres.2018.08.096.

2. Vazquez-Olivo, G., Criollo-Mendoza, M.S., Gutierrez-Grijalva, E.P., Picos-Salas, M.A., and Heredia, J.B. 2019. "Lignin and Its Derivatives as Antioxidant, Antiviral and Antimicrobial Agents: Applicability in Human Health Promotion." In *Lignin: Biosynthesis, Functions and Economic Significance*, edited by Fachuang Lu and Fengxia Yue, 261-280. New York, USA: Nova Science Publishers, Inc.
3. Gutiérrez-Grijalva, E.P., Contreras-Angulo, L.A., Emus-Medina, A., Vázquez-Olivo, G., and Heredia, J.B. 2019. "Anticancer Properties of Medicinal Plants Listed in the Herbal Pharmacopoeia of The United Mexican States." In *Medicinal Plants: Chemistry, Pharmacology, and Therapeutic Applications*, edited by Mallappa Kumara Swamy, Jayanta Kumar Patra and Gudepalya Renukaiah Rudramurthy. CRC Press.
4. Vazquez-Olivo, G., Gutiérrez-Grijalva, E.P., and Heredia, J.B. (2019). Prebiotic compounds from agro-industrial by-products. *Journal of Food Biochemistry* 43 (6):e12711. doi: 10.1111/jfbc.12711.
5. Gutierrez-Grijalva, E.P., López-Martínez, L.X., Leyva-López, N., Vazquez-Olivo, G., Cabanillas-Bojorquez, L.A., and Heredia, J.B. 2019. "Germination of Edible Seeds: Effect on Bioactive Compounds and Their Health-Promoting Benefits." In *Germination: Types, Process and Effects*, edited by Rosalva Mora-Escobedo, Cristina Martinez and Rosalía Reynoso. Nova Science Publishers.
6. Méndez-Zúñiga, S.M., Corrales-García, J.E., Gutiérrez-Grijalva, E.P., García-Mateos, R., Pérez-Rubio, V., and Heredia, J.B. (2019). Fatty Acid Profile, Total Carotenoids, and Free Radical-Scavenging from the Lipophilic Fractions of 12 Native Mexican Avocado Accessions. *Plant Foods for Human Nutrition*. doi: 10.1007/s11130-019-00766-2.
7. Gutiérrez-Grijalva, E.P., Picos-Salas, M.A., Leyva-López, N., Criollo-Mendoza, M.S., Vazquez-Olivo, G., and Heredia, J.B. (2018). Flavonoids and phenolic acids from Oregano: Occurrence, biological activity and health benefits. *Plants* 7 (1). doi: 10.3390/plants7010002.

8. Gutiérrez-Grijalva, E.P., Flores-Acosta, L.G., Vazquez-Olivo, G., and Heredia, J.B. 2018. "By-Products from Grains and Cereals Processing." In *Plant Food By-Products: Industrial Relevance for Food Additives and Nutraceuticals*, edited by Jesús Fernando Ayala-Zavala, Gustavo González-Aguilar and Mohammed Wasim Siddiqui, 327-. Oakville, Canada: Apple Academic Press, Inc.
9. Leyva-López, N., Gutiérrez-Grijalva, E.P., Vazquez-Olivo, G., Contreras-Angulo, L.A., Emus-Medina, A., and Heredia, J.B. 2018. "Terpenes in Oregano: Constituents, Extraction, Analysis and Biological Properties." In *Terpenes: Biosynthesis, Applications And Research*, edited by Albin Bjarke. New York, USA: Nova Science Publishers.
10. Gutiérrez-Grijalva, E.P., Ambriz-Pérez, D.L., and Heredia, J.B. 2018. "Subproductos de Aguacate: Fuente Potencial de Compuestos Bioactivos." In *Aprovechamiento de subproductos de la industria alimentaria para la obtención de compuestos bioactivos*, edited by Gustavo A. González-Aguilar, Adrián Hernández-Mendoza, Jorge Milán-Carrillo, Belinda Vallejo-Cordoba and Aaron F. González-Córdova, 347-363. México, D.F.: AGT Editor, S. A.
11. Gutiérrez-Grijalva, E.P., Angulo-Escalante, M.A., León-Félix, J., and Heredia, J.B. (2017). Effect of In Vitro Digestion on the Total Antioxidant Capacity and Phenolic Content of 3 Species of Oregano (Hedeoma patens, Lippia graveolens, Lippia palmeri). *Journal of Food Science* 82 (12):2832-2839. doi: 10.1111/1750-3841.13954.
12. López-Martínez, L.X., Leyva-López, N., Gutiérrez-Grijalva, E.P., and Heredia, J.B. (2017). Effect of cooking and germination on bioactive compounds in pulses and their health benefits. *Journal of Functional Foods* 38:624-634. doi: 10.1016/j.jff.2017.03.002.
13. Leyva-López, N., Gutiérrez-Grijalva, E.P., Vazquez-Olivo, G., and Heredia, J.B. (2017). Essential oils of oregano: Biological activity beyond their antimicrobial properties. *Molecules* 22 (6). doi: 10.3390/molecules22060989.

14. Castillo-López, R.I., Gutiérrez-Grijalva, E.P., Leyva-Lópe, N., López-Martínez, L.X., and Heredia, J.B. (2017). Natural alternatives to growth-promoting antibiotics (GPA) in animal production. *Journal of Animal and Plant Sciences* 27 (2):349-359.
15. Leyva-López, N., Gutierrez-Grijalva, E.P., Ambriz-Perez, D.L., and Heredia, J.B. (2016). Flavonoids as cytokine modulators: A possible therapy for inflammation-related diseases. *International Journal of Molecular Sciences* 17 (6). doi: 10.3390/ijms17060921.
16. Gutiérrez-Grijalva, E.P., Ambriz-Pérez, D.L., Leyva-López, N., Castillo-López, R.I., and Heredia, J.B. (2016). Bioavailability of dietary phenolic compounds: Review. *Revista Espanola de Nutricion Humana y Dietetica* 20 (2):140-147. doi: 10.14306/renhyd.20.2.184.
17. Ambriz-Pérez, D.L., Leyva-López, N., Gutierrez-Grijalva, E.P., Heredia, J.B., and Yildiz, F. (2016). Phenolic compounds: Natural alternative in inflammation treatment. A Review. *Cogent Food & Agriculture* 2 (1):1131412. doi: 10.1080/23311932.2015.1131412.
18. Gutiérrez-Grijalva, E.P., Ambriz-Pérez, D.L., Leyva-López, N., Castillo-López, R.I., and Heredia, J.B. (2016). Review: dietary phenolic compounds, health benefits and bioaccessibility. *Archivos Latinoamericanos de Nutrición* 66 (2):87-100.
19. Ambriz-Pérez, D.L., Young Bang, W., Nair, V., Angulo-Escalante, M.A., Cisneros-Zevallos, L., Heredia, J.B. (2016). Protective Role of Flavonoids and Lipophilic Compounds from *Jatropha platyphylla* on the Suppression of Lipopolysaccharide (LPS)-Induced Inflammation in Macrophage Cells. *Journal of Agricultural and Food Chemistry*. 64:1899–1909.
20. Leyva-López, N., Nair, V., Young Bang, W., Cisneros-Zevallos, L., Heredia, J.B. (2016). Protective role of terpenes and polyphenols from three species of Oregano (*Lippia graveolens, Lippia palmeri* and *Hedeoma* patens) on the suppression of lipopolysaccharide-induced inflammation in RAW264.7macrophage cells. *Journal of Ethnopharmacology*. 187:302–312.

Erick Paul Gutiérrez Grijalva, PhD
Cátedras CONACYT (CONACYT Research Fellow)
Cátedras CONACYT - Centro de Investigación en Alimentación y Desarrollo, A. C., Culiacán, Sinaloa, México
Email: erick.gutierrez@ciad.mx

Dr. Erick Paul Gutiérrez-Grijalva is a CONACYT Research Fellow at the Centro de Investigación en Alimentación y Desarrollo, A. C. Dr. Gutiérrez-Grijalva has a degree in Biochemical Engineering, and a Doctorate in Sciences, with specialization in the study of metabolomics, bioavailability and bioactivity of phytochemicals from medicinal plants from CIAD. His research is focused on the metabolomics bioactivity and bioavailability of phytochemicals from plants of the Northwestern region of Mexico.

Publications within the past 3 years:

1. Gutiérrez-Grijalva, E.P., Antunes-Ricardo, M., Acosta-Estrada, B.A., Gutiérrez-Uribe, J.A., and Basilio Heredia, J. (2019). Cellular antioxidant activity and in vitro inhibition of α-glucosidase, α-amylase and pancreatic lipase of oregano polyphenols under simulated gastrointestinal digestion. *Food Research International* 116:676-686. doi: 10.1016/j.foodres.2018.08.096.
2. Vazquez-Olivo, G., Criollo-Mendoza, M.S., Gutierrez-Grijalva, E.P., Picos-Salas, M.A., and Heredia, J.B. 2019. "Lignin and Its Derivatives as Antioxidant, Antiviral and Antimicrobial Agents: Applicability in Human Health Promotion." In *Lignin: Biosynthesis, Functions and Economic Significance*, edited by Fachuang Lu and Fengxia Yue, 261-280. New York, USA: Nova Science Publishers, Inc.
3. Gutiérrez-Grijalva, E.P., Contreras-Angulo, L.A., Emus-Medina, A., Vázquez-Olivo, G., and Heredia, J.B. 2019. "Anticancer Properties of Medicinal Plants Listed in the Herbal Pharmacopoeia of The United Mexican States." In *Medicinal Plants: Chemistry, Pharmacology, and Therapeutic Applications*, edited by Mallappa Kumara Swamy, Jayanta Kumar Patra and Gudepalya Renukaiah Rudramurthy. CRC Press.

4. Vazquez-Olivo, G., Gutiérrez-Grijalva, E.P., and Heredia, J.B. (2019). Prebiotic compounds from agro-industrial by-products. *Journal of Food Biochemistry* 43 (6):e12711. doi: 10.1111/jfbc.12711.
5. Gutierrez-Grijalva, E.P., López-Martínez, L.X., Leyva-López, N., Vazquez-Olivo, G., Cabanillas-Bojorquez, L.A., and Heredia, J.B. 2019. "Germination of Edible Seeds: Effect on Bioactive Compounds and Their Health-Promoting Benefits." In *Germination: Types, Process and Effects*, edited by Rosalva Mora-Escobedo, Cristina Martinez and Rosalía Reynoso. Nova Science Publishers.
6. Méndez-Zúñiga, S.M., Corrales-García, J.E., Gutiérrez-Grijalva, E.P., García-Mateos, R., Pérez-Rubio, V., and Heredia, J.B. (2019). Fatty Acid Profile, Total Carotenoids, and Free Radical-Scavenging from the Lipophilic Fractions of 12 Native Mexican Avocado Accessions. *Plant Foods for Human Nutrition*. doi: 10.1007/s11130-019-00766-2.
7. Gutiérrez-Grijalva, E.P., Picos-Salas, M.A., Leyva-López, N., Criollo-Mendoza, M.S., Vazquez-Olivo, G., and Heredia, J.B. (2018). Flavonoids and phenolic acids from Oregano: Occurrence, biological activity and health benefits. *Plants* 7 (1). doi: 10.3390/plants7010002.
8. Gutiérrez-Grijalva, E.P., Flores-Acosta, L.G., Vazquez-Olivo, G., and Heredia, J.B. 2018. "By-Products from Grains and Cereals Processing." In *Plant Food By-Products: Industrial Relevance for Food Additives and Nutraceuticals*, edited by Jesús Fernando Ayala-Zavala, Gustavo González-Aguilar and Mohammed Wasim Siddiqui, 327-. Oakville, Canada: Apple Academic Press, Inc.
9. Leyva-López, N., Gutiérrez-Grijalva, E.P., Vazquez-Olivo, G., Contreras-Angulo, L.A., Emus-Medina, A., and Heredia, J.B. 2018. "Terpenes in Oregano: Constituents, Extraction, Analysis and Biological Properties." In *TERPENES: BIOSYNTHESIS, APPLICATIONS AND RESEARCH*, edited by Albin Bjarke. New York, USA: Nova Science Publishers.
10. Gutiérrez-Grijalva, E.P., Ambriz-Pérez, D.L., and Heredia, J.B. 2018. "Subproductos de Aguacate: Fuente Potencial de Compuestos Bioactivos." In *Aprovechamiento de subproductos de la industria alimentaria para la obtención de compuestos bioactivos*, edited by

Gustavo A. González-Aguilar, Adrián Hernández-Mendoza, Jorge Milán-Carrillo, Belinda Vallejo-Cordoba and Aaron F. González-Córdova, 347-363. México, D.F.: AGT Editor, S. A.

11. Gutiérrez-Grijalva, E.P., Angulo-Escalante, M.A., León-Félix, J., and Heredia, J.B. (2017). Effect of In Vitro Digestion on the Total Antioxidant Capacity and Phenolic Content of 3 Species of Oregano (Hedeoma patens, Lippia graveolens, Lippia palmeri). *Journal of Food Science* 82 (12):2832-2839. doi: 10.1111/1750-3841.13954.
12. López-Martínez, L.X., Leyva-López, N., Gutiérrez-Grijalva, E.P., and Heredia, J.B. (2017). Effect of cooking and germination on bioactive compounds in pulses and their health benefits. *Journal of Functional Foods* 38:624-634. doi: 10.1016/j.jff.2017.03.002.
13. Leyva-López, N., Gutiérrez-Grijalva, E.P., Vazquez-Olivo, G., and Heredia, J.B. (2017). Essential oils of oregano: Biological activity beyond their antimicrobial properties. *Molecules* 22 (6). doi: 10.3390/molecules22060989.
14. Castillo-López, R.I., Gutiérrez-Grijalva, E.P., Leyva-Lópe, N., López-Martínez, L.X., and Heredia, J.B. (2017). Natural alternatives to growth-promoting antibiotics (GPA) in animal production. *Journal of Animal and Plant Sciences* 27 (2):349-359.
15. Leyva-López, N., Gutierrez-Grijalva, E.P., Ambriz-Perez, D.L., and Basilio Heredia, J. (2016). Flavonoids as cytokine modulators: A possible therapy for inflammation-related diseases. *International Journal of Molecular Sciences* 17 (6). doi: 10.3390/ijms17060921.
16. Gutiérrez-Grijalva, E.P., Ambriz-Pérez, D.L., Leyva-López, N., Castillo-López, R.I., and Heredia, J.B. (2016). Bioavailability of dietary phenolic compounds: Review. *Revista Espanola de Nutricion Humana y Dietetica* 20 (2):140-147. doi: 10.14306/renhyd.20.2.184.
17. Ambriz-Pérez, D.L., Leyva-López, N., Gutierrez-Grijalva, E.P., Heredia, J.B., and Yildiz, F. (2016). Phenolic compounds: Natural alternative in inflammation treatment. A Review. *Cogent Food & Agriculture* 2 (1):1131412. doi: 10.1080/23311932.2015.1131412.
18. Gutiérrez-Grijalva, E.P., Ambriz-Pére, D.L., Leyva-López, N., Castillo-López, R.I., and Heredia, J.B. (2016). Review: dietary

phenolic compounds, health benefits and bioaccessibility. *Archivos Latinoamericanos de Nutrición* 66 (2):87-100.

Leticia Xochitl Lopez-Martinez, PhD
Cátedras CONACYT-Centro de Investigación en Alimentación y Desarrollo, A.C., Culiacán, Hermosillo, Sonora, Mexico
Email: leticia.lopez@ciad.mx
www.ciad.edu.mx

Dr. Leticia Xochitl López-Martínez, Researcher Fellow at CONACYT-CIAD. She has a degree in Biochemical Engineering in Food from Instituto Tecnologico de Veracruz, has a Master's Degree in Biochemical Engineering and a Doctorate in Food Sciences from the Food Research and Development Unit of the Instituto Tecnologico de Veracruz. He is currently studying the bioavailability, antioxidant and anti-inflammatory potential of the bioactive compounds associated with the dietary fiber of tropical fruits, as well as their processing products and their by-products.

Publications within the past 3 years:

1. Alejandra Nabil Alvarado-López, Leobardo Manuel Gómez-Oliván, Ramiro Baeza-Jiménez[2], Hugo Sergio Garcia-Galindo, J. Basilio Heredia, Leticia Xochitl Lopez-Martinez. (2019). Nutritional and bioactive characteristics of Ayocote bean (*Phaseolus coccienus* L.) an underutilized legume harvested in Mexico. *CyTA - Journal of Food*, ISSN: 1947-6337.
2. Guadalupe López-García, Ramiro Baeza-Jiménez, Hugo Sergio Garcia-Galindo, Octavio Dublán-García & Leticia X. Lopez-Martinez. (2018). Cooking treatments effect on bioactive compounds and antioxidant activity of quintonil (*Amaranthus hybridus*) harvested in spring and fall seasons. *CyTA - Journal of Food*, 16:1, 707-714. ISSN:1947-6337.

3. Paola Fabila-Garcia, Octavio Dublán-García, Leobardo M. Gómez-Oliván, R. Baeza-Jimenez, Leticia X. López-Martínez. (2017). *In vitro* antioxidant and bioactive properties of corn (*Zea mays* L.). *Archivos Latinoamericanos de la Nutrición*, 67 (4), 301-307.
4. López-García, L.X. López-Martínez, O. Dublán-García, R. Baeza-Jimenez (2017). Extracción y caracterización del perfil de ácidos grasos del quintonil (*Amaranthus hybridus*). *Revista Mexicana de Ingeniería Química.* 16 (3): 835-844.
5. 5.- López-Martínez, L. X., Leyva-López, N., Gutiérrez-Grijalva, E. P., & Heredia, J. B. (2017). Effect of cooking and germination on bioactive compounds in pulses and their health benefits. *Journal of Functional Foods*. 38(B):624-634
6. astillo-López, R. I., Gutiérrez-Grijalva, E. P., Leyva-López, N., López-Martínez, L. X., Heredia, J. B. (2017). Natural Alternatives to Growth-Promoting Antibiotics (GPA) In Animal Production. *JAPS: Journal of Animal & Plant Sciences*, *27*(2), 349-359. ISSN:
7. Sánchez-Aceves, L. M., Dublán-García, O., López-Martínez, L. X., Novoa-Luna, K. A., Islas-Flores, H., Galar-Martínez, M., ... & Gómez-Oliván, L. M. (2017). Reduction of the Oxidative Stress Status Using Steviol Glycosides in a Fish Model (Cyprinus carpio). *BioMed Research International*. 1-9.
8. Vazquez-Olivo, G., López-Martínez, L. X., Contreras-Angulo, L., & Heredia, J. B. (2017). Antioxidant Capacity of Lignin and Phenolic Compounds from Corn Stover. *Waste and Biomass Valorization*, 1-8.

Chapters

1. Ramiro Baeza-Jiménez, Leticia X. López-Martínez, Rebeca García-Varela, Hugo S. García. 2018. "Lipids in fruits and vegetables: chemistry and biological activities" in *Fruit and Vegetable Phytochemicals: Chemistry and Human Health, 2 Volumes 2nd edition.* John Wiley & Sons. pp 423-449.
2. Leticia Xochitl López-Martínez, Alberto Alonso Arrollo Collantes, Alejandra N. Alvarado-López y Henri Marquez-Pacheco. 2017. "Propiedades bioactivas de hidrolizados de ayocote (*Phaseolus*

coccineus L.)" in *Avances de la Ciencia en México. 1ª. Ed.* Centro de Investigaciones en Óptica, A.C. (607-95228) ISBN. 978-607-95228-8-9.

3. Leticia X. López-Martínez, Miguel A. Angulo-Escalante, Laura A. Contreras-Angulo, J. Basilio Heredia. 2018. Propiedades de compuestos bioactivos de subproductos de especies de *Jatropha spp.* IN Aprovechamiento de subproductos de la industria alimentaria para la obtención de compuestos bioactivos. 1ª. Ed. AGT Editor. ISBN:978-607-7551-46-1.

INDEX

B

C

D

E

F

G

J

K

L

M

N

O